Physics and Partial Differential Equations
Volume II

Physics and Partial Differential Equations
Volume II

Tatsien Li
Fudan University
Shanghai, People's Republic of China

Tiehu Qin
Fudan University
Shanghai, People's Republic of China

Translated by Yachun Li
Shanghai Jiao Tong University
Shanghai, People's Republic of China

siam.
Society for Industrial and Applied Mathematics
Philadelphia

高等教育出版社
HIGHER EDUCATION PRESS
Beijing, People's Republic of China

Co-published by Higher Education Press, People's Republic of China and the Society for Industrial and Applied Mathematics, Philadelphia, Pennsylvania.

10 9 8 7 6 5 4 3 2 1

Library of Congress Cataloging-in-Publication Data

880-01 Li, Daqian.
[880-01 Wu li xue yu pian wei fen fang cheng. English]
Physics and partial differential equations / Tatsien Li, Tiehu Qin ; translated by Yachun Li.
p. cm. -- (Other titles in applied mathematics)
Includes bibliographical references and index.
ISBN 978-1-611972-26-9 (Volume I)
ISBN 978-1-611973-31-0 (Volume II)
1. Mathematical physics. 2. Differential equations, Partial. I. 880-01 Qin, Tiehu. II. Title.
QC20.7.D5L5313 2012
530.15'5353--dc23
2012010830

高等教育出版社 HIGHER EDUCATION PRESS

Contents

Preface to the English Edition

The first volume of the Chinese edition of this book was published in July 1997, and the second volume was published in June 2000. In July 2000, upon the readers' request, we corrected several typographical errors and republished the first volume.

In this edition, only minor typographical errors are corrected.

We would like to take this opportunity to express our sincere thanks to our teachers, friends, and readers for their encouragement and support.

Tatsien Li
Tiehu Qin
Shanghai
September 2003

Preface to the Chinese Edition

Partial differential equations are fundamental in many important physical and mechanical disciplines. Although the names of these equations are well known, and they have been the focus of a considerable amount of research, gaining a comprehensive and deep understanding of the related physical and mechanical background is not an easy task. The purpose of this book is to offer some help not only to teachers, graduate students, and senior undergraduate students engaged in studying, researching, and teaching applied partial differential equations, but also to scholars and researchers in other disciplines and application areas. Readers will find the tools to become proficient in the use of important fundamental equations in modern physics, to gain familiarity with the whys, wherefores, and derivation of these equations, and to understand some commonly used mathematical models more easily. They will be able to study and use partial differential equations more consciously, and also learn to grasp some significant problems in order to properly carry out their research. Therefore, our purpose in writing this book is to build a bridge between physics and partial differential equations.

In this book, starting with the most basic concepts of physics, we focus on the whole process of establishing the fundamental equations for physical and mechanical disciplines such as electrodynamics, fluid dynamics, magnetohydrodynamics, reacting fluid dynamics, elastic mechanics, thermoelastic mechanics, viscoelastic mechanics, kinetic theory of gases, special relativity, and quantum mechanics. At the same time, we give a brief description of the mathematical structures and features of these equations, including their types and basic characteristics, the behavior of solutions, and some approaches commonly used to solve these equations. We selectively introduce some worldwide research results from recent years, including those of the authors and their research group. We hope that readers who are unfamiliar with the related physical and mechanical disciplines can gain access to the core of these disciplines in an easy-to-digest way so as to complete as soon as possible their transition from physics to mathematics and from related physical and mechanical fields to their mathematical models described by partial differential equations. On the other hand, for readers who are more familiar with the related physical and mechanical disciplines, we hope that an in-depth understanding of the mathematical structures and features of the fundamental equations will ultimately reveal the advantages of effective mathematical tools and expressions to more clearly present the basic contents of physics; consequently, readers who acquire such an understanding will be able to use modern mathematical concepts, methods, and tools more purposefully to solve related physical and mechanical problems.

This book is divided into two volumes, each consisting of five chapters. The contents of each chapter are relatively independent; however, all of the chapters echo and relate with each other to a certain extent. Exercises and a bibliography are included in each chapter. Most of the chapters are not meant to be difficult for those readers who have taken basic undergraduate courses in mathematics and physics. This book can be used as a textbook for graduate courses or elective senior undergraduate courses, as well as a reference book or extracurricular reading material.

Since the second half of 1987, the contents of this book have been continuously and successfully taught in Fudan University as both an elective senior undergraduate course and a required graduate degree course. The lecture notes have been constantly supplemented and revised, and it is on these that the final version of this book is based.

The authors would like to thank Higher Education Press for its enthusiastic support of the publication of this book, and Professor Sixu Guo for careful and meticulous typesetting. Thanks also go to Dr. Zhijie Cai for his responsible and proficient typing of the entire manuscript, and to Dr. Yingqiu Gu for his assistance in ensuring all the physical units in this book conform to the international system of units (SI units). In particular, we are grateful to Minyou Qi, Professor of Mathematics in the Department of Mathematics at Wuhan University, and Guangjiong Ni, Professor of Physics in the Department of Physics at Fudan University. They have carefully reviewed the manuscript and supplied many helpful comments and suggestions. Their hard work enriched this book.

As mathematicians, the authors may have a superficial understanding of physics to a certain extent, so errors and omissions are inevitable. We hope that readers will not spare their comments and corrections.

Tatsien Li
Tiehu Qin
November 10, 1996

Chapter 6

Thermoelasticity

6.1 Introduction

In Chapter 5, we discussed the deformation and stress of an elastic body under mechanical loads without considering the impact of temperature change. In fact, a temperature change in the elastic body will generate additional strain and stress. In this chapter, we will focus our attention on the impact of temperature change on the deformation of an elastic body; i.e., we will discuss the laws obeyed by the deformation of a thermoelastic body and its interior temperature distribution.

What is a thermoelastic body? A *thermoelastic body* is an object with the following characteristics: for an object that was originally in the natural state (there was neither deformation nor a temperature gradient inside), the deformation and the change of temperature field affected by mechanical loads and thermal environment will disappear immediately on being withdrawn from the mechanical loads and thermal environment, so that the object is restored to its original natural state with neither permanent deformation nor permanent temperature gradient.

To obtain the law of deformation and temperature distribution of a thermoelastic body, i.e., to establish the mathematical model of thermoelasticity—the system of thermoelasticity—the corresponding conservation laws and constitutive relations must be given, as was done for the elastic body in Chapter 5. Since the conservation laws of mass and momentum have nothing to do with the temperature change of an elastic body, the two conservation laws obtained in Chapter 5 are still valid for thermoelastic bodies. But since temperature changes must be taken into account in the discussion of thermoelastic bodies, and there is heat transfer among different parts of the thermoelastic body, in addition to the above two conservation laws, we should establish the corresponding *conservation law of energy*. Moreover, from the thermodynamical point of view, heat conduction is an irreversible process, and the conservation law of energy itself is not enough to determine whether a process can be carried on or not, so we have to establish the corresponding *entropy inequality*. Meanwhile, the quantities involved in the constitutive relations in elasticity are expressed as functions of only the deformation gradient tensor $\boldsymbol{F}$, while in thermoelasticity they are functions of the deformation gradient tensor, the absolute temperature, and the temperature gradient, since the deformation and heat conduction caused by the temperature and the temperature gradient have to be taken into consideration.

In this chapter, we will adopt relevant notation as in Chapter 5.

6.2 The Conservation Law of Energy and the Entropy Inequality

6.2.1 The Conservation Law of Energy

Let $\Omega \subset \mathbb{R}^3$ be the reference configuration of the thermoelastic body, i.e., the domain occupied by the thermoelastic body before deformation (assumed to be time $t = 0$). Denote by $\boldsymbol{x} = (x_1, x_2, x_3)$ a point in this domain. At time t (> 0), assume that the domain occupied by the thermoelastic body turns into Ω_t from Ω. This deformation is described by

$$\boldsymbol{y} = \boldsymbol{y}(t, \boldsymbol{x}),$$

where $\boldsymbol{y} = (y_1, y_2, y_3) \in \Omega_t$.

For any given domain $G_t \subset \Omega_t$, let us examine the changes in the total energy in G_t. For the elastic body, the total energy is the sum of the kinetic energy and the strain energy. However, for the thermoelastic body, the strain energy should be replaced by the internal energy, due to the heat transfer caused by the temperature change among different parts. Thus, the total energy in G_t should be the sum of the total kinetic energy and the total internal energy. Here the internal energy, of course, includes both the thermodynamic internal energy and the strain energy.

The total kinetic energy in G_t is

$$\int_{G_t} \frac{1}{2}\rho|\boldsymbol{v}|^2 \mathrm{d}y,$$

where ρ is the mass density and $\boldsymbol{v}$ is the velocity vector. Suppose that the internal energy per unit mass of the thermoelastic body is e. Then the total internal energy in G_t is

$$\int_{G_t} \rho e \mathrm{d}y.$$

From the conservation law of energy, the rate of change of the total energy in G_t should be equal to the sum of the following four parts: the work done by the volume force in G_t, the work done by the stress on the boundary S_t of G_t, the heat produced by the heat source in G_t, and the heat flowing into G_t across S_t in unit time.

Suppose that the volume force density, i.e., the volume force per unit mass, is $\boldsymbol{b} = (b_1, b_2, b_3)$. Then the work done by the volume force in unit time is

$$\int_{G_t} \rho \boldsymbol{b} \cdot \boldsymbol{v} \mathrm{d}y.$$

Moreover, the stress on the boundary S_t of G_t received from the exterior of G_t in Ω_t is $\boldsymbol{T}\boldsymbol{\nu}$, where $\boldsymbol{T}$ is the Cauchy stress tensor (see section 5.3.5 in Chapter 5) and $\boldsymbol{\nu}$ is the unit outward normal vector on S_t. Thus, the work done by the stress on S_t in unit time is

$$\int_{S_t} (\boldsymbol{T}\boldsymbol{\nu}) \cdot \boldsymbol{v} \mathrm{d}S_t.$$

Suppose that the heat source density of the thermoelastic body, i.e., the heat generated per unit mass in unit time, is γ; then the heat produced by the heat source in G_t in unit time is

$$\int_{G_t} \rho\gamma \mathrm{d}y.$$

Suppose furthermore that $\boldsymbol{q} = (q_1, q_2, q_3)^{\mathrm{T}}$ is the heat flux density vector: its direction is that of the heat flux, i.e., that of the heat conduction, while its length stands for the heat across a unit area perpendicular to the direction of heat flux in unit time. Thus, for any given area element $\mathrm{d}S_t$ on S_t, the heat flowing across $\mathrm{d}S_t$ along the normal vector $\boldsymbol{\nu}$ in unit time is given by

$$\boldsymbol{q} \cdot \boldsymbol{\nu} \mathrm{d}S_t.$$

Therefore, the heat flowing into G_t across S_t in unit time is

$$-\int_{S_t} \boldsymbol{q} \cdot \boldsymbol{\nu} \mathrm{d}S_t.$$

In summary, from the conservation law of energy we obtain

$$\begin{aligned} &\frac{\mathrm{d}}{\mathrm{d}t}\left(\int_{G_t} \frac{1}{2}\rho|\boldsymbol{v}|^2 \mathrm{d}y + \int_{G_t} \rho e \mathrm{d}y\right) \\ &= \int_{S_t} (\boldsymbol{T}\boldsymbol{\nu}) \cdot \boldsymbol{v} \mathrm{d}S_t + \int_{G_t} \rho \boldsymbol{b} \cdot \boldsymbol{v} \mathrm{d}y + \int_{G_t} \rho\gamma \mathrm{d}y - \int_{S_t} \boldsymbol{q} \cdot \boldsymbol{\nu} \mathrm{d}S_t. \end{aligned} \tag{6.1}$$

From Lemma 5.2 in Chapter 5, we have

$$\frac{\mathrm{d}}{\mathrm{d}t}\int_{G_t}\left(\frac{1}{2}\rho|\boldsymbol{v}|^2 + \rho e\right)\mathrm{d}y = \int_{G_t} \rho \frac{\mathrm{d}}{\mathrm{d}t}\left(\frac{1}{2}|\boldsymbol{v}|^2 + e\right)\mathrm{d}y,$$

where $\frac{\mathrm{d}}{\mathrm{d}t} = \frac{\partial}{\partial t} + (\boldsymbol{v} \cdot \nabla_y)$ and $\nabla_y = \left(\frac{\partial}{\partial y_1}, \frac{\partial}{\partial y_2}, \frac{\partial}{\partial y_3}\right)$. Then (6.1) can be rewritten as

$$\begin{aligned} &\int_{G_t} \rho \frac{\mathrm{d}}{\mathrm{d}t}\left(\frac{1}{2}|\boldsymbol{v}|^2 + e\right)\mathrm{d}y \\ &= \int_{S_t} (\boldsymbol{T}\boldsymbol{\nu}) \cdot \boldsymbol{v} \mathrm{d}S_t + \int_{G_t} \rho \boldsymbol{b} \cdot \boldsymbol{v} \mathrm{d}y + \int_{G_t} \rho\gamma \mathrm{d}y - \int_{S_t} \boldsymbol{q} \cdot \boldsymbol{\nu} \mathrm{d}S_t. \end{aligned} \tag{6.2}$$

This is the integral form of the conservation law of energy under the spatial description (see section 5.1 in Chapter 5).

To obtain the differential form of (6.2), similarly to Chapter 5, we need to reduce the surface integral on its right-hand side to the volume integral on the domain enclosed by this surface. Assume that the integrands under discussion are smooth. Then by Green's formula we have

$$\int_{S_t} (\boldsymbol{T}\boldsymbol{\nu}) \cdot \boldsymbol{v} \mathrm{d}S_t = \int_{G_t} \mathrm{div}_y(\boldsymbol{T}\boldsymbol{v}) \mathrm{d}y, \tag{6.3}$$

$$\int_{S_t} \boldsymbol{q} \cdot \boldsymbol{\nu} \mathrm{d}S_t = \int_{G_t} \mathrm{div}_y \boldsymbol{q} \mathrm{d}y, \tag{6.4}$$

where div_y stands for the divergence with respect to $\boldsymbol{y} = (y_1, y_2, y_3)$. We also used the symmetry of the Cauchy stress tensor $\boldsymbol{T}$ when deriving (6.3) (see Theorem 5.2 in Chapter 5).

Using (6.3) and (6.4), it follows from (6.2) that

$$\begin{aligned} &\int_{G_t} \rho \frac{\mathrm{d}}{\mathrm{d}t} \left(\frac{1}{2}|\boldsymbol{v}|^2 + e \right) \mathrm{d}y \\ = &\int_{G_t} (\mathrm{div}_y(\boldsymbol{T}\boldsymbol{v}) - \mathrm{div}_y \boldsymbol{q} + \rho \boldsymbol{b} \cdot \boldsymbol{v} + \rho \gamma) \mathrm{d}y. \end{aligned} \tag{6.5}$$

Since (6.5) holds for any given domain $G_t \subset \Omega_t$, we have

$$\rho \frac{\mathrm{d}}{\mathrm{d}t} \left(\frac{1}{2}|\boldsymbol{v}|^2 + e \right) = \mathrm{div}_y(\boldsymbol{T}\boldsymbol{v}) - \mathrm{div}_y \boldsymbol{q} + \rho \boldsymbol{b} \cdot \boldsymbol{v} + \rho \gamma. \tag{6.6}$$

This is exactly the differential form of the conservation law of energy under the spatial description.

Obviously, we have

$$\frac{\mathrm{d}}{\mathrm{d}t} \left(\frac{1}{2}|\boldsymbol{v}|^2 \right) = \frac{\mathrm{d}\boldsymbol{v}}{\mathrm{d}t} \cdot \boldsymbol{v}$$

and

$$\mathrm{div}_y(\boldsymbol{T}\boldsymbol{v}) = \mathrm{div}_y \boldsymbol{T} \cdot \boldsymbol{v} + \sum_{i,j=1}^{3} t_{ij} \frac{\partial v_i}{\partial y_j}, \tag{6.7}$$

where t_{ij} are components of $\boldsymbol{T}$ and $\mathrm{div}_y \boldsymbol{T} = \left(\sum_{j=1}^{3} \frac{\partial t_{ij}}{\partial y_j} \right)$ is a vector (see Appendix A in Volume I). We used the symmetry of $\boldsymbol{T}$ again when obtaining (6.7). Thus, using the differential form of the conservation laws of momentum obtained in Chapter 5 (see (5.57) in Chapter 5),

$$\rho \frac{\mathrm{d}\boldsymbol{v}}{\mathrm{d}t} - \mathrm{div}_y \boldsymbol{T} - \rho \boldsymbol{b} = 0, \tag{6.8}$$

equation (6.6) can also be rewritten in the following simpler form:

$$\rho \frac{\mathrm{d}e}{\mathrm{d}t} = \sum_{i,j=1}^{3} t_{ij} \frac{\partial v_i}{\partial y_j} - \mathrm{div}_y \boldsymbol{q} + \rho \gamma. \tag{6.9}$$

This is another expression of the differential form of the conservation law of energy under the spatial description.

Just as in elasticity, the form of the above equations under the material description (see section 5.1 in Chapter 5) is the most common one used in thermoelasticity. Let the domain $G_0 \subset \Omega$ in the reference configuration correspond to $G_t \subset \Omega_t$. In order to obtain the conservation equation of energy under the material description, we need to reduce the volume integral with respect to variables (y_1, y_2, y_3) over G_t in (6.2) to the volume integral with respect to variables (x_1, x_2, x_3) over G_0, and the surface integral over S_t into the surface integral over the boundary S_0 of G_0. The former can easily be done by the usual substitution of integral variables and by noticing that

$$\rho J = \rho_0 \tag{6.10}$$

(see (5.34) in Chapter 5), where $J = \det \boldsymbol{F}$, and $\boldsymbol{F} = (f_{ij})$, $f_{ij} = \frac{\partial y_i}{\partial x_j}$, while ρ_0 represents the mass density of the thermoelastic body before the deformation (depending only on x, and independent of t). For example,

$$\begin{aligned}
&\int_{G_t} \rho \frac{\mathrm{d}}{\mathrm{d}t}\left(\frac{1}{2}|\boldsymbol{v}|^2 + e\right)\mathrm{d}y \\
&= \int_{G_0} \rho \frac{\mathrm{d}}{\mathrm{d}t}\left(\frac{1}{2}|\boldsymbol{v}|^2 + e\right) J\mathrm{d}x \\
&= \int_{G_0} \rho_0 \frac{\mathrm{d}}{\mathrm{d}t}\left(\frac{1}{2}|\boldsymbol{v}|^2 + e\right)\mathrm{d}x \\
&= \int_{G_0} \frac{\partial}{\partial t}\left(\rho_0 e + \frac{1}{2}\rho_0|\boldsymbol{v}|^2\right)\mathrm{d}x.
\end{aligned} \tag{6.11}$$

Similarly, we have

$$\int_{G_t} \rho \boldsymbol{b}\cdot\boldsymbol{v}\mathrm{d}y = \int_{G_0} \rho_0 \boldsymbol{b}\cdot\boldsymbol{v}\mathrm{d}x, \tag{6.12}$$

$$\int_{G_t} \rho\gamma\mathrm{d}y = \int_{G_0} \rho_0\gamma\mathrm{d}x. \tag{6.13}$$

Now we inspect the corresponding change of the surface integral over S_t in (6.2). Noting that

$$\boldsymbol{\nu}\mathrm{d}S_t = J\boldsymbol{F}^{-\mathrm{T}}\boldsymbol{n}\mathrm{d}S_0 \tag{6.14}$$

(see (5.63) of Lemma 5.3 in Chapter 5), where $\boldsymbol{n} = (n_1, n_2, n_3)$ is the unit outward normal vector of S_0, and using the symmetry of $\boldsymbol{T}$, we have

$$\begin{aligned}
\int_{S_t} (\boldsymbol{T}\boldsymbol{\nu})\cdot\boldsymbol{v}\mathrm{d}S_t &= \int_{S_t} (\boldsymbol{T}\boldsymbol{v})\cdot\boldsymbol{\nu}\mathrm{d}S_t \\
&= \int_{S_0} J(\boldsymbol{T}\boldsymbol{v})\cdot(\boldsymbol{F}^{-\mathrm{T}}\boldsymbol{n})\mathrm{d}S_0 \\
&= \int_{S_0} \sum_{i,j=1}^{3} p_{ij} v_i n_j \mathrm{d}S_0,
\end{aligned} \tag{6.15}$$

where p_{ij} are the components of the Piola stress tensor

$$\boldsymbol{P} = J\boldsymbol{T}\boldsymbol{F}^{-\mathrm{T}} \tag{6.16}$$

(see (5.69) in Chapter 5). Similarly, it is easy to see that

$$\int_{S_t} \boldsymbol{q}\cdot\boldsymbol{\nu}\mathrm{d}S_t = \int_{S_0} \boldsymbol{h}\cdot\boldsymbol{n}\mathrm{d}S_0, \tag{6.17}$$

where

$$\boldsymbol{h} = J\boldsymbol{F}^{-1}\boldsymbol{q}. \tag{6.18}$$

It is not hard to see that $\boldsymbol{q}\cdot\boldsymbol{\nu}$ stands for the heat across unit area on S_t along $\boldsymbol{\nu}$ in unit time, while $\boldsymbol{h}\cdot\boldsymbol{n}$ is the heat flux in unit time measured by unit area before deformation (i.e., on S_0). Plugging (6.11)–(6.13), (6.15), and (6.17) into (6.2), we then get

$$\begin{aligned}&\int_{G_0} \frac{\partial}{\partial t}\left(\rho_0 e + \frac{1}{2}\rho_0|\boldsymbol{v}|^2\right)\mathrm{d}x \\ &= \int_{S_0}\sum_{i,j=1}^{3} p_{ij}v_i n_j \mathrm{d}S_0 + \int_{G_0}\rho_0\boldsymbol{b}\cdot\boldsymbol{v}\mathrm{d}x \\ &+ \int_{G_0}\rho_0\gamma\mathrm{d}x - \int_{S_0}\boldsymbol{h}\cdot\boldsymbol{n}\mathrm{d}S_0.\end{aligned} \tag{6.19}$$

This is just the integral form of the conservation law of energy under the material description.

To obtain the differential form of (6.19), under the assumption that the integrands are smooth, we reduce the surface integrals on its right-hand side to the volume integrals over G_0 by using Green's formula:

$$\int_{S_0}\sum_{i,j=1}^{3} p_{ij}v_i n_j \mathrm{d}S_0 = \int_{G_0}\sum_{i,j=1}^{3}\frac{\partial}{\partial x_j}(p_{ij}v_i)\mathrm{d}x, \tag{6.20}$$

$$\int_{S_0}\boldsymbol{h}\cdot\boldsymbol{n}\mathrm{d}S_0 = \int_{G_0}\operatorname{div}\boldsymbol{h}\mathrm{d}x, \tag{6.21}$$

where div denotes the divergence with respect to $\boldsymbol{x} = (x_1, x_2, x_3)$. Plugging (6.20) and (6.21) into (6.19), we obtain

$$\begin{aligned}&\int_{G_0} \frac{\partial}{\partial t}\left(\rho_0 e + \frac{1}{2}\rho_0|\boldsymbol{v}|^2\right)\mathrm{d}x \\ &= \int_{G_0}\left(\sum_{i,j=1}^{3}\frac{\partial}{\partial x_j}(p_{ij}v_i) - \operatorname{div}\boldsymbol{h} + \rho_0\boldsymbol{b}\cdot\boldsymbol{v} + \rho_0\gamma\right)\mathrm{d}x.\end{aligned} \tag{6.22}$$

Since (6.22) holds for any given domain $G_0 \subset \Omega$, we have

$$\frac{\partial}{\partial t}\left(\rho_0 e + \frac{1}{2}\rho_0|\boldsymbol{v}|^2\right) = \sum_{i,j=1}^{3} \frac{\partial}{\partial x_j}(p_{ij}v_i) - \operatorname{div}\boldsymbol{h} + \rho_0\boldsymbol{b}\cdot\boldsymbol{v} + \rho_0\gamma. \tag{6.23}$$

This is exactly the differential form of the conservation law of energy under the material description. It is a conservation equation of divergence form.

Using the conservation equations of momentum,

$$\rho_0\frac{\partial v_i}{\partial t} = \sum_{j=1}^{3}\frac{\partial p_{ij}}{\partial x_j} + \rho_0 b_i \quad (i = 1,2,3), \tag{6.24}$$

obtained in Chapter 5 (see (5.74) in Chapter 5), it is easy to see that (6.23) can also be written in the following simpler equivalent form:

$$\rho_0\frac{\partial e}{\partial t} = \sum_{i,j=1}^{3} p_{ij}\frac{\partial v_i}{\partial x_j} - \operatorname{div}\boldsymbol{h} + \rho_0\gamma. \tag{6.25}$$

This is another expression of the conservation law of energy under the material description.

6.2.2 Entropy Inequality

Let η be the entropy density, i.e., the entropy per unit mass. Now we consider the change of entropy in any given domain $G_t \subset \Omega_t$. From the second law of thermodynamics (see Appendix B in Volume I) we know that the increment of entropy in G_t in unit time is not less than the sum of the entropy supplied by the entropy source in G_t and the entropy flowing into G_t across the boundary S_t of G_t during this time period.

The total entropy in G_t is

$$\int_{G_t} \rho\eta \mathrm{d}y.$$

The entropy source in G_t is exactly the heat source. It is known that the heat produced by the volume element $\mathrm{d}y = \mathrm{d}y_1\mathrm{d}y_2\mathrm{d}y_3$ in G_t in unit time is $\rho\gamma\mathrm{d}y$, the entropy supplied by which is $\rho\gamma\mathrm{d}y/\theta$, where θ is the absolute temperature of the element $\mathrm{d}y$ (see (B.4) in Appendix B in Volume I). Then the entropy supplied by the entropy source in G_t in unit time is

$$\int_{G_t} \frac{\rho\gamma}{\theta}\mathrm{d}y.$$

Moreover, it is known that the heat flowing into G_t across the boundary area element $\mathrm{d}S_t$ in unit time is $-\boldsymbol{q}\cdot\boldsymbol{\nu}\mathrm{d}S_t$, and thus the entropy is $-\boldsymbol{q}\cdot\boldsymbol{\nu}\mathrm{d}S_t/\theta$. Hence, the entropy flowing into G_t across S_t in unit time is

$$-\int_{S_t}\frac{\boldsymbol{q}\cdot\boldsymbol{\nu}}{\theta}\mathrm{d}S_t.$$

From the above analysis and the second law of thermodynamics, we obtain

$$\frac{\mathrm{d}}{\mathrm{d}t}\int_{G_t}\rho\eta\mathrm{d}y \geqslant \int_{G_t}\frac{\rho\gamma}{\theta}\mathrm{d}y - \int_{S_t}\frac{\boldsymbol{q}\cdot\boldsymbol{\nu}}{\theta}\mathrm{d}S_t. \tag{6.26}$$

Applying Lemma 5.2 in Chapter 5 to the left-hand side of the above formula, (6.26) can be rewritten as

$$\int_{G_t}\rho\frac{\mathrm{d}\eta}{\mathrm{d}t}\mathrm{d}y \geqslant \int_{G_t}\frac{\rho\gamma}{\theta}\mathrm{d}y - \int_{S_t}\frac{\boldsymbol{q}\cdot\boldsymbol{\nu}}{\theta}\mathrm{d}S_t. \tag{6.27}$$

This is exactly the integral form of the entropy inequality under the spatial description.

Under the assumption that the integrand is smooth, reducing the surface integral on the right-hand side of (6.27) into the volume integral over G_t by Green's formula, (6.27) can then be written in the following form:

$$\int_{G_t}\rho\frac{\mathrm{d}\eta}{\mathrm{d}t}\mathrm{d}y \geqslant \int_{G_t}\left(\frac{\rho\gamma}{\theta} - \mathrm{div}_y\left(\frac{\boldsymbol{q}}{\theta}\right)\right)\mathrm{d}y. \tag{6.28}$$

Since (6.28) holds for any given domain $G_t \subset \Omega_t$, we then have

$$\rho\frac{\mathrm{d}\eta}{\mathrm{d}t} \geqslant \frac{\rho\gamma}{\theta} - \mathrm{div}_y\left(\frac{\boldsymbol{q}}{\theta}\right). \tag{6.29}$$

This is exactly the differential form of the entropy inequality under the spatial description.

To obtain the entropy inequality under the material description, we need to reduce the volume integral and the surface integral in (6.27) to the volume integral over G_0 and the surface integral on S_0, respectively. Similarly to what we have done to the conservation equation of energy, (6.27) can be reduced to

$$\int_{G_0}\rho_0\frac{\partial\eta}{\partial t}\mathrm{d}x \geqslant \int_{G_0}\frac{\rho_0\gamma}{\theta}\mathrm{d}x - \int_{S_0}\frac{\boldsymbol{h}\cdot\boldsymbol{n}}{\theta}\mathrm{d}S_0, \tag{6.30}$$

where $\boldsymbol{h}$ is defined by (6.18). This is just the integral form of the entropy inequality under the material description.

Reducing the surface integral on S_0 on the right-hand side of (6.30) to the volume integral over G_0 by Green's formula, (6.30) can also be rewritten as

$$\int_{G_0}\rho_0\frac{\partial\eta}{\partial t}\mathrm{d}x \geqslant \int_{G_0}\left(\frac{\rho_0\gamma}{\theta} - \mathrm{div}\left(\frac{\boldsymbol{h}}{\theta}\right)\right)\mathrm{d}x. \tag{6.31}$$

Since (6.31) holds for any given domain $G_0 \subset \Omega$, we then have

$$\rho_0\frac{\partial\eta}{\partial t} \geqslant \frac{\rho_0\gamma}{\theta} - \mathrm{div}\left(\frac{\boldsymbol{h}}{\theta}\right). \tag{6.32}$$

This is just the differential form of the second law of thermodynamics under the material description—the *entropy inequality*, also called the *Clausius–Duhem inequality*.

6.3 Constitutive Relations in Thermoelasticity

6.3.1 Constitutive Relations; Free Energy

For the thermoelastic body, the volume force density $\boldsymbol{b}$ and the heat source density γ are given as known quantities. However, according to the characteristics of the thermoelastic body described in section 6.1, the Cauchy stress tensor $\boldsymbol{T}$ (and consequently the Piola stress tensor $\boldsymbol{P}$), the internal energy density e, the heat flux density vector $\boldsymbol{q}$ (and consequently $\boldsymbol{h}$), and the entropy density η should be expressed as functions of deformation (depicted by the deformation gradient tensor $\boldsymbol{F}$), absolute temperature θ, and temperature gradient $\nabla\theta = \left(\frac{\partial\theta}{\partial x_1}, \frac{\partial\theta}{\partial x_2}, \frac{\partial\theta}{\partial x_3}\right)$, i.e.,

$$\boldsymbol{T} = \widehat{\boldsymbol{T}}(\boldsymbol{F}, \theta, \nabla\theta), \tag{6.33}$$

$$e = \widehat{e}(\boldsymbol{F}, \theta, \nabla\theta), \tag{6.34}$$

$$\boldsymbol{q} = \widehat{\boldsymbol{q}}(\boldsymbol{F}, \theta, \nabla\theta), \tag{6.35}$$

$$\eta = \widehat{\eta}(\boldsymbol{F}, \theta, \nabla\theta). \tag{6.36}$$

These relationships are called the *constitutive relations* or *constitutive equations* of the thermoelastic body. The specific forms of functions $\widehat{\boldsymbol{T}}$, $\widehat{e}$, $\widehat{\boldsymbol{q}}$, and $\widehat{\eta}$ in the above constitutive equations are completely determined by the material under discussion. If the functions $\widehat{\boldsymbol{T}}$, $\widehat{e}$, $\widehat{\boldsymbol{q}}$, and $\widehat{\eta}$ on the right-hand side of the constitutive equations (6.33)–(6.36) do not depend explicitly on $\boldsymbol{x}$, then this material is called *homogeneous*; otherwise it is *inhomogeneous*. For narrative simplicity, hereafter we assume that all materials under discussion are homogeneous, although most of the following discussions are still valid for inhomogeneous thermoelastic materials in principle.

When we discussed elasticity in Chapter 5, in addition to expressing the constitutive relations by using the stress as a function of the deformation gradient, the constitutive relations of hyperelastic material could also be given by using the stored-energy function $W = \widehat{W}(\boldsymbol{F})$. When we derived the conservation equation of energy for the thermoelastic body, we replaced the stored-energy function W by the internal energy density e (actually, by ρe). So, for thermoelastic material, is it possible to obtain the constitutive relations only by giving the internal energy density e? Through careful study, we find that this idea is at least incomplete. In fact, the internal energy of a thermoelastic body includes the heat energy. Thermodynamics tells us that it is impossible for the heat energy to be completely transformed into mechanical work, and the thermoelastic body can be deformed only by the part of the energy involving mechanical work. From thermodynamics (see Appendix B in Volume I) we know that in the case of constant temperature, the Helmholtz free energy measures how much mechanical work the medium can do. Therefore, instead of the internal energy density e, we can take the *Helmholtz free energy* per unit mass

$$\psi = e - \theta\eta \tag{6.37}$$

to give the constitutive relations. Although here e is not just the thermodynamic internal energy, the above Helmholtz free energy can still be used as a measurement for the ability of the medium (at constant temperature) to do mechanical work since the strain energy can be completely converted into mechanical work.

Now we show that if the constitutive equation of the Helmholtz free energy is given by

$$\psi = \widehat{\psi}(\boldsymbol{F}, \theta, \nabla\theta), \tag{6.38}$$

we can accordingly derive equations for $\boldsymbol{P}$ (and thus $\boldsymbol{T}$), η, and e in the constitutive relations.

Substituting $e = \psi + \theta\eta$ into the conservation equation (6.25) of energy, and noticing that $v_i = \frac{\partial y_i}{\partial t}$ and $f_{ij} = \frac{\partial y_i}{\partial x_j}$, we can get

$$\rho_0 \frac{\partial \psi}{\partial t} + \rho_0 \eta \frac{\partial \theta}{\partial t} + \rho_0 \theta \frac{\partial \eta}{\partial t} = \sum_{i,j=1}^{3} p_{ij} \frac{\partial f_{ij}}{\partial t} - \operatorname{div} \boldsymbol{h} + \rho_0 \gamma. \tag{6.39}$$

Differentiating (6.38) with respect to t gives

$$\frac{\partial \psi}{\partial t} = \sum_{i,j=1}^{3} \frac{\partial \widehat{\psi}}{\partial f_{ij}} \frac{\partial f_{ij}}{\partial t} + \frac{\partial \widehat{\psi}}{\partial \theta} \frac{\partial \theta}{\partial t} + \sum_{k=1}^{3} \frac{\partial \widehat{\psi}}{\partial (\frac{\partial \theta}{\partial x_k})} \frac{\partial}{\partial t} \left(\frac{\partial \theta}{\partial x_k} \right). \tag{6.40}$$

Eliminating $\frac{\partial \psi}{\partial t}$ in (6.39) and (6.40), we get

$$\begin{aligned} &\sum_{i,j=1}^{3} \left(p_{ij} - \rho_0 \frac{\partial \widehat{\psi}}{\partial f_{ij}} \right) \frac{\partial f_{ij}}{\partial t} - \rho_0 \left(\eta + \frac{\partial \widehat{\psi}}{\partial \theta} \right) \frac{\partial \theta}{\partial t} \\ &- \rho_0 \sum_{k=1}^{3} \frac{\partial \widehat{\psi}}{\partial (\frac{\partial \theta}{\partial x_k})} \frac{\partial}{\partial t} \left(\frac{\partial \theta}{\partial x_k} \right) + \rho_0 \left(\gamma - \theta \frac{\partial \eta}{\partial t} \right) - \operatorname{div} \boldsymbol{h} = 0. \end{aligned} \tag{6.41}$$

Equation (6.41) is true for any $\frac{\partial f_{ij}}{\partial t}$, $\frac{\partial \theta}{\partial t}$, and $\frac{\partial}{\partial t}\left(\frac{\partial \theta}{\partial x_k}\right)$, so the corresponding coefficients should be zero. Thus we have

$$p_{ij} = \rho_0 \frac{\partial \widehat{\psi}}{\partial f_{ij}}, \tag{6.42}$$

$$\eta = -\frac{\partial \widehat{\psi}}{\partial \theta}, \tag{6.43}$$

$$\frac{\partial \widehat{\psi}}{\partial (\frac{\partial \theta}{\partial x_k})} = 0, \tag{6.44}$$

and

$$\rho_0 \left(\gamma - \theta \frac{\partial \eta}{\partial t} \right) - \operatorname{div} \boldsymbol{h} = 0. \tag{6.45}$$

Equations (6.42) and (6.43) are exactly the constitutive equations for the Piola stress tensor $\boldsymbol{P}$ and the entropy density η, given by the free energy function $\widehat{\psi}$. Thus, the constitutive equation of e follows from $e = \psi + \theta\eta$ as well as (6.38) and (6.43). And (6.44) implies that the free energy function $\widehat{\psi}$ is independent of the temperature gradient $\nabla\theta$; that is, $\widehat{\psi}$ is a function of only $\boldsymbol{F}$ and θ. Consequently, the Piola stress tensor $\boldsymbol{P}$ and the entropy density η, as well as the internal energy density e, are also independent of the temperature gradient $\nabla\theta$ and depend only on $\boldsymbol{F}$ and θ. Equation (6.45) describes the heat conduction phenomenon, which is actually another expression of the conservation equation of energy.

From the above discussion we can see that it is enough to give the constitutive equations for the free energy ψ and the heat flux density vector $\boldsymbol{q}$ in the constitutive relations of the thermoelastic body. Furthermore, the constitutive equation for the free energy ψ must be of the following form:

$$\psi = \widehat{\psi}(\boldsymbol{F},\theta). \tag{6.46}$$

Of course, the thermoelastic constitutive relations should also satisfy the objectivity hypothesis (see section 5.4 in Chapter 5). Just like the stored-energy function W, the Helmholtz free energy ψ is a scalar whose objectivity hypothesis takes the form

$$\widehat{\psi}(\boldsymbol{QF},\theta) = \widehat{\psi}(\boldsymbol{F},\theta) \tag{6.47}$$

for any given orthogonal matrix $\boldsymbol{Q}$ (see (5.93) in Chapter 5). Regarding the objectivity hypothesis for the constitutive relation (6.35) of the heat flux density vector $\boldsymbol{q}$, we omit the details here.

Using the polar decomposition theorem (see Lemma 5.1 in Chapter 5), $\boldsymbol{F}$ can be represented in the form $\boldsymbol{F} = \boldsymbol{RU}$, where $\boldsymbol{R}$ is an orthogonal matrix and $\boldsymbol{U}$ is a positive definite symmetric matrix. Taking $\boldsymbol{Q} = \boldsymbol{R}^{\mathrm{T}}$ in (6.47), we have

$$\widehat{\psi}(\boldsymbol{F},\theta) = \widehat{\psi}(\boldsymbol{U},\theta) = \overline{\psi}(\boldsymbol{C},\theta), \tag{6.48}$$

where $\boldsymbol{C} = \boldsymbol{F}^{\mathrm{T}}\boldsymbol{F} = \boldsymbol{U}^2$ is the right Cauchy–Green strain tensor and $\overline{\psi}(\boldsymbol{C},\theta) = \widehat{\psi}(\boldsymbol{C}^{\frac{1}{2}},\theta)$. This shows that the Helmholtz free energy ψ is actually a function of only the right Cauchy–Green strain tensor $\boldsymbol{C}$ and the absolute temperature θ. Using (6.42) and (6.48), it is not difficult to directly verify that the Piola stress tensor $\boldsymbol{P}$ can be represented in the form

$$\boldsymbol{P} = \boldsymbol{F}\overline{\boldsymbol{P}}(\boldsymbol{C},\theta), \tag{6.49}$$

where $\overline{\boldsymbol{P}}$ is a second-order tensor function depending only on $\boldsymbol{C}$ and θ. Equation (6.49), like (5.89) in Chapter 5, can also be obtained from the expression of the Cauchy stress tensor $\boldsymbol{T}$.

In addition, if the constitutive relations (6.46) and (6.35) of the Helmholtz free energy and the heat flux density vector satisfy

$$\widehat{\psi}(\boldsymbol{FQ},\theta) = \widehat{\psi}(\boldsymbol{F},\theta), \tag{6.50}$$

$$\widehat{q}(\boldsymbol{FQ},\theta,\boldsymbol{Q}^{\mathrm{T}}\nabla\theta) = \widehat{q}(\boldsymbol{F},\theta,\nabla\theta), \tag{6.51}$$

where $\boldsymbol{Q}$ is any given orthogonal matrix, then the corresponding thermoelastic material is said to be *isotropic*; otherwise it is *anisotropic*.

6.3.2 Directionality of Heat Conduction

Next we will use the entropy inequality (6.32) to discuss the directionality of heat conduction.

From (6.45), the entropy inequality (6.32) can be rewritten as

$$\operatorname{div} \boldsymbol{h} - \theta \operatorname{div}\left(\frac{\boldsymbol{h}}{\theta}\right) \leqslant 0. \tag{6.52}$$

Noticing that $\operatorname{div}\left(\frac{\boldsymbol{h}}{\theta}\right) = \frac{1}{\theta}\operatorname{div} \boldsymbol{h} + \boldsymbol{h}\cdot\nabla\left(\frac{1}{\theta}\right)$ and $\nabla\left(\frac{1}{\theta}\right) = -\frac{1}{\theta^2}\nabla\theta$, the above formula implies

$$\boldsymbol{h}\cdot\nabla\theta \leqslant 0. \tag{6.53}$$

This is another expression for the entropy inequality. It depicts the restrictions of the second law of thermodynamics on the direction of heat conduction: the angle between the direction of heat flow and the opposite direction of temperature gradient does not exceed $\pi/2$. This is intuitively consistent with the heat conductivity phenomenon: heat always flows from a place with higher temperature to a place with lower temperature. In fact, the genuine direction of heat flow in a thermoelastic body is $\boldsymbol{q}$, while $\nabla_y\theta$ is the temperature gradient of a thermoelastic body under deformation. But from the definition (6.18) of $\boldsymbol{h}$, and noticing that $\nabla\theta = (\nabla_y\theta)\boldsymbol{F}$ and $J > 0$, it is easy to see that (6.53) is equivalent to the formula

$$\boldsymbol{q}\cdot\nabla_y\theta \leqslant 0. \tag{6.54}$$

This inequality can also be derived from the conservation equation (6.9) of energy and the entropy inequality (6.29) under the spatial description (see Exercise 2).

In addition, from the definition (6.18) of $\boldsymbol{h}$ and the constitutive equation (6.35) of $\boldsymbol{q}$, we may assume that

$$\boldsymbol{h} = \widehat{\boldsymbol{h}}(\boldsymbol{F},\theta,\nabla\theta). \tag{6.55}$$

Thus, from (6.53) we can also get

$$\widehat{\boldsymbol{h}}(\boldsymbol{F},\theta,0) = 0 \tag{6.56}$$

for any given $\boldsymbol{F}$ and θ. In fact, for any given real number λ, denote

$$f(\lambda) = \widehat{\boldsymbol{h}}(\boldsymbol{F},\theta,\lambda\nabla\theta)\cdot\nabla\theta.$$

From (6.53) we have

$$\lambda f(\lambda) \leqslant 0, \quad \forall\lambda \in \mathbb{R}. \tag{6.57}$$

This shows that $f(\lambda)$ changes sign at $\lambda = 0$, and thus $f(0) = 0$, that is,

$$\widehat{\boldsymbol{h}}(\boldsymbol{F},\theta,0)\cdot\nabla\theta = 0.$$

Since the above formula is true for any given $\nabla\theta$, we have (6.56) accordingly.

6.3.3 Linear Thermoelasticity

Suppose that the reference configuration is in the natural state at some constant temperature $\theta = \theta_0$ (> 0), i.e.,

$$\widehat{\boldsymbol{T}}(\boldsymbol{I},\theta_0) = 0, \tag{6.58}$$

where $\widehat{\boldsymbol{T}}$ is the tensor function on the right-hand side of (6.33). Now we examine the very small deformation and the slight change of temperature near the natural state of the thermoelastic body. We always assume

$$|\nabla \boldsymbol{u}| \stackrel{\mathrm{d}}{=} \left(\sum_{i,j=1}^{3} \left| \frac{\partial u_i}{\partial x_j} \right|^2 \right)^{\frac{1}{2}}, \quad \left| \frac{\Theta}{\theta_0} \right|, |\nabla \theta| \leqslant \varepsilon \ll 1, \tag{6.59}$$

where $\boldsymbol{u} = \boldsymbol{y} - \boldsymbol{x}$ and $\Theta = \theta - \theta_0$.

From (6.58) and (6.16), we have

$$\overline{\boldsymbol{P}}(\boldsymbol{I},\theta_0) = 0, \tag{6.60}$$

where $\overline{\boldsymbol{P}}$ is the tensor function on the right-hand side of (6.49). Expanding $\overline{\boldsymbol{P}}(\boldsymbol{C},\theta)$ around $\boldsymbol{C} = \boldsymbol{I}$ and $\theta = \theta_0$, and taking into account (6.59)–(6.60), we get

$$\overline{\boldsymbol{P}}(\boldsymbol{C},\theta) = \frac{1}{2}\boldsymbol{A}(\boldsymbol{C} - \boldsymbol{I}) - \Theta \boldsymbol{G} + O(\varepsilon^2), \tag{6.61}$$

where $\boldsymbol{A} = (a_{ijkl})$ and $\boldsymbol{G} = (g_{ij})$ are fourth-order and second-order tensors, respectively, with components

$$a_{ijkl} = 2\frac{\partial \overline{p}_{ij}}{\partial c_{kl}}\bigg|_{\substack{\boldsymbol{C}=\boldsymbol{I}\\ \theta=\theta_0}}, \tag{6.62}$$

$$g_{ij} = -\frac{\partial \overline{p}_{ij}}{\partial \theta}\bigg|_{\substack{\boldsymbol{C}=\boldsymbol{I}\\ \theta=\theta_0}}, \tag{6.63}$$

where $\overline{p}_{ij}$ and c_{kl} are the components of $\overline{\boldsymbol{P}}$ and $\overline{\boldsymbol{C}}$, respectively, and $\boldsymbol{A}(\boldsymbol{C} - \boldsymbol{I})$ stands for the second-order tensor with components

$$\sum_{k,l=1}^{3} a_{ijkl}(c_{kl} - \delta_{kl}).$$

It is known that $\boldsymbol{F} = \boldsymbol{I} + \nabla \boldsymbol{u}$ and $\boldsymbol{C} - \boldsymbol{I} = 2\boldsymbol{E} + (\nabla \boldsymbol{u})^{\mathrm{T}} \nabla \boldsymbol{u}$, where $\boldsymbol{E}$ is the infinitesimal strain tensor with components $e_{ij} = \frac{1}{2}\left(\frac{\partial u_i}{\partial x_j} + \frac{\partial u_j}{\partial x_i}\right)$ (see section 5.2.3 in Chapter 5). From (6.49) and (6.61), and taking into account (6.59), it is easy to get

$$\boldsymbol{P} = \boldsymbol{A}\boldsymbol{E} - \Theta \boldsymbol{G} + O(\varepsilon^2). \tag{6.64}$$

Expanding the function $\widehat{\eta}$ on the right-hand side of the constitutive equation (6.36) of the entropy density η (from (6.43) and (6.46), $\widehat{\eta}$ is independent of $\nabla\theta$) around $\boldsymbol{F}=\boldsymbol{I}$ and $\theta=\theta_0$, we get

$$\begin{aligned}\eta = &\widehat{\eta}(\boldsymbol{I},\theta_0)+\widehat{\eta}_F(\boldsymbol{I},\theta_0)\cdot(\boldsymbol{F}-\boldsymbol{I})\\ &+\widehat{\eta}_\theta(\boldsymbol{I},\theta_0)\Theta+O(\varepsilon^2),\end{aligned} \tag{6.65}$$

where $\widehat{\eta}_F=\left(\frac{\partial\widehat{\eta}}{\partial f_{ij}}\right)$, and

$$\widehat{\eta}_F(\boldsymbol{I},\theta_0)\cdot(\boldsymbol{F}-\boldsymbol{I})=\sum_{i,j=1}^{3}\frac{\partial\widehat{\eta}}{\partial f_{ij}}(\boldsymbol{I},\theta_0)(f_{ij}-\delta_{ij}). \tag{6.66}$$

For simplicity and without loss of generality, we now take

$$\widehat{\eta}(\boldsymbol{I},\theta_0)=0. \tag{6.67}$$

The right-hand side of the constitutive equation of the heat flux density vector $\boldsymbol{h}$ (see (6.18) and (6.35)) depends on $\boldsymbol{F}$, θ, and $\nabla\theta$; then, expanding it around $\boldsymbol{F}=\boldsymbol{I}$, $\theta=\theta_0$, and $\nabla\theta=0$ and taking into account (6.56), we have

$$\boldsymbol{h}=-\boldsymbol{K}\nabla\theta+O(\varepsilon^2), \tag{6.68}$$

where $\boldsymbol{K}=(k_{ij})$ is a second-order constant tensor.

From (6.42) and (6.43), we have

$$\widehat{\eta}_F=-\frac{\partial\widehat{\psi}_F}{\partial\theta}=-\frac{1}{\rho_0}\frac{\partial\widehat{\boldsymbol{P}}}{\partial\theta}, \tag{6.69}$$

where $\widehat{\psi}_F=\left(\frac{\partial\widehat{\psi}}{\partial f_{ij}}\right)$ and $\widehat{\boldsymbol{P}}$ is the function on the right-hand side of the constitutive equation of the Piola stress tensor $\boldsymbol{P}$: $\boldsymbol{P}=\widehat{\boldsymbol{P}}(\boldsymbol{F},\theta)$. Substituting the expression (6.64) of $\boldsymbol{P}$ into (6.69), and letting $\boldsymbol{F}=\boldsymbol{I}$ and $\theta=\theta_0$, we get

$$\widehat{\eta}_F(\boldsymbol{I},\theta_0)=\frac{1}{\rho_0}\boldsymbol{G}. \tag{6.70}$$

Furthermore, from (6.43) and the definition (6.37) of ψ, we have

$$-\widehat{\eta}=\frac{\partial\widehat{\psi}}{\partial\theta}=\frac{\partial\widehat{e}}{\partial\theta}-\widehat{\eta}-\theta\frac{\partial\widehat{\eta}}{\partial\theta};$$

therefore

$$\widehat{\eta}_\theta=\frac{1}{\theta}\widehat{e}_\theta. \tag{6.71}$$

Substituting (6.67), (6.70), and (6.71) into (6.65), we have

$$\eta=\frac{1}{\rho_0}\boldsymbol{G}\cdot(\boldsymbol{F}-\boldsymbol{I})+\frac{1}{\theta_0}\widehat{e}_\theta(\boldsymbol{I},\theta_0)\Theta+O(\varepsilon^2). \tag{6.72}$$

Ignoring the higher-order terms, i.e., after linearization in (6.64), (6.72), and (6.68), we get

$$p_{ij} = \sum_{k,l=1}^{3} a_{ijkl}e_{kl} - g_{ij}\Theta \quad (i,j = 1,2,3), \tag{6.73}$$

$$\eta = \sum_{i,j=1}^{3} \frac{1}{\rho_0} g_{ij}\frac{\partial u_i}{\partial x_j} + \frac{a}{\rho_0}\Theta, \tag{6.74}$$

and

$$h_i = -\sum_{j=1}^{3} k_{ij}\frac{\partial \Theta}{\partial x_j} \quad (i = 1,2,3), \tag{6.75}$$

where $a = \frac{\rho_0}{\theta_0}\widehat{e}_\theta(\boldsymbol{I},\theta_0)$. Just like the case of a purely elastic body, the fourth-order tensor $A = (a_{ijkl})$ has the following symmetry:

$$a_{ijkl} = a_{klij} = a_{ijlk} = a_{jikl} \tag{6.76}$$

(see (5.133) in Chapter 5). $\boldsymbol{P}$ and $\boldsymbol{T}$ are the same after linearization, so they also have this symmetry, and the second-order tensor $\boldsymbol{G} = (g_{ij})$ on the right-hand side of (6.73) is also symmetric:

$$g_{ij} = g_{ji} \quad (i,j = 1,2,3). \tag{6.77}$$

Moreover, we can usually assume physically that

$$\widehat{e}_\theta(\boldsymbol{I},\theta_0) > 0, \tag{6.78}$$

and then we have $a > 0$.

Equations (6.73), (6.74), and (6.75) are just the constitutive relations of linear thermoelasticity. Equation (6.75) is a more general form of the Fourier experimental law for heat conduction, where $\boldsymbol{K} = (k_{ij})$ is called the *heat conduction tensor* or *heat conduction matrix*.

The constraints for the heat conduction tensor $\boldsymbol{K}$ can be obtained from the second law of thermodynamics. Substituting (6.75) into (6.53), we have

$$\sum_{i,j=1}^{3} k_{ij}\frac{\partial \Theta}{\partial x_i}\frac{\partial \Theta}{\partial x_j} \geqslant 0. \tag{6.79}$$

This shows that the *heat conduction matrix* $\boldsymbol{K}$ is positive semidefinite, namely, $\boldsymbol{K} + \boldsymbol{K}^{\mathrm{T}}$ is a positive semidefinite symmetric matrix.

For the isothermal case, $\Theta \equiv 0$, the linear thermoelasticity is reduced to the purely elastic problem depending only on the constant temperature parameter θ_0. On this occasion, the constitutive equation (6.73) is reduced to the generalized Hooke's law (see (5.132) in Chapter 5)

$$p_{ij} = \sum_{k,l=1}^{3} a_{ijkl}e_{kl}. \tag{6.80}$$

Therefore, if the material is isotropic, the first term on the right-hand side of the constitutive equation (6.73) takes the following form:

$$\sum_{k,l=1}^{3} a_{ijkl}e_{kl} = \lambda(e_{11}+e_{22}+e_{33})\delta_{ij} + 2\mu e_{ij} \qquad (i,j=1,2,3), \tag{6.81}$$

where λ and μ are Lamé constants (see (5.134) in Chapter 5).

For the case of full constraint (i.e., $e_{ij} \equiv 0,\ i,j=1,2,3$), the constitutive equation (6.73) is reduced to

$$p_{ij} = -g_{ij}\Theta \quad (i,j=1,2,3). \tag{6.82}$$

Therefore, for an isotropic material, $\boldsymbol{G}=(g_{ij})$ should be an isotropic second-order tensor. From the general form of isotropic second-order tensors (see Appendix A in Volume I), we have

$$g_{ij} = \beta\delta_{ij} \quad (i,j=1,2,3), \tag{6.83}$$

where β is a scalar.

Moreover, for an isotropic material, the heat conduction tensor $\boldsymbol{K}=(k_{ij})$ should also be an isotropic second-order tensor, taking the form

$$k_{ij} = k\delta_{ij} \quad (i,j=1,2,3), \tag{6.84}$$

where $k>0$ is a scalar.

Combining (6.81), (6.83), and (6.84), for isotropic materials, the constitutive equations (6.73), (6.74), and (6.75) of linear thermoelasticity should take the following forms:

$$p_{ij} = \lambda(e_{11}+e_{22}+e_{33})\delta_{ij} + 2\mu e_{ij} - \beta\Theta\delta_{ij} \qquad (i,j=1,2,3) \tag{6.85}$$

or

$$e_{ij} = \frac{1}{2\mu}p_{ij} - \frac{\lambda}{2\mu(3\lambda+2\mu)}(p_{11}+p_{22}+p_{33})\delta_{ij} + \alpha\Theta\delta_{ij} \qquad (i,j=1,2,3), \tag{6.86}$$

$$\eta = \frac{\beta}{\rho_0}(e_{11}+e_{22}+e_{33}) + \frac{a}{\rho_0}\Theta, \tag{6.87}$$

and

$$h_i = -k\frac{\partial\Theta}{\partial x_i} \quad (i=1,2,3), \tag{6.88}$$

where $\alpha = \frac{\beta}{3\lambda+2\mu}$. Equation (6.88) is just the ordinary Fourier experimental law of heat conduction.

It can be seen from (6.86) that α represents the increment of normal strain, caused by a 1-degree increase of temperature when the stress remains unchanged, which is called the *linear expansion coefficient*. Most thermoelastic materials expand when heated and contract when cooled (but there are exceptions, e.g., rubber); for such materials, the linear expansion coefficient is positive, i.e., $\alpha>0$. According to section 5.4 in Chapter 5, the Lamé constants λ and μ should satisfy $\kappa = \lambda+\frac{2}{3}\mu>0$. Therefore, when $\alpha>0$, the β appearing on the right-hand side of (6.85) is also positive: $\beta>0$.

6.4 System of Thermoelastodynamics and Its Mathematical Structure

6.4.1 System of Linear Thermoelastodynamics

Substituting the constitutive equations (6.73) of linear thermoelasticity into the conservation equations (6.24) of momentum, and using (6.76), we get

$$\rho_0 \frac{\partial^2 u_i}{\partial t^2} = \sum_{j,k,l=1}^{3} a_{ijkl} \frac{\partial^2 u_k}{\partial x_j \partial x_l} - \sum_{j=1}^{3} g_{ij} \frac{\partial \Theta}{\partial x_j} + \rho_0 b_i \quad (i = 1,2,3). \tag{6.89}$$

Substituting the constitutive equations (6.74) and (6.75) into the conservation equation (6.45) of energy, we have

$$a\theta \frac{\partial \Theta}{\partial t} - \sum_{i,j=1}^{3} k_{ij} \frac{\partial^2 \Theta}{\partial x_i \partial x_j} + \theta \sum_{i,j=1}^{3} g_{ij} \frac{\partial^2 u_i}{\partial t \partial x_j} = \rho_0 \gamma. \tag{6.90}$$

This is still a nonlinear equation which, by ignoring the higher-order infinitesimals, i.e., replacing with θ_0 those θ appearing in the first and third terms on the right-hand side, can be reduced to the following linear equation:

$$a\theta_0 \frac{\partial \Theta}{\partial t} - \sum_{i,j=1}^{3} k_{ij} \frac{\partial^2 \Theta}{\partial x_i \partial x_j} + \theta_0 \sum_{i,j=1}^{3} g_{ij} \frac{\partial^2 u_i}{\partial t \partial x_j} = \rho_0 \gamma. \tag{6.91}$$

Equations (6.89) and (6.91) constitute the *system of linear thermoelastodynamics*. It is a closed second-order linear system of partial differential equations involving four unknown functions (u_1, u_2, u_3, Θ). It should be pointed out that we assumed only that $|\nabla \boldsymbol{u}|$, $|\Theta|$, and $|\nabla \theta|$ are small (see (6.59)) when deriving the constitutive relations (6.73)–(6.75) of linear thermoelasticity in the last section, but we have further assumed that $\left|\nabla \frac{\partial \boldsymbol{u}}{\partial t}\right|$ and $\left|\frac{\partial \theta}{\partial t}\right|$ are also small when deriving equation (6.91) for the dynamics problem.

Now we discuss the mathematical structure of this system. We assume that the elastic tensor $\boldsymbol{A} = (a_{ijkl})$ satisfies the strong ellipticity condition (see Definition 5.1 in Chapter 5), that is, there exists a constant $\alpha > 0$ such that

$$\sum_{i,j,k,l=1}^{3} a_{ijkl} \xi_i \xi_k \eta_j \eta_l \geqslant \alpha |\boldsymbol{\xi}|^2 |\boldsymbol{\eta}|^2, \quad \forall \boldsymbol{\xi}, \boldsymbol{\eta} \in \mathbb{R}^3. \tag{6.92}$$

Thus, system (6.89) is hyperbolic with respect to the unknown functions $\boldsymbol{u} = (u_1, u_2, u_3)$. We also assume that the heat conduction matrix $\boldsymbol{K} = (k_{ij})$ contained in equation (6.91) is positive definite, i.e., there exists a constant $k > 0$ such that

$$\sum_{i,j=1}^{3} k_{ij} \xi_i \xi_j \geqslant k |\boldsymbol{\xi}|^2, \quad \forall \boldsymbol{\xi} \in \mathbb{R}^3. \tag{6.93}$$

This requirement is a little bit stronger than the positive semidefiniteness obtained in the last section (see (6.79)); however, it is still reasonable for the heat transfer process of many common thermoelastic materials. Under the condition that (6.93) is satisfied, equation (6.91) is parabolic with respect to the unknown function Θ.

System (6.89) is hyperbolic with respect to the unknown functions (u_1, u_2, u_3), while equation (6.91) is parabolic with respect to the unknown function Θ; they constitute a *second-order linear hyperbolic-parabolic coupled system* (6.89) and (6.91). We have seen the hyperbolic-parabolic coupled system when discussing the system of viscous heat-conducting fluid dynamics in Chapter 2, but the hyperbolic system there is a first-order symmetric hyperbolic system instead of a second-order hyperbolic system. Now the well-posed initial and boundary conditions for this kind of hyperbolic-parabolic coupled system can be assigned to the unknown functions $\boldsymbol{u} = (u_1, u_2, u_3)$ and Θ according to the requirements for the second-order hyperbolic system and the second-order parabolic equation, respectively.

The well-posed problems for system (6.89) and (6.91) of thermoelastodynamics can be divided into the initial value problem (Cauchy problem) and the mixed initial-boundary value problem, which can be put forward as follows:

Initial value problem. Find the solution $\boldsymbol{u}$ and Θ of system (6.89) and (6.91) in $t > 0$, $\boldsymbol{x} \in \mathbb{R}^3$, such that it satisfies at $t = 0$ the initial conditions

$$\begin{aligned} \boldsymbol{u}(0,\boldsymbol{x}) &= \boldsymbol{u}^0(\boldsymbol{x}), \\ \frac{\partial \boldsymbol{u}}{\partial t}(0,\boldsymbol{x}) &= \boldsymbol{u}^1(\boldsymbol{x}), \\ \Theta(0,\boldsymbol{x}) &= \Theta^0(\boldsymbol{x}), \end{aligned} \tag{6.94}$$

where $\boldsymbol{u}^0(\boldsymbol{x})$, $\boldsymbol{u}^1(\boldsymbol{x})$, and $\Theta^0(\boldsymbol{x})$ are given functions.

Mixed initial-boundary value problem. For a domain Ω in $\mathbb{R}^3$, find the solution $\boldsymbol{u}$ and Θ of system (6.89) and (6.91) in $t > 0$, $\boldsymbol{x} \in \Omega$, such that it satisfies the same initial conditions as (6.94) at $t = 0$ and the following boundary conditions on the boundary $\partial\Omega$ of Ω:

$$\boldsymbol{u}|_{\Gamma_1} = \boldsymbol{d}, \boldsymbol{P}\boldsymbol{n}|_{\Gamma_2} = \boldsymbol{\sigma}, \tag{6.95}$$

$$\Theta|_{S_1} = f, \boldsymbol{h}\cdot\boldsymbol{n}|_{S_2} = g, \tag{6.96}$$

where $\Gamma_1 \cup \Gamma_2 = S_1 \cup S_2 = \partial\Omega$ and $\Gamma_1 \cap \Gamma_2 = S_1 \cap S_2 = \emptyset$, $\boldsymbol{P}$ and $\boldsymbol{h}$ are given by (6.73) and (6.75), respectively, while $\boldsymbol{d}$, $\boldsymbol{\sigma}$, f, and g are all given functions.

The boundary condition (6.95) shows that the displacement of the thermoelastic body is given on the part Γ_1 of the boundary $\partial\Omega$, while the stress is given on the other part Γ_2 (either Γ_1 or Γ_2 may be an empty set; then only one condition in (6.95) is given). The boundary condition (6.96) indicates that the temperature of the thermoelastic body is given on the part S_1 of the boundary $\partial\Omega$, while the heat flux is given on the other part S_2 (either S_1 or S_2 may be an empty set).

There is no great difficulty in dealing with the existence and uniqueness of solutions of the above problems with respect to the system of linear thermoelastodynamics. For instance, the existence of solutions can be handled by using the semigroup theory of operators. However, the asymptotic behavior of solutions to these problems as $t \to +\infty$ is still an intriguing problem.

Heat conduction implies a dissipation mechanism. In thermoelasticity, solutions to well-posed problems should present some kind of dissipative properties due to the heat effect.

For simplicity, supposing $\boldsymbol{b} \equiv 0$, $\gamma \equiv 0$, and taking the homogeneous boundary conditions

$$\boldsymbol{u}|_{\partial\Omega} = 0, \quad \Theta|_{\partial\Omega} = 0 \tag{6.97}$$

as an example, we roughly investigate the asymptotic behavior of the solution to system (6.89) and (6.91) as $t \to +\infty$.

Multiplying both sides of equation (6.89) by $\frac{\partial u_i}{\partial t}$ and summing over 1 to 3 with respect to i, and multiplying both sides of equation (6.91) by $\frac{\Theta}{\theta_0}$, then adding the results together to integrate over Ω, we can obtain, by using Green's formula and the boundary condition (6.97), that

$$\begin{aligned}&\frac{\mathrm{d}}{\mathrm{d}t}\int_\Omega \frac{1}{2}\left(\rho_0\left|\frac{\partial \boldsymbol{u}}{\partial t}\right|^2 + \sum_{i,j,k,l=1}^{3} a_{ijkl}\frac{\partial u_i}{\partial x_j}\frac{\partial u_k}{\partial x_l} + a\Theta^2\right)\mathrm{d}x\\ &= -\int_\Omega \sum_{i,j=1}^{3}\frac{1}{\theta_0}k_{ij}\frac{\partial\Theta}{\partial x_i}\frac{\partial\Theta}{\partial x_j}\mathrm{d}x.\end{aligned} \tag{6.98}$$

So, if the heat conduction matrix $\boldsymbol{K}$ is positive definite, i.e., condition (6.93) is satisfied, then we have

$$\begin{aligned}&\frac{\mathrm{d}}{\mathrm{d}t}\int_\Omega \frac{1}{2}\left(\rho_0\left|\frac{\partial \boldsymbol{u}}{\partial t}\right|^2 + \sum_{i,j,k,l=1}^{3} a_{ijkl}\frac{\partial u_i}{\partial x_j}\frac{\partial u_k}{\partial x_l} + a\Theta^2\right)\mathrm{d}x\\ &\leqslant -\frac{k}{\theta_0}\int_\Omega |\nabla\theta|^2\mathrm{d}x.\end{aligned} \tag{6.99}$$

If the elastic tensor $\boldsymbol{A} = (a_{ijkl})$ still satisfies the stability condition, i.e., there exists a positive constant $\widetilde{\alpha}$ such that

$$\sum_{i,j,k,l=1}^{3} a_{ijkl}e_{ij}e_{kl} \geqslant \widetilde{\alpha}|\boldsymbol{E}|^2 \tag{6.100}$$

holds for any given second-order symmetric matrix $\boldsymbol{E} = (e_{ij})$ (see Definition 5.6 in Chapter 5), then from Korn's inequality (see Theorem 5.7 in Chapter 5) we have

$$\int_\Omega \sum_{i,j,k,l=1}^{3} a_{ijkl}e_{ij}e_{kl}\mathrm{d}x \geqslant \widetilde{\alpha}C_0\left(\|\boldsymbol{u}\|^2_{(L^2(\Omega))^3} + \sum_{i,j=1}^{3}\left\|\frac{\partial u_i}{\partial x_j}\right\|^2_{L^2(\Omega)}\right). \tag{6.101}$$

Noting

$$\sum_{i,j,k,l=1}^{3} a_{ijkl}\frac{\partial u_i}{\partial x_j}\frac{\partial u_k}{\partial x_l} = \sum_{i,j,k,l=1}^{3} a_{ijkl}e_{ij}e_{kl}, \tag{6.102}$$

we know that the integral

$$L(t)=\int_{\Omega}\frac{1}{2}\left(\rho_0\left|\frac{\partial u}{\partial t}\right|^2+\sum_{i,j,k,l=1}^{3}a_{ijkl}\frac{\partial u_i}{\partial x_j}\frac{\partial u_k}{\partial x_l}+a\Theta^2\right)\mathrm{d}x,$$

which is usually called the *energy integral*, is positive. Inequality (6.99) shows that the energy integral $L(t)$ is decreasing with increasing time t. There are many detailed results about the decay property of the energy integral. For example, in the one-dimensional case, it can be proved that the energy integral decays to zero as $t\to+\infty$ (see [3]), and further the exponential decay property can also be proved.

We also point out that both the coefficients of $\frac{\partial\Theta}{\partial x_j}$ in equation (6.89) and the coefficients of $\frac{\partial^2 u_i}{\partial t\partial x_j}$ in equation (6.91) are g_{ij} (up to a constant factor θ_0 only), which provides a guarantee for the effectiveness of the coupling of these two equations.

6.4.2 System of Nonlinear Thermoelastodynamics

Substituting the constitutive equation

$$\boldsymbol{P}=\widehat{\boldsymbol{P}}(\boldsymbol{F},\theta)=(\widehat{p}_{ij}(\boldsymbol{F},\theta)) \tag{6.103}$$

into the conservation equations (6.24) of momentum, and noticing $\boldsymbol{F}=(f_{ij})$, where $f_{ij}=\frac{\partial y_i}{\partial x_j}$, we have

$$\rho_0\frac{\partial^2 y_i}{\partial t^2}=\sum_{j,k,l=1}^{3}a_{ijkl}(\nabla\boldsymbol{y},\theta)\frac{\partial^2 y_k}{\partial x_j\partial x_l}-\sum_{j=1}^{3}g_{ij}(\nabla\boldsymbol{y},\theta)\frac{\partial\theta}{\partial x_j}+\rho_0 b_i \quad (i=1,2,3), \tag{6.104}$$

where

$$a_{ijkl}(\nabla\boldsymbol{y},\theta)=\frac{\partial\widehat{p}_{ij}}{\partial f_{kl}} \tag{6.105}$$

and

$$g_{ij}(\nabla\boldsymbol{y},\theta)=-\frac{\partial\widehat{p}_{ij}}{\partial\theta}. \tag{6.106}$$

Substituting the constitutive equations

$$\eta=\widehat{\eta}(\boldsymbol{F},\theta),\quad \boldsymbol{h}=\widehat{\boldsymbol{h}}(\boldsymbol{F},\theta,\nabla\theta) \tag{6.107}$$

into the conservation equation (6.45) of energy, and noticing that it follows from (6.42)–(6.43) and (6.106) that

$$\rho_0 \frac{\partial \widehat{\eta}}{\partial f_{ij}} = -\rho_0 \frac{\partial^2 \widehat{\psi}}{\partial f_{ij} \partial \theta} = -\frac{\partial \widehat{p}_{ij}}{\partial \theta} = g_{ij}, \tag{6.108}$$

we get

$$\begin{aligned} &\rho_0 \theta c(\nabla \boldsymbol{y}, \theta) \frac{\partial \theta}{\partial t} - \sum_{i,j=1}^{3} k_{ij}(\nabla \boldsymbol{y}, \theta, \nabla \theta) \frac{\partial^2 \theta}{\partial x_i \partial x_j} \\ &- \sum_{i,j,k=1}^{3} k_{ijk}(\nabla \boldsymbol{y}, \theta, \nabla \theta) \frac{\partial^2 y_j}{\partial x_i \partial x_k} \\ &+ \sum_{i,j=1}^{3} \theta g_{ij}(\nabla \boldsymbol{y}, \theta) \frac{\partial^2 y_i}{\partial t \partial x_j} \\ &- \sum_{i=1}^{3} k_i(\nabla \boldsymbol{y}, \theta, \nabla \theta) \frac{\partial \theta}{\partial x_i} = \rho_0 \gamma, \end{aligned} \tag{6.109}$$

where

$$c(\nabla \boldsymbol{y}, \theta) = \frac{\partial \widehat{\eta}}{\partial \theta} > 0 \tag{6.110}$$

and

$$k_{ijk} = -\frac{\partial \widehat{h}_i}{\partial f_{jk}}, \quad k_{ij} = -\frac{\partial \widehat{h}_i}{\partial (\frac{\partial \theta}{\partial x_j})}, \quad k_i = -\frac{\partial \widehat{h}_i}{\partial \theta}. \tag{6.111}$$

Formula (6.110) is actually an assumption. Noting (6.71), (6.110) is equivalent to $\widehat{e}_\theta > 0$, which is always assumed to be true in physics (see (6.78)).

Then (6.104) and (6.109) constitute the *system of nonlinear thermoelastodynamics*, where a_{ijkl} and g_{ij} defined by (6.105)–(6.106) are generally known as the *elasticity rate*, while k_{ijk}, k_{ij}, and k_i defined by (6.111) are called the *heat conductivity*.

$\boldsymbol{A} = (a_{ijkl}(\nabla \boldsymbol{y}, \theta))$ is also called the *elastic tensor*. From (6.105) and (6.42), it is easy to see that tensor $\boldsymbol{A}$ has the following symmetry:

$$a_{ijkl} = a_{klij}. \tag{6.112}$$

In addition, we also assume that tensor $\boldsymbol{A}$ satisfies the strong ellipticity condition (see Definition 5.7 in Chapter 5), i.e.,

$$\begin{aligned} &\sum_{i,j,k,l=1}^{3} a_{ijkl}(\boldsymbol{F}, \theta) \xi_i \xi_k \eta_j \eta_l > 0, \\ &\forall \boldsymbol{F} \in \mathbb{R}^{3\times 3},\ \det \boldsymbol{F} > 0,\ \theta > 0,\ \boldsymbol{\xi}, \boldsymbol{\eta} \in \mathbb{R}^3 \backslash \{0\}. \end{aligned} \tag{6.113}$$

Now we try to find the conditions that should be satisfied by the heat conduction tensor $\boldsymbol{K}=(k_{ij})$. Noting (6.56), it is easy to see from inequality (6.53) that

$$\begin{aligned}\boldsymbol{h}\cdot\nabla\theta &= \sum_{i=1}^{3}(\widehat{h}_i(\boldsymbol{F},\theta,\nabla\theta)-\widehat{h}_i(\boldsymbol{F},\theta,0))\frac{\partial\theta}{\partial x_i}\\ &= \int_0^1\sum_{i,j=1}^{3}\frac{\partial\widehat{h}_i}{\partial(\frac{\partial\theta}{\partial x_j})}(\boldsymbol{F},\theta,\tau\nabla\theta)\frac{\partial\theta}{\partial x_j}\frac{\partial\theta}{\partial x_i}\mathrm{d}\tau\\ &= -\int_0^1\sum_{i,j=1}^{3}k_{ij}(\boldsymbol{F},\theta,\tau\nabla\theta)\mathrm{d}\tau\frac{\partial\theta}{\partial x_i}\frac{\partial\theta}{\partial x_j}\leqslant 0.\end{aligned} \tag{6.114}$$

According to this, we can assume that tensor $\boldsymbol{K}=(k_{ij})$ is positive, i.e.,

$$\begin{aligned}&\sum_{i,j=1}^{3}k_{ij}(\boldsymbol{F},\theta,\nabla\theta)\xi_i\xi_j>0,\\ &\forall\boldsymbol{F}\in\mathbb{R}^{3\times3},\ \det\boldsymbol{F}>0,\ \theta>0,\ \nabla\theta\in\mathbb{R}^3,\ \boldsymbol{\xi}\in\mathbb{R}^3\backslash\{0\}.\end{aligned} \tag{6.115}$$

Under the assumption of the strong ellipticity condition (6.113), system (6.104) is a second-order quasi-linear hyperbolic system with respect to unknown variables $\boldsymbol{y}=(y_1,y_2,y_3)$, while under assumption (6.115), equation (6.109) is a second-order quasi-linear parabolic equation with respect to unknown variable θ. Hence, system (6.104) and (6.109) of nonlinear thermoelastodynamics is a *second-order quasi-linear hyperbolic-parabolic coupled system*. The formulation of the well-posed problems of this system is similar to that of the linear case (see section 6.4.1), which is not restated here. As for the features and research situation of these well-posed problems, we will explain when discussing the acceleration wave in the next section.

Under many circumstances, especially when studying discontinuous solutions, it is better to write the system in the form of conservation laws. To this end, instead of (6.45), we represent the conservation law of energy by equation (6.23) of the conservation law, while for the conservation system (6.24) of momentum, by introducing new unknown variables

$$f_{ij}=\frac{\partial y_i}{\partial x_j}\quad(i,j=1,2,3),\quad v_i=\frac{\partial y_i}{\partial t}\quad(i=1,2,3), \tag{6.116}$$

it can be written as a system of first-order partial differential equations of conservation laws. Hence, the system of nonlinear thermoelastodynamics can be written as the following system of conservation laws:

$$\frac{\partial f_{ij}}{\partial t}-\frac{\partial v_i}{\partial x_j}=0\quad(i,j=1,2,3), \tag{6.117}$$

$$\frac{\partial}{\partial t}(\rho_0 v_k)=\sum_{l=1}^{3}\frac{\partial p_{kl}}{\partial x_l}+\rho_0 b_k\quad(k=1,2,3), \tag{6.118}$$

$$\frac{\partial}{\partial t}\left(\rho_0 e+\frac{1}{2}\rho_0|\boldsymbol{v}|^2\right)=\mathrm{div}(\boldsymbol{P}^{\mathrm{T}}\boldsymbol{v})-\mathrm{div}\,\boldsymbol{h}+\rho_0\boldsymbol{b}\cdot\boldsymbol{v}+\rho_0\gamma. \tag{6.119}$$

The first two systems, (6.117) and (6.118), constitute a first-order quasi-linear (nonstrictly, nonsymmetric) hyperbolic system (see section 5.5.3 in Chapter 5) with respect to unknown functions $(f_{11}, f_{12}, \dots, f_{33}, v_1, v_2, v_3)$, while (6.119) (which is equivalent to (6.45)) is essentially a quasi-linear parabolic equation. Therefore, system (6.117)–(6.119) is still a quasi-linear hyperbolic-parabolic coupled system.

6.4.3 System of One-Dimensional Nonlinear Thermoelastodynamics

Now we discuss two kinds of problems of one-dimensional nonlinear thermoelastodynamics.

1. Pure axial deformation of a thermoelastic rod. Assume that the deformation of a thermoelastic rod is given by

$$y_1 = x_1 + u(t, x_1), \quad y_2 = x_2, \quad y_3 = x_3. \tag{6.120}$$

Assume furthermore that the temperature of each point on any cross section of the rod is the same at a given time, that is, the distribution of temperature depends only on t and x_1:

$$\theta = \theta(t, x_1). \tag{6.121}$$

For this kind of deformation, we have

$$\boldsymbol{F} = \begin{pmatrix} 1 + \dfrac{\partial u}{\partial x_1} & 0 & 0 \\ 0 & 1 & 0 \\ 0 & 0 & 1 \end{pmatrix}. \tag{6.122}$$

Then, from the constitutive relations of thermoelasticity, the components of Piola stress tensor $\boldsymbol{P}$ and vector $\boldsymbol{h}$ are finally functions of only t and x_1, denoted by $p_{ij}(u_{x_1}, \theta)$ and $h_i(u_{x_1}, \theta, \theta_{x_1})$, respectively.

Thus, the first equation of conservation system (6.24) or (6.104) of momentum is reduced to

$$\rho_0 \frac{\partial^2 u}{\partial t^2} = \frac{\partial}{\partial x_1}(p_{11}(u_{x_1}, \theta)) + \rho_0 b_1 \tag{6.123}$$

or

$$\rho_0 \frac{\partial^2 u}{\partial t^2} = a_{1111}(u_{x_1}, \theta) \frac{\partial^2 u}{\partial x_1^2} - g_{11}(u_{x_1}, \theta) \frac{\partial \theta}{\partial x_1} + \rho_0 b_1, \tag{6.124}$$

where

$$a_{1111} = \frac{\partial p_{11}(u_{x_1}, \theta)}{\partial u_{x_1}}, \quad g_{11} = -\frac{\partial p_{11}(u_{x_1}, \theta)}{\partial \theta}. \tag{6.125}$$

Similarly, conservation equation (6.23) or (6.109) of energy can be reduced to

$$\begin{aligned} &\frac{\partial}{\partial t}\left(\rho_0 e + \frac{1}{2}\rho_0 u_t^2\right) - \frac{\partial}{\partial x_1}\left(p_{11}(u_{x_1}, \theta) \frac{\partial u}{\partial t}\right) \\ &\quad + \frac{\partial}{\partial x_1} h_1(u_{x_1}, \theta, \theta_{x_1}) = \rho_0 b_1 u_t + \rho_0 \gamma \end{aligned} \tag{6.126}$$

or

$$\rho_0\theta c(u_{x_1},\theta)\frac{\partial\theta}{\partial t}-k_{11}(u_{x_1},\theta,\theta_{x_1})\frac{\partial^2\theta}{\partial x_1^2}-k_{111}(u_{x_1},\theta,\theta_{x_1})\frac{\partial^2 u}{\partial x_1^2}$$

$$+\theta g_{11}(u_{x_1},\theta)\frac{\partial^2 u}{\partial t\partial x_1}-k_1(u_{x_1},\theta,\theta_{x_1})\frac{\partial\theta}{\partial x_1}=\rho_0\gamma, \tag{6.127}$$

where

$$k_{111}=-\frac{\partial h_1}{\partial u_{x_1}},\quad k_{11}=-\frac{\partial h_1}{\partial\theta_{x_1}},\quad k_1=-\frac{\partial h_1}{\partial\theta}. \tag{6.128}$$

The combination of equations (6.124) (or (6.123)) and (6.127) (or (6.126)) then constitutes the corresponding system of one-dimensional nonlinear thermoelastodynamics.

Taking $\xi_1=\eta_1=1$, $\xi_2=\xi_3=\eta_2=\eta_3=0$ in the strong ellipticity condition (6.113), we get

$$a_{1111}(u_{x_1},\theta)>0. \tag{6.129}$$

Similarly, from the positiveness condition (6.115) of heat conduction tensor $\boldsymbol{K}$, we obtain

$$k_{11}(u_{x_1},\theta,\theta_{x_1})>0. \tag{6.130}$$

Thus, equation (6.124) is a one-dimensional nonlinear wave equation with respect to u, while equation (6.127) is a one-dimensional nonlinear heat equation with respect to θ. Therefore, the system (6.124) and (6.127) is a *one-dimensional quasi-linear hyperbolic-parabolic coupled system.*

2. Pure shear deformation of a thermoelastic body. Assume that the deformation of the thermoelastic rod is given by

$$y_1=x_1+u(t,x_2),\quad y_2=x_2,\quad y_3=x_3. \tag{6.131}$$

Assume furthermore that under this deformation, the distribution of temperature θ is only a function of t and x_2:

$$\theta=\theta(t,x_2). \tag{6.132}$$

For the above deformation, we have

$$\boldsymbol{F}=\begin{pmatrix}1 & \frac{\partial u}{\partial x_2} & 0\\ 0 & 1 & 0\\ 0 & 0 & 1\end{pmatrix}. \tag{6.133}$$

Therefore, the components of $\boldsymbol{P}$ and $\boldsymbol{h}$ given by the constitutive relations are finally functions of only t and x_2, denoted by $p_{ij}(u_{x_2},\theta)$ and $h_i(u_{x_2},\theta,\theta_{x_2})$, respectively.

Hence, the first equation of conservation system (6.24) or (6.104) of momentum can be reduced to

$$\rho_0 \frac{\partial^2 u}{\partial t^2} = \frac{\partial}{\partial x_2} p_{12}(u_{x_2}, \theta) + \rho_0 b_1 \tag{6.134}$$

or

$$\rho_0 \frac{\partial^2 u}{\partial t^2} = a_{1212}(u_{x_2}, \theta) \frac{\partial^2 u}{\partial x_2^2} - g_{12}(u_{x_2}, \theta) \frac{\partial \theta}{\partial x_2} + \rho_0 b_1, \tag{6.135}$$

where

$$a_{1212} = \frac{\partial p_{12}(u_{x_2}, \theta)}{\partial u_{x_2}} > 0, \quad g_{12} = -\frac{\partial p_{12}(u_{x_2}, \theta)}{\partial \theta}. \tag{6.136}$$

Here $a_{1212} > 0$ is a corollary of the strong ellipticity condition (6.113). Moreover, the conservation equation (6.23) or (6.109) of energy is reduced to

$$\begin{aligned} &\frac{\partial}{\partial t}\left(\rho_0 e + \frac{1}{2}\rho_0 u_t^2\right) - \frac{\partial}{\partial x_2}\left(p_{12}(u_{x_2}, \theta) u_t\right) \\ &+ \frac{\partial}{\partial x_2} h_2(u_{x_2}, \theta, \theta_{x_2}) = \rho_0 b_1 u_t + \rho_0 \gamma \end{aligned} \tag{6.137}$$

or

$$\begin{aligned} &\rho_0 \theta c(u_{x_2}, \theta) \frac{\partial \theta}{\partial t} - k_{22}(u_{x_2}, \theta, \theta_{x_2}) \frac{\partial^2 \theta}{\partial x_2^2} \\ &- k_{212}(u_{x_2}, \theta, \theta_{x_2}) \frac{\partial^2 u}{\partial x_2^2} + \theta g_{12}(u_{x_2}, \theta) \frac{\partial^2 u}{\partial t \partial x_2} \\ &- k_2(u_{x_2}, \theta, \theta_{x_2}) \frac{\partial \theta}{\partial x_2} = \rho_0 \gamma, \end{aligned} \tag{6.138}$$

where

$$\begin{aligned} k_{212} &= -\frac{\partial h_2(u_{x_2}, \theta, \theta_{x_2})}{\partial u_{x_2}}, \\ k_{22} &= -\frac{\partial h_2(u_{x_2}, \theta, \theta_{x_2})}{\partial \theta_{x_2}} > 0, \\ k_2 &= -\frac{\partial h_2(u_{x_2}, \theta, \theta_{x_2})}{\partial \theta}. \end{aligned} \tag{6.139}$$

Here $k_{22} > 0$ is a corollary of the positiveness assumption (6.115) of $\boldsymbol{K}$.

The system (6.135) (or (6.134)) and (6.138) (or (6.137)) is also a *one-dimensional quasi-linear hyperbolic-parabolic coupled system.*

The above systems obtained under two different circumstances are formally identical. Therefore, the system of one-dimensional thermoelastodynamics can be unified and written as

$$\rho_0 \frac{\partial^2 u}{\partial t^2} - \frac{\partial}{\partial x} p(u_x, \theta) = \rho_0 b, \tag{6.140}$$

$$\frac{\partial}{\partial t}\left(\rho_0 e + \frac{1}{2}\rho_0 u_t^2\right) - \frac{\partial}{\partial x}(p(u_x,\theta)u_t) + \frac{\partial}{\partial x} h(u_x,\theta,\theta_x) = \rho_0 b u_t + \rho\gamma \tag{6.141}$$

or

$$\rho_0 \theta c(u_x,\theta)\frac{\partial \theta}{\partial t} + h_{\theta_x}(u_x,\theta,\theta_x)\frac{\partial^2 \theta}{\partial x^2} + h_{u_x}(u_x,\theta,\theta_x)\frac{\partial^2 u}{\partial x^2} - \theta p_\theta(u_x,\theta)\frac{\partial^2 u}{\partial t \partial x} + h_\theta(u_x,\theta,\theta_x)\frac{\partial \theta}{\partial x} = \rho_0 \gamma, \tag{6.142}$$

where

$$p_{u_x}(u_x,\theta) > 0, \quad h_{\theta_x}(u_x,\theta,\theta_x) < 0. \tag{6.143}$$

Under many circumstances, especially when studying discontinuous solutions, it is more convenient to write the first equation in the system (6.140) and (6.141) as a form of first-order conservation law. For this, setting

$$v = u_x, \qquad w = u_t, \tag{6.144}$$

the system (6.140) and (6.141) is written in the following equivalent form:

$$v_t - w_x = 0, \tag{6.145}$$

$$\rho_0 w_t - p(v,\theta)_x = \rho_0 b, \tag{6.146}$$

$$\left(\rho_0 e + \frac{1}{2}\rho_0 w^2\right)_t - (p(v,\theta)w)_x + (h(v,\theta,\theta_x))_x = \rho_0 b w + \rho_0 \gamma, \tag{6.147}$$

where

$$p_v(v,\theta) > 0, \quad h_{\theta_x}(v,\theta,\theta_x) < 0. \tag{6.148}$$

In addition, the entropy inequality (6.32) can now be taken in the following form:

$$\rho_0 \eta_t + \left(\frac{h}{\theta}\right)_x \geqslant \frac{\rho_0 \gamma}{\theta}. \tag{6.149}$$

There are already quite a number of research results about system (6.145)–(6.147) (see the next section and the references cited therein). The propagation of the acceleration wave will provide important ideas on the study of the global existence and blow-up phenomenon of smooth solutions to its well-posed problems, which will be discussed in the next section.

6.5 Propagation of Acceleration Waves

The discussion of the propagation of acceleration waves will be helpful in illuminating under what situation the smooth solution of a well-posed problem to the system of thermoelastodynamics will exist globally or blow up in finite time. Now we take the system (6.140) and (6.141) of one-dimensional thermoelastodynamics as an example to answer the following questions: What is the acceleration wave? What are the characteristics of its propagation? What inspiration will these characteristics give to the research on the well-posed problems? For simplicity, we assume that $\rho_0 \equiv 1$, $b \equiv \gamma \equiv 0$.

Let Γ be a curve on the (t,x) plane:

$$x = x(t). \tag{6.150}$$

If u and θ are suitably smooth and satisfy the system outside of Γ, u and its first-order partial derivatives as well as θ are continuous on Γ, while the second-order partial derivatives of u and the first-order partial derivatives of θ have discontinuity of the first type, then Γ is called an *acceleration wave* curve (named because the acceleration u_{tt} is discontinuous on it).

Now we prove that the first-order partial derivatives of θ are also continuous on the acceleration wave curve Γ. So the acceleration wave curve is actually the *weak discontinuity curve* of the one-dimensional nonlinear thermoelastodynamics system. In fact, since θ is a continuous function of (t,x), it is easy to see that

$$\oint \theta_t \mathrm{d}t + \theta_x \mathrm{d}x = 0 \tag{6.151}$$

is true for any given closed curve on the (t,x) plane. Then, by the deduction of (2.179) in Chapter 2, on Γ we have

$$[\theta_t] = -V[\theta_x], \tag{6.152}$$

where V is the propagation speed of the acceleration wave,

$$V(t) = x'(t), \tag{6.153}$$

and $x'(t) = \frac{\mathrm{d}x(t)}{\mathrm{d}t}$. Here $[\cdot]$ stands for the jump across Γ. Similarly, applying (2.179) in Chapter 2 to the corresponding integral form of the equation (6.141) of the conservation law (see Exercise 8), and noting that e, u_t, and $p(u_x,\theta)$ are all continuous on Γ, then we get

$$[h] = 0. \tag{6.154}$$

Moreover, from the second formula in (6.143), we have that h is a strictly monotonic function of θ_x, and thus

$$[\theta_x] = 0. \tag{6.155}$$

Then from (6.152) we also have

$$[\theta_t] = 0. \tag{6.156}$$

This proves the continuity of the first-order partial derivatives of θ on Γ.

Now we examine the change of jump

$$a(t) = [u_{tt}] \tag{6.157}$$

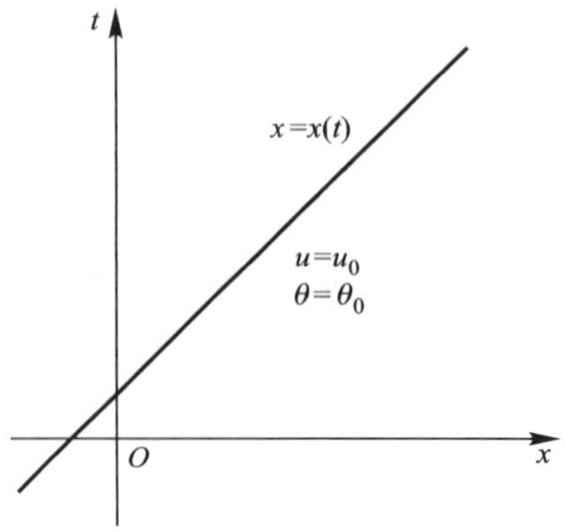

Figure 6.1

of acceleration u_{tt} across Γ with respect to time t. For simplicity, suppose that the thermoelastic body is in a fixed state under uniform deformation and constant temperature before the acceleration wave propagates into the medium, i.e.,

$$u \equiv u_0, \quad \theta \equiv \theta_0 \quad \text{as } x > x(t) \tag{6.158}$$

(see Figure 6.1), where u_0 and θ_0 are constants. According to the knowledge about the mathematical physics equations (see [5]), the weak discontinuity curve Γ of the solution to equation (6.140) must be its characteristic curve; thus $V = \pm\sqrt{p_{u_x}(0,\theta_0)}$. Assume that the acceleration wave propagates from left to right, i.e.,

$$x'(t) > 0. \tag{6.159}$$

Then we have

$$V = \sqrt{p_{u_x}(0,\theta_0)}. \tag{6.160}$$

Differentiating equation (6.140) with respect to t once, we get

$$\begin{aligned} u_{ttt} &= C_1(u_x,\theta)u_{xxt} + C_{11}(u_x,\theta)u_{xx}u_{xt} + C_{12}(u_x,\theta)u_{xx}\theta_t \\ &\quad + C_2(u_x,\theta)\theta_{xt} + C_{12}(u_x,\theta)u_{xt}\theta_x + C_{22}(u_x,\theta)\theta_x\theta_t, \end{aligned} \tag{6.161}$$

where

$$\begin{aligned} &C_1 = p_{u_x}, \quad C_2 = p_\theta, \quad C_{11} = p_{u_x u_x}, \\ &C_{12} = p_{u_x\theta}, \quad C_{22} = p_{\theta\theta}. \end{aligned} \tag{6.162}$$

Since the acceleration wave is a weak discontinuity, noting (6.158), we have

$$[u_{xx}u_{xt}] = [u_{xx}][u_{xt}], \quad [u_{xx}\theta_t] = [\theta_{xt}\theta_x] = 0. \tag{6.163}$$

From (6.161), it is easy to obtain

$$[u_{ttt}] = C_1(0,\theta_0)[u_{xxt}] + C_{11}(0,\theta_0)[u_{xx}][u_{xt}] + C_2(0,\theta_0)[\theta_{xt}]. \tag{6.164}$$

Differentiating both sides of

$$[u_x] = 0$$

with respect to t once, and noting (6.160), we obtain

$$\frac{\mathrm{d}}{\mathrm{d}t}[u_x] = \left[\frac{\mathrm{d}}{\mathrm{d}_V t}u_x\right] = [u_{xt}] + V[u_{xx}] = 0,$$

where

$$\frac{\mathrm{d}}{\mathrm{d}_V t} = \frac{\partial}{\partial t} + V\frac{\partial}{\partial x} \tag{6.165}$$

is the directional derivative with respect to t along curve Γ. Thus, we have

$$[u_{xt}] = -V[u_{xx}]. \tag{6.166}$$

Similarly,

$$[u_{tt}] = -V[u_{xt}], \tag{6.167}$$

$$[\theta_{xt}] = -V[\theta_{xx}]. \tag{6.168}$$

From (6.166) and (6.167), we have

$$[u_{xx}] = \frac{1}{V^2}[u_{tt}], \tag{6.169}$$

$$[u_{xt}] = -\frac{1}{V}[u_{tt}]. \tag{6.170}$$

In addition, it is not hard to verify that

$$\frac{\mathrm{d}}{\mathrm{d}t}[u_{tt}] = \left[\frac{\mathrm{d}u_{tt}}{\mathrm{d}_V t}\right] = [u_{ttt}] + V[u_{xtt}]. \tag{6.171}$$

Differentiating (6.167) with respect to t once, and noting (6.160), we also have

$$\begin{aligned}\frac{\mathrm{d}}{\mathrm{d}t}[u_{tt}] &= -V\frac{\mathrm{d}}{\mathrm{d}t}[u_{xt}] = -V\left[\frac{\mathrm{d}u_{xt}}{\mathrm{d}_V t}\right]\\ &= -V[u_{xtt}] - V^2[u_{xxt}].\end{aligned} \tag{6.172}$$

Adding (6.171) and (6.172) together, we get

$$2\frac{\mathrm{d}}{\mathrm{d}t}[u_{tt}] = [u_{ttt}] - V^2[u_{xxt}]. \tag{6.173}$$

Plugging (6.168)–(6.170) and (6.173) into (6.164), and noticing $C_1(0,\theta_0) = V^2$, we then obtain

$$2\frac{\mathrm{d}}{\mathrm{d}t}[u_{tt}] + \frac{C_{11}(0,\theta_0)}{V^3}[u_{tt}]^2 + VC_2(0,\theta_0)[\theta_{xx}] = 0. \tag{6.174}$$

Finally, we need to represent $[\theta_{xx}]$ with $[u_{tt}]$. For this, taking the jump of both sides of equation (6.142) across Γ, we get

$$h_{\theta_x}(0,\theta_0,0)[\theta_{xx}] + h_{u_x}(0,\theta_0,0)[u_{xx}] - \theta_0 p_\theta(0,\theta_0)[u_{xt}] = 0. \tag{6.175}$$

Noticing that (6.56) implies $h_{u_x}(0,\theta_0,0)=0$, from (6.175) and (6.170) we obtain

$$[\theta_{xx}] = -\frac{\theta_0 p_\theta(0,\theta_0)}{V h_{\theta_x}(0,\theta_0,0)}[u_{tt}]. \tag{6.176}$$

Substituting this into (6.174), we arrive at the ordinary differential equation

$$a'(t)+\alpha a^2(t)+\beta a(t)=0 \tag{6.177}$$

satisfied by $a(t)=[u_{tt}]$, where

$$\alpha = \frac{p_{u_x u_x}(0,\theta_0)}{2V^3}, \quad \beta = -\frac{\theta_0 p_\theta^2(0,\theta_0)}{2h_{\theta_x}(0,\theta_0,0)}. \tag{6.178}$$

About this ordinary differential equation, we have the following theorem.

Theorem 6.1. *For the system* (6.140)–(6.141) *of one-dimensional nonlinear thermoelastodynamics, the amplitude of the acceleration wave propagating from left to right into the state of uniform deformation and constant temperature determined by* (6.158) *is given by*

$$a(t) = \frac{\beta}{\left(\frac{\beta}{a(0)}+\alpha\right)e^{\beta t}-\alpha}, \tag{6.179}$$

where α *and* β *are given by* (6.178). *Moreover, we have the following:*

(1) *Suppose that*

$$p_\theta(0,\theta_0)\neq 0 \tag{6.180}$$

if

$$|a(0)| < \left|\frac{\beta}{\alpha}\right| \tag{6.181}$$

or

$$\operatorname{sign} a(0) = \operatorname{sign} p_{u_x u_x}(0,\theta_0). \tag{6.182}$$

Then $a(t)$ *decays exponentially to zero at the rate* $e^{-\beta t}$ *as* $t\to+\infty$.

(2) *Suppose that*

$$p_{u_x u_x}(0,\theta_0)\neq 0 \tag{6.183}$$

if

$$|a(0)| > \left|\frac{\beta}{\alpha}\right| \tag{6.184}$$

and

$$\operatorname{sign} a(0) = -\operatorname{sign} p_{u_x u_x}(0,\theta_0). \tag{6.185}$$

Then $a(t)\to\infty$ *as* $t\to t_\infty$, *where*

$$t_\infty = -\frac{1}{\beta}\ln\left(1+\frac{\beta}{\alpha a(0)}\right). \tag{6.186}$$

Proof. The ordinary differential equation (6.177) is of the type of J. Bernoulli; by solving its Cauchy problem with the initial value $a(0)$, (6.179) can be obtained immediately.

We assume (6.180). Then $h_{\theta_x}(0,\theta_0,0) < 0$ from the second formula of (6.143); hence we always have $\beta > 0$. On this occasion, if $p_{u_x u_x}(0,\theta_0) = 0$, then $\alpha = 0$, and the conclusion of (1) is obvious. If $p_{u_x u_x}(0,\theta_0) \neq 0$, then $\alpha \neq 0$, and we further assume (6.181); then from the expression (6.179) of $a(t)$, it is easy to see that $a(t)$ decays exponentially to zero at the rate $e^{-\beta t}$ as $t \to +\infty$. If $p_{u_x u_x}(0,\theta_0) \neq 0$ and (6.182) is satisfied, then $\operatorname{sign}\alpha = \operatorname{sign} a(0)$, and the same conclusion holds.

Under the assumptions (6.183)–(6.185), $\alpha \neq 0$ and $\operatorname{sign}\alpha = -\operatorname{sign} a(0)$. In this circumstance, if $\beta = 0$, using L'Hôpital's rule or solving the equation (6.177) directly, it is easy to see that the expression (6.179) of $a(t)$ is reduced to

$$a(t) = \frac{a(0)}{1 + a(0)\alpha t}, \tag{6.187}$$

and $a(0) \neq 0$ due to (6.184). Then $a(t) \to \infty$ as $t \to t_\infty$, where

$$t_\infty = -\frac{1}{\alpha a(0)}. \tag{6.188}$$

By L'Hôpital's rule, this is just the required conclusion of (2). If $\beta \neq 0$, then from (6.179), $a(t) \to \infty$ as

$$\left(\frac{\beta}{a(0)} + \alpha\right) e^{\beta t} \to \alpha,$$

namely,

$$t \to -\frac{1}{\beta} \ln\left(1 + \frac{\beta}{\alpha a(0)}\right).$$

The theorem is proved. □

For nonlinear wave equations of the form

$$u_{tt} - p(u_x)_x = 0, \tag{6.189}$$

if $p'' \neq 0$ in a certain domain, then we say that it is *genuinely nonlinear* in this domain. Condition (6.183) implies that equation (6.140) is genuinely nonlinear with respect to the unknown variable u in a neighborhood of the origin. The conclusion (2) in Theorem 6.1 tells us that the nonlinear elastic response results in the instability of the acceleration wave. In fact, when there is no effect of heat conduction, i.e., $\beta = 0$, no matter how small the initial amplitude $a(0)$ of acceleration wave is restricted to be, there always exists a finite value t_∞ given by (6.188) such that $a(t)$ goes to infinity as t goes to t_∞. However, when there is a heat conduction effect, i.e., $\beta > 0$, but no nonlinear elastic response, i.e., $\alpha = 0$, from conclusion (1) in Theorem 6.1, it can be seen that the propagation of the acceleration wave is stable and $a(t)$ decays exponentially to zero as $t \to +\infty$.

Under the general framework of nonlinear thermoelastodynamics, there are both nonlinear elastic response and heat conduction effects, i.e., both α and β are not zero. These two kinds of effects interact with each other; sometimes one dominates, sometimes the other. Theorem 6.1 tells us that, on this occasion, when the initial amplitude $a(0)$ is relatively small, the propagation of the acceleration wave $a(t)$ is stable and $a(t)$ decays exponentially to zero as $t \to +\infty$, while when the initial amplitude $a(0)$ is relatively large, the acceleration wave $a(t)$ may go to infinity, i.e., blow up in finite time.

The above discussion inspires us to think that, for the well-posed problem of the system of nonlinear thermoelastodynamics, there may exist global smooth solutions when the initial value is sufficiently small, while for a large initial value, its smooth solutions may blow up in finite time. The above discussions about the acceleration wave of thermoelasticity were given by B. D. Coleman and M. E. Gurtin in 1965 (see [2]). Shortly thereafter, subsequent research results showed that the above conjecture is true for the system of one-dimensional nonlinear thermoelastodynamics (see [4], [6], [7]). For the existence of global smooth solutions of the well-posed problem to the system of multidimensional nonlinear thermoelastodynamics, there are similar results when the initial value is sufficiently small.

Exercises

1. Prove: If the Helmholtz free energy function $\psi = \widehat{\psi}(\boldsymbol{F},\theta)$ satisfies the objectivity hypothesis (6.47), and $\boldsymbol{P} = \widehat{\boldsymbol{P}}(\boldsymbol{F},\theta)$ is given by (6.42), then the Cauchy stress tensor $\boldsymbol{T} = \widehat{\boldsymbol{T}}(\boldsymbol{F},\theta)$ satisfies the relation
$$\widehat{\boldsymbol{T}}(\boldsymbol{Q}\boldsymbol{F},\theta) = \boldsymbol{Q}\widehat{\boldsymbol{T}}(\boldsymbol{F},\theta)\boldsymbol{Q}^{\mathrm{T}},$$
where $\boldsymbol{Q}$ is any given orthogonal matrix.

2. Using directly the conservation equation (6.9) of energy and the entropy inequality (6.29) under the spatial description, prove that
$$\boldsymbol{q} \cdot \nabla_y \theta \leqslant 0.$$

3. Suppose that the boundary $\partial\Omega$ of the linear thermoelastic body Ω is free (i.e., there is no surface force), the temperature of the surrounding medium on the body surface is known as $\theta_1(t,\boldsymbol{x})$, and heat is exchanged over the boundary between the thermoelastic body and the external medium according to *Newton's experimental law*: the heat flowing into the surrounding medium across the area element $\mathrm{d}S$ on $\partial\Omega$ in unit time is
$$k_1(\theta - \theta_1)|_{\partial\Omega}\mathrm{d}S,$$
where k_1 is a positive constant called the *heat exchange coefficient*. Formulate the well-posed problem satisfied by the deformation and temperature of this linear thermoelastic body.

4. Suppose that $\boldsymbol{u} = \boldsymbol{u}(t,\boldsymbol{x})$ and $\Theta = \Theta(t,\boldsymbol{x})$ constitute the solution to system (6.89) and (6.91) of linear thermoelastodynamics in the domain $(0,+\infty)\times\Omega$ ($\Omega \subset \mathbb{R}^3$), and the following boundary conditions are satisfied on the boundary $\partial\Omega$ of Ω:
$$\left.\sum_{j,k,l=1}^{3} a_{ijkl}\frac{\partial u_k}{\partial x_l}n_j - \sum_{j=1}^{3} g_{ij}n_j\right|_{\partial\Omega} = 0 \quad (i = 1,2,3),$$
$$\left.\sum_{i,j=1}^{3} k_{ij}\frac{\partial \Theta}{\partial x_j}n_i\right|_{\partial\Omega} = 0,$$

where $\boldsymbol{n} = (n_1, n_2, n_3)$ is the unit outward normal vector on $\partial\Omega$, and $\boldsymbol{b} \equiv 0$, $\gamma \equiv 0$. Under the assumption that the heat conduction matrix (k_{ij}) is positive definite, prove that

$$\frac{\mathrm{d}L(t)}{\mathrm{d}t} \leqslant 0,$$

where

$$L(t) = \frac{1}{2}\int_\Omega \left(\rho_0 \left| \frac{\partial \boldsymbol{u}}{\partial t} \right|^2 + \sum_{i,j,k,l=1}^{3} a_{ijkl} \frac{\partial u_i}{\partial x_j}\frac{\partial u_k}{\partial x_l} + a\Theta^2 \right) \mathrm{d}x.$$

5. Prove: For the thermoelastic body under the isothermal state (i.e., $\theta \equiv$ constant), system (6.104) reduces to the system of nonlinear elastodynamics with constant parameter θ.
6. When there is only rigid motion for the thermoelastic body, i.e., for the rigid heat conduction state, prove that equation (6.109) reduces to a nonlinear heat equation of θ (second-order quasi-linear parabolic equation).
7. Suppose that $\boldsymbol{y} = \boldsymbol{y}(t, \boldsymbol{x})$ and $\theta = \theta(t, \boldsymbol{x})$ constitute the solution to system (6.104) and (6.109) of nonlinear thermoelastodynamics in the domain $(0, +\infty) \times \Omega$ ($\Omega \subset \mathbb{R}^3$), suppose the fixed (namely, the displacement is zero) and adiabatic boundary conditions are satisfied on the boundary $\partial\Omega$ of Ω, and suppose furthermore that $\boldsymbol{b} \equiv 0$, $\gamma \equiv 0$. Prove that

$$\frac{\mathrm{d}L(t)}{\mathrm{d}t} = 0,$$

where

$$L(t) = \int_\Omega \rho_0 \left(\frac{1}{2}\left| \frac{\partial \boldsymbol{u}}{\partial t} \right|^2 + \psi + \theta\eta \right) \mathrm{d}x,$$

and give the physical explanation of this exercise.
8. Give the integral form of the conservation law to the conservation equation (6.141) of energy for one-dimensional nonlinear thermoelastodynamics, and according to this, prove that on the acceleration wave curve $x = x(t)$ we have

$$[\theta_t] = -V(t)[\theta_x],$$

where $V(t) = x'(t)$ is the propagation speed of the acceleration wave.
9. Discuss the propagation of the acceleration wave of the nonlinear wave equation

$$u_{tt} - (p(u_x))_x + \alpha u_t = 0$$

with dissipation term, where α is a positive constant and p satisfies the following conditions:

$$p'(0) > 0, \quad p''(0) \neq 0.$$

Here it is assumed that u takes constant value u_0 on the right-hand side of the acceleration wave curve $x = x(t)$, i.e.,

$$u \equiv u_0 \quad \text{as } x > x(t)\,.$$

Bibliography

[1] Carlson D. E. *Linear Thermoelasticity*. Handbuch der Physik, VI a/2, 297–345. Berlin: Springer–Verlag, 1972.

[2] Coleman B. D., Gurtin M. E. *Waves in materials with memory* III, *Thermodynamic influences on growth and decay of acceleration waves*. Arch. Rat. Mech. Anal., 19 (1965), 266–298.

[3] Dafermos C. M. *On the existence and the asymptotic stability of solutions to the equations of linear thermoelasticity*. Arch. Rat. Mech. Anal., 29 (1968), 241–271.

[4] Dafermos C. M., Hsiao L. *Development of singularities in solutions of the equations of nonlinear thermoelasticity*. Quart. Appl. Math., 44 (1986), 463–474.

[5] Gu C., Li T., et al. *Mathematical Physics Equations* (in Chinese). Shanghai: Shanghai Science and Technology Press, 1987.

[6] Slemrod M. *Global existence, uniqueness and asymptotic stability of classical smooth solutions in one-dimensional, nonlinear thermoelasticity*. Arch. Rat. Mech. Anal., 76 (1981), 97–133.

[7] Zheng S., Shen W. *Global solutions to the Cauchy problem of quasilinear hyperbolic–parabolic coupled system*. Scientia Sinica, Ser. A 10 (1987), 1133–1149.

Chapter 7

Viscoelasticity

7.1 Introduction

7.1.1 Viscoelastic Materials

In Chapter 5 of Volume I, we explored the deformation law of pure elastic bodies, in which we ignored internal friction as well as the effect of heat conduction. In Chapter 6, we inspected the coupling of the heat conduction effect and the elastic response for thermoelastic bodies; however, the effect of internal friction was still not taken into account. In this chapter, we will study *viscoelastic materials* with internal friction. This type of material has both an instantaneous elastic response and a continuous internal friction effect under loads. Studies on laws of stress and deformation satisfied by *viscoelastic bodies* (solid or fluid, or an object between a solid and a fluid) under loads constitute the basic content of the subject of viscoelasticity. To further explain the concept of a viscoelastic body, we first examine the difference of response characteristics among different types of materials under loads.

For an elastic body, a sudden stress $\sigma = \sigma_0$ produces a sudden strain $\varepsilon = \varepsilon_0$, and the strain disappears immediately ($\varepsilon = 0$) (see Figure 7.1) when the load is discharged ($\sigma = 0$).

For a viscous fluid (see (2.111) in Chapter 2), a sudden constant stress $\sigma = \sigma_0$ generates a sudden rate of strain (i.e., the derivative of strain with respect to time t) $\frac{\mathrm{d}\varepsilon}{\mathrm{d}t}$, and then causes a continuous flow of fluid; moreover, the strain does not disappear when the load is discharged (see Figure 7.2).

There is also a type of material that will produce, under a sudden load, not only a sudden elastic response like an elastic solid, but also a continuous strain like a viscous fluid, and the corresponding stress response is between that of an elastic solid and that of a viscous fluid. This is an irreversible process because there is an internal friction effect, so a thermodynamic loss exists and the strain cannot completely disappear after having been unloaded (see Figure 7.3). This type of object is called a *viscoelastic body*; the material constituting this type of object is called *viscoelastic material*. Materials like plastics, wood, natural or synthetic fibers, heated concrete and metal, and asphalt all belong to the category of viscoelastic materials.

The study of viscoelastic bodies started with the work of Kelvin in 1865. When he prepared to write the entry "second law of thermodynamics" for the *Encyclopædia*

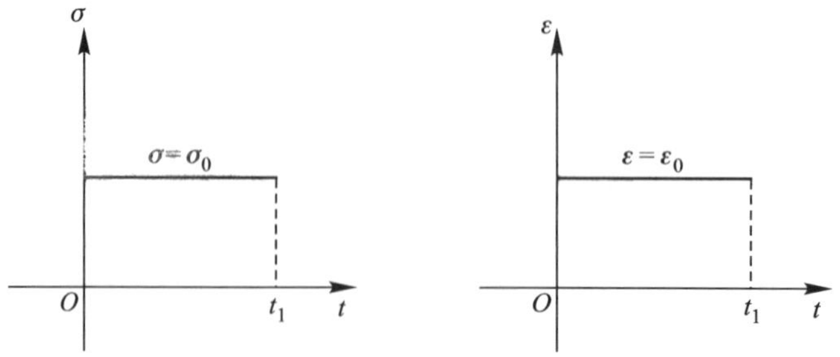

Figure 7.1

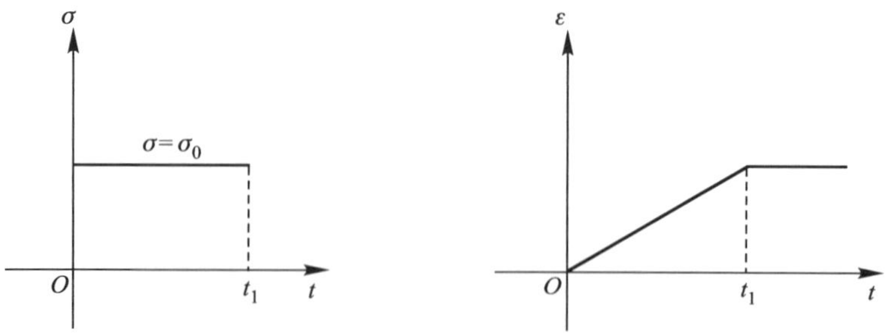

Figure 7.2

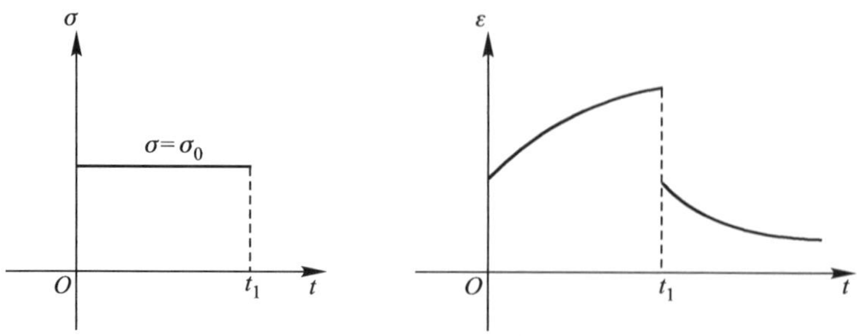

Figure 7.3

Britannica, he conducted a torsional pendulum experiment with metal, wood, cotton, and so on, and found that these materials did not completely obey Hooke's law and the viscous effect must be introduced to consider the internal friction of the material; then he proposed the parallel spring–damper model (see Example 7.4 below), which is now named after him, to simulate this phenomenon. Soon after, in 1867, Maxwell proposed the series spring–damper model (see Example 7.3 below). In 1874 Boltzmann established the three-dimensional theory of the isotropic linear viscoelastic body by using the superposition principle. In 1909, Volterra extended Boltzmann's results to anisotropic material. But the study of viscoelastic bodies did not draw the widespread attention of the scientific and engineering communities during the early stage, and its development was slow. In recent years, people have become compelled to pay more attention to the theory of viscoelasticity thanks to research on macromolecular polymers and composite materials; analysis of the mechanical response and stability of some critical engineering structures (such as nuclear power engineering, and underground engineering); and the appearance of some new disciplines (such as biomechanics). During the late 1950s and early 1960s, Coleman and Noll and others developed the constitutive theory of material with memory. In particular, Coleman's paper, entitled "Thermodynamics of Materials with memory" (see [2]), published in 1964, played an important role in the development of this discipline. During the past 40 years or so, viscoelasticity, along with its mathematical theory, has developed rapidly, some monographs have been published (e.g., [12]), and it has become an active research topic in the mathematical theory of material science, a frontier that is drawing much attention in the worldwide community of applied mathematics.

7.1.2 Simple Viscoelastic Elements

The simplest model to describe both "elastic" and "viscous" features is a spring plus a damper. Now we introduce a few simple models, i.e., viscoelastic elements, to describe the characteristics of viscoelasticity. For completeness, we first discuss the spring and the damper.

Example 7.1 (spring). Let ε be the elongation and σ be the tension of the spring. Then Hooke's law gives

$$\sigma = k\varepsilon \tag{7.1}$$

or

$$\varepsilon = \frac{1}{k}\sigma, \tag{7.2}$$

where k is a positive constant called the *elasticity coefficient*. The above two formulas are the constitutive equations of the spring. ∎

Example 7.2 (damper). The damper is shown in Figure 7.4, and ε and σ are the same as in Example 7.1. For the damper, the damping force is proportional to the velocity of the piston. Then

$$\sigma = \mu\frac{\mathrm{d}\varepsilon}{\mathrm{d}t}, \tag{7.3}$$

where μ is called the *damping coefficient*. Suppose that $\varepsilon(t) \equiv 0$ when t is sufficiently negative; then from the above formula we get

$$\varepsilon(t) = \frac{1}{\mu}\int_{-\infty}^{t} \sigma(\tau)\mathrm{d}\tau. \tag{7.4}$$

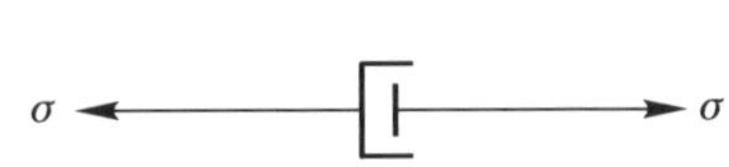

Figure 7.4

Figure 7.5

The above two equations are the constitutive equations of the damper. From (7.4), we know that the value of the damper elongation ε at time t depends on the whole history of the tension σ before time t. ■

Example 7.3 (Maxwell model). Consider a viscoelastic element made up of a spring and a damper in series (see Figure 7.5). Denote by k and μ the elastic coefficient of the spring and the damping coefficient of the damper, respectively, and by ε_1 and ε_2 their elongations, respectively. In the case of a series connection, the tensions received by the spring and the damper are the same, denoted by σ. From (7.1) and (7.3) we get

$$\sigma = k\varepsilon_1, \quad \sigma = \mu\frac{\mathrm{d}\varepsilon_2}{\mathrm{d}t}. \tag{7.5}$$

Noting that the total elongation of the element is $\varepsilon = \varepsilon_1 + \varepsilon_2$, we have

$$\frac{\mathrm{d}\varepsilon}{\mathrm{d}t} = \frac{1}{k}\frac{\mathrm{d}\sigma}{\mathrm{d}t} + \frac{1}{\mu}\sigma. \tag{7.6}$$

Solving the constitutive relation expressed by the differential equation (7.6), under the assumptions that $\sigma(t)$ and $\varepsilon(t)$ are continuously differentiable and $\varepsilon(t) \equiv \sigma(t) \equiv 0$ for sufficiently negative t, we get

$$\varepsilon(t) = \frac{1}{k}\sigma(t) + \frac{1}{\mu}\int_{-\infty}^{t} \sigma(\tau)\mathrm{d}\tau \tag{7.7}$$

or

$$\sigma(t) = k\varepsilon(t) - \frac{k^2}{\mu}\int_{-\infty}^{t} \mathrm{e}^{-\frac{k}{\mu}(t-\tau)}\varepsilon(\tau)\mathrm{d}\tau. \tag{7.8}$$

Equations (7.7) and (7.8) are the constitutive equations of the Maxwell model. From (7.8), we know that the tension σ at time t also depends on the whole history of the elongation ε before time t. ■

Example 7.4 (Kelvin model). Consider a viscoelastic element made up of a spring and a damper in parallel (see Figure 7.6). In the case of a parallel connection, the elongations of the spring and the damper are the same, denoted by ε. Denote by σ_1 and σ_2 their tensions, respectively; then the total tension is

$$\sigma = \sigma_1 + \sigma_2. \tag{7.9}$$

From (7.1) and (7.3) we get

$$\sigma_1 = k\varepsilon, \quad \sigma_2 = \mu\frac{\mathrm{d}\varepsilon}{\mathrm{d}t}. \tag{7.10}$$

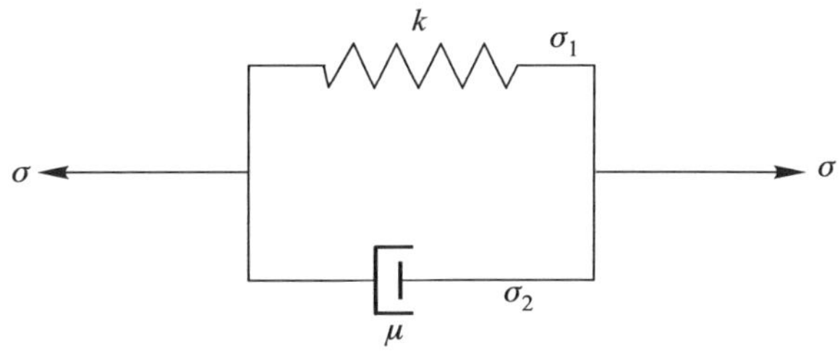

Figure 7.6

Substituting this into (7.9) gives

$$\sigma = k\varepsilon + \mu\frac{\mathrm{d}\varepsilon}{\mathrm{d}t}. \tag{7.11}$$

Assuming that $\varepsilon(t) \equiv 0$ for sufficiently negative t and solving the above differential equation, we obtain

$$\varepsilon(t) = \frac{1}{\mu}\int_{-\infty}^{t} \mathrm{e}^{-\frac{k}{\mu}(t-\tau)}\sigma(\tau)\mathrm{d}\tau. \tag{7.12}$$

Equations (7.11) and (7.12) are the constitutive equations of the Kelvin model. ∎

In Example 7.1, the deformation ε is a linear function of the tension σ. This is a purely elastic model. In Example 7.2, from (7.3), the tension σ is a linear function of the deformation rate $\frac{\mathrm{d}\varepsilon}{\mathrm{d}t}$. This is a purely viscous model. Examples 7.3 and 7.4 are two typical viscoelastic models. For the Maxwell model, if the tension disappears from a certain moment (i.e., there exists a time t_1 such that $\sigma(t) \equiv 0$ when $t \geqslant t_1$), then it is easy to see from (7.7) that $\varepsilon(t)$ tends to a nonzero limit as $t \to +\infty$. That is, the viscoelastic element will have a permanent deformation when the tension disappears, so it behaves like a fluid. Therefore, the Maxwell model is a viscoelastic fluid model. For the Kelvin model, if the tension σ disappears from a certain moment, then it is easy to see from (7.12) that $\varepsilon(t) \to 0$ as $t \to +\infty$. That is, the deformation of the viscoelastic element will eventually disappear when the tension disappears (but unlike the purely elastic case, the deformation does not disappear immediately at the moment when the tension disappears), so it behaves like a solid. Therefore, the Kelvin model is a viscoelastic solid model. Here we point out that it is always very difficult to definitely classify many specific viscoelastic materials as liquid or solid.

Using some basic elements, like the spring and the damper, we can construct a variety of viscoelastic models. For actual viscoelastic materials, we usually use a proper combination of springs and dampers to simulate their stress response and then give the corresponding constitutive relations. For example, there is a model that regards a macromolecular polymer as small beads linked up by springs, and these small beads receive friction from the surrounding medium. The corresponding constitutive relations of this model can be derived accordingly.

The conservation equation(s) of mass and momentum in Chapter 5 of Volume I are still valid for viscoelastic bodies. Next we will discuss the constitutive equations and some related properties of viscoelastic materials.

7.2 Constitutive Equations of Viscoelastic Materials and the Dissipation Inequality

7.2.1 Constitutive Equations of Single-Integral Form

We continue to use the notation in Chapter 5 of Volume I. Suppose that the viscoelastic body occupies a domain $\Omega \subset \mathbb{R}^3$ in space before deformation (assumed to be a certain moment, say, $t = 0$). This domain is called the reference configuration. A particle in this domain is represented by its coordinate vector $\boldsymbol{x} = (x_1, x_2, x_3)$. At time t, Ω deforms to Ω_t, in which the particle is represented by the coordinate vector $\boldsymbol{y} = (y_1, y_2, y_3)$. Then the deformation law is described by

$$\boldsymbol{y} = \boldsymbol{y}(t, \boldsymbol{x}). \tag{7.13}$$

For a purely elastic body, its Cauchy stress tensor $\boldsymbol{T}$ is a function of the deformation gradient tensor $\boldsymbol{F} = \left(\frac{\partial y_i}{\partial x_j}\right)$. From the last section we know that viscoelastic material has memory in general: the Cauchy stress tensor $\boldsymbol{T}$ at time t and particle $\boldsymbol{x}$ should depend on the whole deformation history of the material before time t and around the particle $\boldsymbol{x}$. That is, if we denote $\boldsymbol{F}(t-s, \boldsymbol{x}) = \nabla_x \boldsymbol{y}(t-s, \boldsymbol{x})$ for short by $\boldsymbol{F}^t(s) = \boldsymbol{F}(t-s)$, where $s \in [0, +\infty)$, then $\boldsymbol{T}(t, \boldsymbol{y})$ should be a functional of $\boldsymbol{F}^t(s) = \boldsymbol{F}(t-s)$ $(s \in [0, +\infty))$:

$$\boldsymbol{T}(t, \boldsymbol{y}) = \mathcal{T}(\boldsymbol{F}^t(\cdot)). \tag{7.14}$$

Here, the material is assumed to be homogeneous; i.e., the right-hand side of (7.14) does not depend on $\boldsymbol{x}$ explicitly. Equation (7.14) is the general form of the constitutive equations of a viscoelastic material with memory.

The constitutive equation (7.14) should still be required to satisfy the objectivity hypothesis (see section 5.4 in Chapter 5 of Volume I). In the current situation, the objectivity hypothesis should take the form

$$\boldsymbol{Q}(t)\mathcal{T}(\boldsymbol{F}^t(\cdot))\boldsymbol{Q}^{\mathrm{T}}(t) = \mathcal{T}(\boldsymbol{Q}^t(\cdot)\boldsymbol{F}^t(\cdot)), \tag{7.15}$$

where $\boldsymbol{Q}(t)$ is any given matrix function that assumes values in orthogonal matrices and $\boldsymbol{Q}^t(s) = \boldsymbol{Q}(t-s)$, $s \in [0, +\infty)$.

The discussion starting from the general constitutive equations in the form of (7.14) is extremely complicated. In this chapter, we only discuss a special case that is very important in applications and relatively easy to handle; that is, we consider the following constitutive equation in *single-integral form*:

$$\boldsymbol{P}(\boldsymbol{F}^t) = \boldsymbol{G}(\boldsymbol{F}(t)) + \int_0^\infty \boldsymbol{H}(\tau, \boldsymbol{F}(t), \boldsymbol{F}(t-\tau))\mathrm{d}\tau, \tag{7.16}$$

where $\boldsymbol{P} = (p_{ij})$ is the Piola stress tensor (see section 5.3 in Chapter 5 of Volume I) and $\boldsymbol{G} = (g_{ij})$ and $\boldsymbol{H} = (h_{ij})$ are both second-order tensor functions. Meanwhile, for simplicity, in (7.16) and later discussion, we no longer specify the dependence on $\boldsymbol{x}$ of the tensor functions $\boldsymbol{F}$ and $\boldsymbol{P}$, and so on. From the relation $\boldsymbol{P} = J\boldsymbol{T}\boldsymbol{F}^{-\mathrm{T}}$ between the Cauchy stress tensor and the Piola stress tensor with $J = \det \boldsymbol{F}$, it is easy to see that (7.16) also gives the constitutive equation of single-integral form representing the Cauchy stress tensor.

Affected by the viscous friction, there must be a part of work to be transferred to the heat energy during the deformation of the viscoelastic material, so the heat conduction effect has to be taken into consideration. Therefore, strictly speaking, the right-hand side of the constitutive equations (7.14) and (7.16) should depend on the whole history of the absolute temperature θ before time t. It is a research topic of thermoviscoelasticity to discuss the deformation law of the viscoelastic body under this general kind of framework. However, in this chapter we only consider the special isothermal case; that is, the absolute temperature θ is a constant that depends on neither the particle $\boldsymbol{x}$ nor the time t. In many practical problems, either the temperature of the viscoelastic medium is not much different from point to point and does not change obviously with time, or the viscoelastic medium has rather good heat conductivity so that the nonuniformity of temperature distribution will disappear rapidly due to the heat conduction effect. Under these situations, the deformation of the viscoelastic body can always be regarded as an isothermal process.

Now we start a preliminary discussion about the properties of the tensor functions $\boldsymbol{G}(\boldsymbol{\xi})$ and $\boldsymbol{H}(\tau,\boldsymbol{\xi},\boldsymbol{\eta})$ on the right-hand side of the constitutive equation (7.16). Further discussion about the conditions to be satisfied by these functions are left to the next section when we consider the dissipation inequality.

Is the single-integral form (7.16) unique? That is, are $\boldsymbol{G}(\boldsymbol{\xi})$ and $\boldsymbol{H}(\tau,\boldsymbol{\xi},\boldsymbol{\eta})$ for the same constitutive relation given by the single-integral form (7.16) uniquely determined? The answer is negative. We can see this by simply examining the Maxwell model in Example 7.3 in the last section. In fact, the constitutive equation (7.8) of the Maxwell model can be written in the following single-integral form:

$$\sigma(t) = k\varepsilon(t) - \frac{k^2}{\mu}\int_0^\infty \mathrm{e}^{-\frac{k}{\mu}\tau}\,\varepsilon(t-\tau)\mathrm{d}\tau, \tag{7.17}$$

which can also be written as

$$\sigma(t) = \frac{k^2}{\mu}\int_0^\infty \mathrm{e}^{-\frac{k}{\mu}\tau}(\varepsilon(t)-\varepsilon(t-\tau))\mathrm{d}\tau. \tag{7.18}$$

To guarantee the uniqueness of the single-integral form, we require that $\boldsymbol{H}(\tau,\boldsymbol{\xi},\boldsymbol{\eta})$ satisfy some normalization conditions. Hereafter, we always assume that

$$(\mathrm{A}_1) \quad \boldsymbol{H}(\tau,\boldsymbol{\xi},\boldsymbol{\xi}) = 0, \quad \forall \tau > 0,\ \boldsymbol{\xi} \in \mathbb{R}^{3\times 3}. \tag{7.19}$$

In addition, we always assume that $\boldsymbol{G}(\boldsymbol{\xi})$ and $\boldsymbol{H}(\tau,\boldsymbol{\xi},\boldsymbol{\eta})$ are sufficiently smooth tensor functions with respect to $\tau > 0$, $\boldsymbol{\xi},\boldsymbol{\eta} \in \mathbb{R}^{3\times 3}$, but the singularity of $\boldsymbol{H}(\tau,\boldsymbol{\xi},\boldsymbol{\eta})$ at $\tau = 0$ is allowed to some extent. This kind of singularity does happen in some practical problems. For instance, in the constitutive relation of the viscoelastic model describing a kind of macromolecule polymer (see [5]), it holds that

$$\boldsymbol{H}(\tau,\boldsymbol{\xi},\boldsymbol{\eta}) = a(\tau)(\boldsymbol{\xi}-\boldsymbol{\eta}), \tag{7.20}$$

where

$$a(\tau) = \sum_{n=1}^{\infty} \mathrm{e}^{-n^\alpha \tau}, \quad \alpha > \frac{1}{2}. \tag{7.21}$$

Now we examine the behavior of $a(\tau)$ given by the above formula as $\tau \to +0$. It is easy to verify that

$$\int_1^{\infty} \mathrm{e}^{-s^\alpha \tau}\,\mathrm{d}s < a(\tau) < \int_0^{\infty} \mathrm{e}^{-s^\alpha \tau}\,\mathrm{d}s, \quad \forall \tau > 0, \tag{7.22}$$

and as $\tau \to +0$,

$$\int_0^{\infty} \mathrm{e}^{-s^\alpha \tau}\,\mathrm{d}s - \int_1^{\infty} \mathrm{e}^{-s^\alpha \tau}\,\mathrm{d}s = \int_0^{1} \mathrm{e}^{-s^\alpha \tau}\,\mathrm{d}s \to 1. \tag{7.23}$$

Taking the substitution $\delta = \tau^{\frac{1}{\alpha}} s$, we have

$$\int_0^{\infty} \mathrm{e}^{-s^\alpha \tau}\,\mathrm{d}s = \tau^{-\frac{1}{\alpha}} \int_0^{\infty} \mathrm{e}^{-\delta^\alpha}\,\mathrm{d}\delta. \tag{7.24}$$

Formulas (7.22)–(7.24) show that as $\tau \to +0$,

$$a(\tau) \sim \tau^{-\frac{1}{\alpha}}; \tag{7.25}$$

that is, $\boldsymbol{H}(\tau,\boldsymbol{\xi},\boldsymbol{\eta})$ given by (7.20) has singularity like $\tau^{-\frac{1}{\alpha}}$ as $\tau = 0$. When $\alpha > 1$, this singularity is an integrable singularity. For convenience, we assume below that $\boldsymbol{H}(\tau,\boldsymbol{\xi},\boldsymbol{\eta})$ has an integrable singularity at $\tau = 0$. (But in practical problems, this singularity is not always integrable!) To be precise, we make the following assumptions:

(A2) For any given compact set K in $\mathbb{R}^{3\times 3}$, there exists a function $b \in L^1(0,\infty)$ such that

$$|\boldsymbol{H}(\tau,\boldsymbol{\xi},\boldsymbol{\eta})|, \quad |\mathrm{D}_\xi \boldsymbol{H}(\tau,\boldsymbol{\xi},\boldsymbol{\eta})|, \quad |\mathrm{D}_\eta \boldsymbol{H}(\tau,\boldsymbol{\xi},\boldsymbol{\eta})| \leqslant b(\tau), \\ \forall \tau > 0,\ \boldsymbol{\xi},\boldsymbol{\eta} \in K, \tag{7.26}$$

where $\mathrm{D}_\xi \boldsymbol{H}(\tau,\boldsymbol{\xi},\boldsymbol{\eta}) = \left\{\frac{\partial h_{ij}}{\partial \xi_{kl}}\right\}$ is a fourth-order tensor, ξ_{kl} are the components of the second-order tensor $\boldsymbol{\xi}$, and so forth; moreover, the norm $|\cdot|$ of the tensor appearing on the left-hand side of (7.26) stands for the maximum of the absolute values of the components of this tensor.

(A3) For any given compact set K in $\mathbb{R}^{3\times 3}$, there exists a function $a(\tau)$ satisfying the following conditions:

$$a \in L^1(\rho,\infty), \quad \forall \rho > 0, \tag{7.27}$$

$$\text{as } \rho \to 0, \quad \rho \int_\rho^{\infty} a(\tau)\mathrm{d}\tau \to 0, \tag{7.28}$$

such that

$$|\partial_\tau \boldsymbol{H}(\tau,\boldsymbol{\xi},\boldsymbol{\eta})|, \quad |\partial_\tau \mathrm{D}_\eta \boldsymbol{H}(\tau,\boldsymbol{\xi},\boldsymbol{\eta})| \leqslant a(\tau), \quad \forall \tau > 0,\ \boldsymbol{\xi},\boldsymbol{\eta} \in K, \tag{7.29}$$

where $\partial_\tau \boldsymbol{H}(\tau,\boldsymbol{\xi},\boldsymbol{\eta}) = \dfrac{\partial}{\partial \tau}\boldsymbol{H}(\tau,\boldsymbol{\xi},\boldsymbol{\eta})$, and so forth.

Denote

$$C_b^1(\mathbb{R};\mathbb{R}^{3\times 3}) = \left\{\boldsymbol{\xi}(t) \in C^1(\mathbb{R};\mathbb{R}^{3\times 3}), \boldsymbol{\xi}(t) \text{ and } \frac{\mathrm{d}\boldsymbol{\xi}(t)}{\mathrm{d}t} \text{ are bounded as } t \to -\infty\right\}.$$

It is easy to see that, if $\boldsymbol{H}(\tau,\boldsymbol{\xi},\boldsymbol{\eta})$ satisfies the above hypotheses (A_1)–(A_3), then for any given $\boldsymbol{\xi}(t) \in C_b^1(\mathbb{R};\mathbb{R}^{3\times 3})$, the integral on the right-hand side of the single-integral form

$$\boldsymbol{P}(\boldsymbol{\xi}^t) = \boldsymbol{G}(\boldsymbol{\xi}(t)) + \int_0^\infty \boldsymbol{H}(\tau,\boldsymbol{\xi}(t),\boldsymbol{\xi}(t-\tau))\mathrm{d}\tau \tag{7.30}$$

makes sense. $\boldsymbol{H}(\tau,\boldsymbol{\xi},\boldsymbol{\eta})$ in (7.30) is called the *kernel of the single-integral form.* From hypothesis (A_1), when $\boldsymbol{\xi}(t)$ takes constant value, we always have $\boldsymbol{P}(\boldsymbol{\xi}^t) = \boldsymbol{G}(\boldsymbol{\xi}(t))$, so $\boldsymbol{G}(\boldsymbol{\xi})$ is called the *equilibrium response function.* For the tensor $\boldsymbol{G}(\boldsymbol{\xi})$, we always assume that

$$\boldsymbol{G}(\boldsymbol{I}) = 0, \tag{7.31}$$

where $\boldsymbol{I}$ is the unit tensor.

Lemma 7.1. *Assume that $\boldsymbol{H}(\tau,\boldsymbol{\xi},\boldsymbol{\eta})$ satisfies the following conditions: for any given compact set K in $\mathbb{R}^{3\times 3}$, there exists a function $a \in L^1(\rho,\infty)$ $\forall \rho > 0$ such that*

$$|\boldsymbol{H}(\tau,\boldsymbol{\xi},\boldsymbol{\eta})| \leqslant a(\tau), \quad \forall \tau > 0,\ \boldsymbol{\xi},\boldsymbol{\eta} \in K \tag{7.32}$$

and

$$\int_{0+}^\infty \boldsymbol{H}(\tau,\boldsymbol{\zeta}(t),\boldsymbol{\zeta}(t-\tau))\mathrm{d}\tau \text{ exists}, \quad \forall \boldsymbol{\zeta} \in C_b^1(\mathbb{R};\mathbb{R}^{3\times 3}). \tag{7.33}$$

Here and hereafter, we adopt the following notation:

$$\int_{0+}^z = \lim_{\rho\to+0}\int_\rho^z .$$

Assume furthermore that $\boldsymbol{\xi} : \mathbb{R} \to \mathbb{R}^{3\times 3}$ is a tensor function with finitely many discontinuous points and piecewise-constant values. Then there exists a function sequence $\boldsymbol{\xi}^n \in C_b^1(\mathbb{R};\mathbb{R}^{3\times 3})$ such that when t is not a discontinuous point, we have

$$\lim_{n\to\infty}\int_0^\infty \boldsymbol{H}(\tau,\boldsymbol{\xi}^n(t),\boldsymbol{\xi}^n(t-\tau))\mathrm{d}\tau = \int_{0+}^\infty \boldsymbol{H}(\tau,\boldsymbol{\xi}(t),\boldsymbol{\xi}(t-\tau))\mathrm{d}\tau. \tag{7.34}$$

Proof. Without loss of generality, we assume that $\boldsymbol{\xi}$ has only one discontinuous point $d \in \mathbb{R}$. Construct $\boldsymbol{\xi}^n \in C_b^1(\mathbb{R};\mathbb{R}^{3\times 3})$ such that

$$\boldsymbol{\xi}^n(\tau) = \boldsymbol{\xi}(\tau) \quad \text{if } \tau \notin \left(d - \frac{1}{n}, d + \frac{1}{n}\right) \tag{7.35}$$

and

$$|\boldsymbol{\xi}^n(\tau)| \leqslant C, \quad \forall \tau \in \mathbb{R}, \forall n, \tag{7.36}$$

where C is a positive constant independent of τ and n.

Noticing that $t \neq d$, there must exist $\rho_0 > 0$ such that when n is sufficiently large,

$$[t-\rho_0, t] \cap \left(d - \frac{1}{n}, d + \frac{1}{n}\right) = \emptyset. \tag{7.37}$$

Then, when $0 < \tau < \rho_0$, $\boldsymbol{\xi}^n(t-\tau) = \boldsymbol{\xi}(t-\tau)$. Hence, for any given $\rho \in (0, \rho_0)$, noting (7.35), when n is suitably large, it is easy to see that

$$\begin{aligned} &\int_\rho^\infty \boldsymbol{H}(\tau, \boldsymbol{\xi}^n(t), \boldsymbol{\xi}^n(t-\tau))\mathrm{d}\tau \\ &= \int_\rho^{\rho_0} \boldsymbol{H}(\tau, \boldsymbol{\xi}(t), \boldsymbol{\xi}(t-\tau))\mathrm{d}\tau + \int_{\rho_0}^\infty \boldsymbol{H}(\tau, \boldsymbol{\xi}^n(t), \boldsymbol{\xi}^n(t-\tau))\mathrm{d}\tau. \end{aligned} \tag{7.38}$$

But from hypothesis (7.33) we know that

$$\lim_{\rho \to 0} \int_\rho^{\rho_0} \boldsymbol{H}(\tau, \boldsymbol{\xi}(t), \boldsymbol{\xi}(t-\tau))\mathrm{d}\tau = \int_{0+}^{\rho_0} \boldsymbol{H}(\tau, \boldsymbol{\xi}(t), \boldsymbol{\xi}(t-\tau))\mathrm{d}\tau; \tag{7.39}$$

moreover, from (7.32), (7.35)–(7.36), and Lebesgue's dominated convergence theorem, we have

$$\lim_{n \to \infty} \int_{\rho_0}^\infty \boldsymbol{H}(\tau, \boldsymbol{\xi}^n(t), \boldsymbol{\xi}^n(t-\tau))\mathrm{d}\tau = \int_{\rho_0}^\infty \boldsymbol{H}(\tau, \boldsymbol{\xi}(t), \boldsymbol{\xi}(t-\tau))\mathrm{d}\tau. \tag{7.40}$$

Thus (7.34) follows. The lemma is proved. □

Theorem 7.1. *For any given single-integral form* (7.30), $\boldsymbol{G}$ *and* $\boldsymbol{H}$ *satisfying hypotheses* (A_1)–(A_3) *are uniquely determined.*

Proof. We need only prove that, if $\boldsymbol{H}(\tau, \boldsymbol{\xi}, \boldsymbol{\eta})$ satisfies hypotheses (A_1)–(A_3) and for any given $\boldsymbol{\xi} \in C_b^1(\mathbb{R}; \mathbb{R}^{3\times 3})$ we have

$$\boldsymbol{G}(\boldsymbol{\xi}(t)) + \int_0^\infty \boldsymbol{H}(\tau, \boldsymbol{\xi}(t), \boldsymbol{\xi}(t-\tau))\mathrm{d}\tau = 0, \quad \forall t \in \mathbb{R}, \tag{7.41}$$

then

$$\boldsymbol{G}(\boldsymbol{\xi}) \equiv 0, \quad \boldsymbol{H}(\tau, \boldsymbol{\xi}, \boldsymbol{\eta}) \equiv 0. \tag{7.42}$$

First, taking $\boldsymbol{\xi}(t)$ to be a constant value $\boldsymbol{\xi}$ in (7.41), it follows immediately from hypothesis (A_1) that

$$\boldsymbol{G}(\boldsymbol{\xi}) = 0, \quad \forall \boldsymbol{\xi} \in \mathbb{R}^{3\times 3}. \tag{7.43}$$

Thus, from (7.41), for any given $\boldsymbol{\xi}(t) \in C_b^1(\mathbb{R};\mathbb{R}^{3\times 3})$ we get

$$\int_0^\infty \boldsymbol{H}(\tau, \boldsymbol{\xi}(t), \boldsymbol{\xi}(t-\tau))\mathrm{d}\tau = 0. \tag{7.44}$$

For any given $\boldsymbol{\xi}, \boldsymbol{\eta} \in \mathbb{R}^{3\times 3}$ and $\delta > 0$, set

$$\zeta(s) = \begin{cases} \boldsymbol{\eta} & \text{if } s \in [-\delta, 0], \\ \boldsymbol{\xi} & \text{otherwise.} \end{cases}$$

Such a $\zeta(s)$ has two discontinuous points and takes piecewise-constant values. By Lemma 7.1, (7.44) still holds for such $\zeta(s)$. From this and noting hypothesis (A_1), for any given $t > 0$, we easily have

$$\int_0^\infty \boldsymbol{H}(\tau, \zeta(t), \zeta(t-\tau))\mathrm{d}\tau = \int_t^{t+\delta} \boldsymbol{H}(\tau, \boldsymbol{\xi}, \boldsymbol{\eta})\mathrm{d}\tau = 0, \quad \forall \delta > 0. \tag{7.45}$$

Dividing the above formula by δ and then letting $\delta \to 0$, we get

$$\boldsymbol{H}(t, \boldsymbol{\xi}, \boldsymbol{\eta}) = 0.$$

Thus Theorem 7.1 is proved. □

Remark 7.1. From the above proof we can see that the conclusion of Theorem 7.1 is still true, provided that $\boldsymbol{H}(\tau, \boldsymbol{\xi}, \boldsymbol{\eta})$ satisfies hypothesis (A_1) and the conditions in Lemma 7.1, and that for any fixed $\boldsymbol{\xi}, \boldsymbol{\eta} \in \mathbb{R}^{3\times 3}$, $\boldsymbol{H}(\tau, \boldsymbol{\xi}, \boldsymbol{\eta})$ is continuous with respect to τ.

7.2.2 The Dissipation Inequality and Its Consequences

Since no heat conduction effect is considered, from (6.25) and (6.32) in the last chapter we know that the conservation equation of energy under the current situation reduces to

$$\rho_0 \frac{\partial e}{\partial t} = \sum_{i,j=1}^{3} p_{ij} \frac{\partial v_i}{\partial x_j} + \rho_0 \gamma, \tag{7.46}$$

while the entropy inequality reduces to

$$\frac{\partial \eta}{\partial t} \geqslant \frac{\gamma}{\theta}. \tag{7.47}$$

Eliminating γ from (7.46) and (7.47), we get

$$\rho_0 \frac{\partial e}{\partial t} - \sum_{i,j=1}^{3} p_{ij} \frac{\partial v_i}{\partial x_j} - \rho_0 \theta \frac{\partial \eta}{\partial t} \leqslant 0. \tag{7.48}$$

In the isothermal (i.e., θ is constant) case, introducing the Helmholtz free energy $\psi = e - \theta\eta$ and noticing that $\frac{\partial v_i}{\partial x_j} = \frac{\partial f_{ij}}{\partial t}$, (7.48) can be rewritten as

$$\rho_0 \frac{\partial \psi}{\partial t} \leqslant \sum_{i,j=1}^{3} p_{ij} \frac{\partial f_{ij}}{\partial t}. \tag{7.49}$$

More generally, for the single-integral form (7.30), if there exists a scalar single-integral form

$$\omega(\boldsymbol{\xi}^t) = l(\boldsymbol{\xi}(t)) + \int_0^{\infty} m(\tau, \boldsymbol{\xi}(t), \boldsymbol{\xi}(t-\tau))\mathrm{d}\tau \tag{7.50}$$

such that

$$\frac{\mathrm{d}\omega(\boldsymbol{\xi}^t)}{\mathrm{d}t} \leqslant \sum_{i,j=1}^{3} p_{ij}(\boldsymbol{\xi}^t) \frac{\mathrm{d}\xi_{ij}(t)}{\mathrm{d}t}, \quad \forall \boldsymbol{\xi} \in C_b^1(\mathbb{R}; \mathbb{R}^{3\times 3}), \tag{7.51}$$

where $p_{ij}(\boldsymbol{\xi}^t)$ are components of the second-order tensor $\boldsymbol{P}(\boldsymbol{\xi}^t)$ given by (7.30), then the single-integral form (7.30) is called *dissipative*, while (7.51) is called the corresponding *dissipation inequality*, and $\omega(\boldsymbol{\xi}^t)$ is called the *dissipation potential* of (7.30). The $l(\boldsymbol{\xi})$ and $m(\tau, \boldsymbol{\xi}, \boldsymbol{\eta})$ on the right-hand side of the scalar single-integral form (7.50) are required to satisfy conditions similar to those satisfied by $\boldsymbol{G}(\xi)$ and $\boldsymbol{H}(\tau, \boldsymbol{\xi}, \boldsymbol{\eta})$, respectively.

From (7.49) we can see that the constitutive relation given by (7.16) is dissipative if the Helmholtz free energy exists and can be expressed by a single-integral form.

Under what conditions satisfied by $\boldsymbol{G}$ and $\boldsymbol{H}$ is the single-integral form (7.30) dissipative? The following theorem (see [10]) will give a sufficient and necessary condition. When $\boldsymbol{G}$ and $\boldsymbol{H}$ satisfy this condition, we can directly construct from them the l and m on the right-hand side of the scalar single-integral form (7.50).

Theorem 7.2. *The single-integral form* (7.30) *is dissipative if and only if, for any given* $\boldsymbol{\xi}, \boldsymbol{\eta} \in \mathbb{R}^{3\times 3}$ *and* $\tau > 0$,

$$\mathrm{D}_{\xi}\, \boldsymbol{G}(\boldsymbol{\xi}), \mathrm{D}_{\xi}\, \boldsymbol{H}(\tau, \boldsymbol{\xi}, \boldsymbol{\eta}) \text{ are symmetric} \tag{7.52}$$

and

$$\int_{\boldsymbol{\eta}}^{\boldsymbol{\xi}} \partial_\tau \boldsymbol{H}(\tau, \boldsymbol{\zeta}, \boldsymbol{\eta}) \cdot \mathrm{d}\boldsymbol{\zeta} \leqslant 0. \tag{7.53}$$

Here and hereafter, $\partial_\tau \boldsymbol{H} \cdot \mathrm{d}\boldsymbol{\zeta} = \sum_{i,j=1}^{3} \partial_\tau h_{ij} \mathrm{d}\zeta_{ij}$, *i.e., it can be regarded as the inner product of two* 3×3*-dimensional vectors.*

Note 7.1. $\mathrm{D}_{\xi}\, \boldsymbol{G}(\boldsymbol{\xi}) = \left(\frac{\partial g_{ij}(\boldsymbol{\xi})}{\partial \xi_{kl}}\right)$ is a fourth-order tensor. The symmetry of $\mathrm{D}_{\xi}\, \boldsymbol{G}(\boldsymbol{\xi})$ in (7.52) means that

$$\frac{\partial g_{ij}(\boldsymbol{\xi})}{\partial \xi_{kl}} = \frac{\partial g_{kl}(\boldsymbol{\xi})}{\partial \xi_{ij}} \quad (i,j,k,l = 1,2,3). \tag{7.54}$$

The symmetry of $\mathrm{D}_{\xi}\, \boldsymbol{H}(\tau, \boldsymbol{\xi}, \boldsymbol{\eta})$ can be similarly understood.

Note 7.2. Suppose that $\mathrm{D}_\xi \boldsymbol{G}(\boldsymbol{\xi})$ and $\mathrm{D}_\xi \boldsymbol{H}(\tau,\boldsymbol{\xi},\boldsymbol{\eta})$ are symmetric; then similar to Lemma 1.2 in Chapter 1 of Volume I, we know that there exist scalar functions $l(\boldsymbol{\xi})$ and $m(\tau,\boldsymbol{\xi},\boldsymbol{\eta})$ such that

$$g_{ij}(\boldsymbol{\xi}) = \frac{\partial l(\boldsymbol{\xi})}{\partial \xi_{ij}}, \qquad h_{ij}(\tau,\boldsymbol{\xi},\boldsymbol{\eta}) = \frac{\partial m(\tau,\boldsymbol{\xi},\boldsymbol{\eta})}{\partial \xi_{ij}},$$
$$\forall \tau > 0,\ \boldsymbol{\xi},\boldsymbol{\eta} \in \mathbb{R}^{3\times 3},\ i,j = 1,2,3.$$

In fact, such $l(\boldsymbol{\xi})$ and $m(\tau,\boldsymbol{\xi},\boldsymbol{\eta})$ can be given, respectively, by the following integrals:

$$l(\boldsymbol{\xi}) = \int_{\boldsymbol{I}}^{\boldsymbol{\xi}} \boldsymbol{G}(\boldsymbol{\zeta})\cdot \mathrm{d}\boldsymbol{\zeta}, \tag{7.55}$$

$$m(\tau,\boldsymbol{\xi},\boldsymbol{\eta}) = \int_{\boldsymbol{\eta}}^{\boldsymbol{\xi}} \boldsymbol{H}(\tau,\boldsymbol{\zeta},\boldsymbol{\eta})\cdot \mathrm{d}\boldsymbol{\zeta}. \tag{7.56}$$

Obviously, under the symmetry condition of (7.52), the above two integrals and the integral on the left-hand side of (7.53) are all independent of the path of integration.

Before proving Theorem 7.2, we first give the following lemmas.

Lemma 7.2. *Suppose that $m(\tau,\boldsymbol{\xi},\boldsymbol{\eta})$ satisfies hypotheses* (A_1)–(A_3)*; then*

$$\int_0^\infty \mathrm{D}_\eta m(\tau,\boldsymbol{\xi}(t),\boldsymbol{\xi}(t-\tau))\cdot \frac{\mathrm{d}\boldsymbol{\xi}(t-\tau)}{\mathrm{d}t}\mathrm{d}\tau$$
$$= \int_{0+}^\infty \partial_\tau m(\tau,\boldsymbol{\xi}(t),\boldsymbol{\xi}(t-\tau))\mathrm{d}\tau, \quad \forall \boldsymbol{\xi} \in C_b^1(\mathbb{R};\mathbb{R}^{3\times 3}), \tag{7.57}$$

where $\mathrm{D}_\eta m(\tau,\boldsymbol{\xi},\boldsymbol{\eta})$ stands for the second-order tensor $\left(\frac{\partial m(\tau,\boldsymbol{\xi},\boldsymbol{\eta})}{\partial \eta_{kl}}\right)$ and η_{kl} are components of the second-order tensor $\boldsymbol{\eta}$.

Proof. Denote

$$\phi(\tau) = \mathrm{D}_\eta m(\tau,\boldsymbol{\xi}(t),\boldsymbol{\xi}(t-\tau))\cdot \frac{\mathrm{d}\boldsymbol{\xi}(t-\tau)}{\mathrm{d}t}. \tag{7.58}$$

It follows from hypothesis (A_2) and $\boldsymbol{\xi} \in C_b^1(\mathbb{R};\mathbb{R}^{3\times 3})$ that $\phi \in L^1(0,\infty)$. Noting that $\frac{\mathrm{d}\boldsymbol{\xi}(t-\tau)}{\mathrm{d}t} = -\frac{\mathrm{d}\boldsymbol{\xi}(t-\tau)}{\mathrm{d}\tau}$, we obviously have

$$\phi(\tau) = \partial_\tau m(\tau,\boldsymbol{\xi}(t),\boldsymbol{\xi}(t-\tau)) - \frac{\mathrm{d}}{\mathrm{d}\tau} m(\tau,\boldsymbol{\xi}(t),\boldsymbol{\xi}(t-\tau)). \tag{7.59}$$

Using hypotheses (A_2) and (A_3), it is easy to verify that

$$\text{as } \tau \to +\infty, \quad m(\tau,\boldsymbol{\xi},\boldsymbol{\eta}) \to 0 \tag{7.60}$$

holds uniformly for $\boldsymbol{\xi},\boldsymbol{\eta} \in K$, where K is any given compact set in $\mathbb{R}^{3\times 3}$. Thus, noticing the boundedness of $\boldsymbol{\xi}(t)$ as $t \to -\infty$, it follows from the above formula that

$$\begin{aligned}
&\int_{\rho}^{\infty} \frac{\mathrm{d}}{\mathrm{d}\tau} m(\tau,\boldsymbol{\xi}(t),\boldsymbol{\xi}(t-\tau))\mathrm{d}\tau \\
&= -m(\rho,\boldsymbol{\xi}(t),\boldsymbol{\xi}(t-\rho)) \\
&= \int_{\rho}^{\infty} \partial_\tau m(\tau,\boldsymbol{\xi}(t),\boldsymbol{\xi}(t-\rho))\mathrm{d}\tau, \quad \forall \rho > 0.
\end{aligned} \tag{7.61}$$

Then from (7.59) we have

$$\begin{aligned}
&\int_0^{\infty} \phi(\tau)\mathrm{d}\tau \\
&= \int_0^{\rho} \phi(\tau)\mathrm{d}\tau + \int_{\rho}^{\infty} \partial_\tau m(\tau,\boldsymbol{\xi}(t),\boldsymbol{\xi}(t-\tau))\mathrm{d}\tau \\
&\quad - \int_{\rho}^{\infty} \partial_\tau m(\tau,\boldsymbol{\xi}(t),\boldsymbol{\xi}(t-\rho))\mathrm{d}\tau, \quad \forall \rho > 0.
\end{aligned} \tag{7.62}$$

Obviously,

$$\text{as } \rho \to +0, \qquad \int_0^{\rho} \phi(\tau)\mathrm{d}\tau \to 0. \tag{7.63}$$

Using hypothesis (A_1) and (7.29) in hypothesis (A_3), we get

$$\begin{aligned}
&|\partial_\tau m(\tau,\boldsymbol{\xi}(t),\boldsymbol{\xi}(t-\rho))| \\
&= |\partial_\tau m(\tau,\boldsymbol{\xi}(t),\boldsymbol{\xi}(t-\rho)) - \partial_\tau m(\tau,\boldsymbol{\xi}(t),\boldsymbol{\xi}(t))| \\
&\leqslant a(\tau)|\boldsymbol{\xi}(t-\rho) - \boldsymbol{\xi}(t)|.
\end{aligned} \tag{7.64}$$

Thus, noting (7.28) in hypothesis (A_3), we can easily see that

$$\text{as } \rho \to +0, \qquad \int_{\rho}^{\infty} \partial_\tau m(\tau,\boldsymbol{\xi}(t),\boldsymbol{\xi}(t-\rho))\mathrm{d}\tau \to 0. \tag{7.65}$$

Then (7.57) follows immediately by letting $\rho \to +0$ in (7.62) and noting (7.63), (7.65), and (7.58). The proof is finished. □

Lemma 7.3. *Suppose that $\boldsymbol{\xi} \in C_b^1(\mathbb{R};\mathbb{R}^{3\times 3})$, $t \in \mathbb{R}$, $\boldsymbol{c} \in \mathbb{R}^{3\times 3}$; then there exists $\boldsymbol{\xi}^n \in C_b^1(\mathbb{R};\mathbb{R}^{3\times 3})$ such that as $n \to \infty$, we have*

$$\begin{aligned}
&\boldsymbol{\xi}^n(\tau) \to \boldsymbol{\xi}(\tau) \text{ uniformly for } \tau \in (-\infty,t); \\
&\frac{\mathrm{d}\boldsymbol{\xi}^n(\tau)}{\mathrm{d}\tau} \to \frac{\mathrm{d}\boldsymbol{\xi}(\tau)}{\mathrm{d}\tau}, \quad \forall \tau \in (-\infty,t); \\
&\left|\frac{\mathrm{d}\boldsymbol{\xi}^n(\tau)}{\mathrm{d}\tau}\right| \text{ is uniformly bounded for } \tau \in (-\infty,t) \text{ and } n;
\end{aligned}$$

and for any given n, we have

$$\boldsymbol{\xi}^n(t) = \boldsymbol{\xi}(t), \qquad \left.\frac{\mathrm{d}\boldsymbol{\xi}^n(\tau)}{\mathrm{d}\tau}\right|_{\tau=t} = \boldsymbol{c}.$$

The proof of this lemma is left as an exercise to the reader (see Exercise 4).

***Proof of Theorem* 7.2.** We first prove the necessity of conditions (7.52)–(7.53). Suppose that there exists a scalar single-integral form (7.50) such that the dissipation inequality (7.51) holds. Differentiating (7.50) once with respect to t, we get

$$\begin{aligned}\frac{\mathrm{d}\omega(\boldsymbol{\xi}^t)}{\mathrm{d}t} &= \mathrm{D}_\xi\,\omega(\boldsymbol{\xi}^t)\cdot\frac{\mathrm{d}\boldsymbol{\xi}(t)}{\mathrm{d}t}\\ &\quad+\int_0^\infty \mathrm{D}_\eta\, m(\tau,\boldsymbol{\xi}(t),\boldsymbol{\xi}(t-\tau))\cdot\frac{\mathrm{d}\boldsymbol{\xi}(t-\tau)}{\mathrm{d}t}\mathrm{d}\tau,\end{aligned} \tag{7.66}$$

where

$$\mathrm{D}_\xi\,\omega(\boldsymbol{\xi}^t)=\mathrm{D}_\xi\, l(\boldsymbol{\xi}(t))+\int_0^\infty \mathrm{D}_\xi\, m(\tau,\boldsymbol{\xi}(t),\boldsymbol{\xi}(t-\tau))\mathrm{d}\tau \tag{7.67}$$

represents the derivative of the single-integral form (7.50) with respect to the current value $\boldsymbol{\xi}(t)$ of $\boldsymbol{\xi}^t$, which is called the *instantaneous derivative* of the single-integral form (7.50). Plugging (7.66) into the dissipation inequality (7.51), we obtain

$$\begin{aligned}&(\mathrm{D}_\xi\,\omega(\boldsymbol{\xi}^t)-P(\boldsymbol{\xi}^t))\cdot\frac{\mathrm{d}\boldsymbol{\xi}(t)}{\mathrm{d}t}+\int_0^\infty \mathrm{D}_\eta\, m(\tau,\boldsymbol{\xi}(t),\boldsymbol{\xi}(t-\tau))\cdot\frac{\mathrm{d}\boldsymbol{\xi}(t-\tau)}{\mathrm{d}t}\mathrm{d}\tau\leqslant 0,\\ &\forall\boldsymbol{\xi}\in C_b^1(\mathbb{R};\mathbb{R}^{3\times 3}).\end{aligned} \tag{7.68}$$

In the above formula, taking $\boldsymbol{\xi}(t)$ as $\boldsymbol{\xi}^n(t)$ in Lemma 7.3 and then sending $n\to\infty$, we get

$$\begin{aligned}&(\mathrm{D}_\xi\,\omega(\boldsymbol{\xi}^t)-P(\boldsymbol{\xi}^t))\cdot\boldsymbol{c}+\int_0^\infty \mathrm{D}_\eta\, m(\tau,\boldsymbol{\xi}(t),\boldsymbol{\xi}(t-\tau))\cdot\frac{\mathrm{d}\boldsymbol{\xi}(t-\tau)}{\mathrm{d}t}\mathrm{d}\tau\leqslant 0,\\ &\forall\boldsymbol{\xi}\in C_b^1(\mathbb{R};\mathbb{R}^{3\times 3}),\ \forall\boldsymbol{c}\in\mathbb{R}^{3\times 3}.\end{aligned} \tag{7.69}$$

This yields immediately that

$$\int_0^\infty \mathrm{D}_\eta\, m(\tau,\boldsymbol{\xi}(t),\boldsymbol{\xi}(t-\tau))\cdot\frac{\mathrm{d}\boldsymbol{\xi}(t-\tau)}{\mathrm{d}t}\mathrm{d}\tau\leqslant 0,\quad\forall\boldsymbol{\xi}\in C_b^1(\mathbb{R};\mathbb{R}^{3\times 3}) \tag{7.70}$$

and

$$\mathrm{D}_\xi\,\omega(\boldsymbol{\xi}^t)-P(\boldsymbol{\xi}^t)=0. \tag{7.71}$$

Noticing (7.67), we see that (7.71) can be also written as

$$P(\boldsymbol{\xi}^t)=\mathrm{D}_\xi\, l(\boldsymbol{\xi}(t))+\int_0^\infty \mathrm{D}_\xi\, m(\tau,\boldsymbol{\xi}(t),\boldsymbol{\xi}(t-\tau))\mathrm{d}\tau. \tag{7.72}$$

In addition, from Lemma 7.2, (7.70) can be written as

$$\int_{0+}^\infty \partial_\tau m(\tau,\boldsymbol{\xi}(t),\boldsymbol{\xi}(t-\tau))\mathrm{d}\tau\leqslant 0,\quad\forall\boldsymbol{\xi}\in C_b^1(\mathbb{R};\mathbb{R}^{3\times 3}). \tag{7.73}$$

Noting that $\partial_\tau m(\tau,\boldsymbol{\xi},\boldsymbol{\eta})$ satisfies the same conditions as $\boldsymbol{H}(\tau,\boldsymbol{\xi},\boldsymbol{\eta})$ in Lemma 7.1, repeating the steps in the proof of Theorem 7.1, from (7.73) we have

$$\partial_\tau m(\tau,\boldsymbol{\xi},\boldsymbol{\eta})\leqslant 0,\quad\forall\tau>0,\ \boldsymbol{\xi},\boldsymbol{\eta}\in\mathbb{R}^{3\times 3}. \tag{7.74}$$

Integrating the above formula with respect to τ from τ to $+\infty$, and noting (7.60), we obtain

$$m(\tau,\boldsymbol{\xi},\boldsymbol{\eta}) \geqslant 0, \quad \forall \tau > 0,\ \boldsymbol{\xi},\boldsymbol{\eta} \in \mathbb{R}^{3\times 3}. \tag{7.75}$$

Hence, by hypothesis (A_1), $m(\tau,\boldsymbol{\xi},\boldsymbol{\eta})$ achieves its minimum at $\boldsymbol{\xi} = \boldsymbol{\eta}$; then

$$\mathrm{D}_{\xi}\, m(\tau,\boldsymbol{\xi},\boldsymbol{\xi}) = 0, \quad \forall \tau > 0,\ \boldsymbol{\xi} \in \mathbb{R}^{3\times 3}, \tag{7.76}$$

i.e., $\mathrm{D}_{\xi}\, m(\tau,\boldsymbol{\xi},\boldsymbol{\eta})$ satisfies hypothesis (A_1) as well. Comparing (7.30) and (7.72), from the uniqueness of the single-integral form ($\mathrm{D}_{\xi}\, m(\tau,\boldsymbol{\xi},\boldsymbol{\eta})$ does not completely satisfy hypotheses (A_1)–(A_3), but from Remark 7.1 we know that Theorem 7.1 is still valid under this situation), we have

$$\begin{aligned} \boldsymbol{G}(\boldsymbol{\xi}) = \mathrm{D}_{\xi}\, l(\boldsymbol{\xi}), \quad & \boldsymbol{H}(\tau,\boldsymbol{\xi},\boldsymbol{\eta}) = \mathrm{D}_{\xi}\, m(\tau,\boldsymbol{\xi},\boldsymbol{\eta}), \\ & \forall \tau > 0,\ \boldsymbol{\xi},\boldsymbol{\eta} \in \mathbb{R}^{3\times 3}. \end{aligned} \tag{7.77}$$

This proves the symmetry of $\mathrm{D}_{\xi}\, \boldsymbol{G}(\boldsymbol{\xi})$ and $\mathrm{D}_{\xi}\, \boldsymbol{H}(\tau,\boldsymbol{\xi},\boldsymbol{\eta})$, and

$$m(\tau,\boldsymbol{\xi},\boldsymbol{\eta}) = \int_{\boldsymbol{\eta}}^{\boldsymbol{\xi}} \boldsymbol{H}(\tau,\boldsymbol{\zeta},\boldsymbol{\eta}) \cdot \mathrm{d}\boldsymbol{\zeta}. \tag{7.78}$$

Inequality (7.53) follows from (7.74). The necessity is proved.

Now we prove the sufficiency of (7.52)–(7.53). Suppose that conditions (7.52) and (7.53) are satisfied; then the integrals on the right-hand side of (7.55) and (7.56) are independent of the path of integration connecting $\boldsymbol{\xi}$ and $\boldsymbol{\eta}$, and then the functions defined by these two integrals are denoted by $l(\boldsymbol{\xi})$ and $m(\tau,\boldsymbol{\xi},\boldsymbol{\eta})$, respectively. Now, the single-integral form $\omega(\boldsymbol{\xi}^t)$ given by (7.50) must satisfy the dissipation inequality (7.51); i.e., the sufficiency is achieved. In fact, from (7.53) and the expression (7.56) of $m(\tau,\boldsymbol{\xi},\boldsymbol{\eta})$, we have

$$\partial_{\tau} m(\tau,\boldsymbol{\xi},\boldsymbol{\eta}) \leqslant 0, \quad \forall \tau > 0,\ \boldsymbol{\xi},\boldsymbol{\eta} \in \mathbb{R}^{3\times 3}. \tag{7.79}$$

Then, starting from (7.74), and deriving in the direction opposite to the proof of necessity, we can prove that $\omega(\boldsymbol{\xi}^t)$ satisfies the dissipation inequality. The theorem is proved. □

Now we consider two special cases.

1. The kernel of the single-integral form (7.30) takes the following form:

$$\boldsymbol{H}(\tau,\boldsymbol{\xi},\boldsymbol{\eta}) = \widehat{\boldsymbol{H}}(\tau,\boldsymbol{\xi}) - \widehat{\boldsymbol{H}}(\tau,\boldsymbol{\eta}). \tag{7.80}$$

To ensure the uniqueness of $\widehat{\boldsymbol{H}}$, it is usually required that, for a specific value $\boldsymbol{\xi}^0 \in \mathbb{R}^{3\times 3}$ (in general, taking $\boldsymbol{\xi}^0 = \boldsymbol{I}$),

$$\widehat{\boldsymbol{H}}(\tau,\boldsymbol{\xi}^0) = 0. \tag{7.81}$$

For the single-integral form (7.30) with this kind of kernel, its instantaneous derivative is

$$\mathrm{D}_{\xi}\, \boldsymbol{P}(\boldsymbol{\xi}^t) = \mathrm{D}_{\xi}\, \boldsymbol{G}(\boldsymbol{\xi}(t)) + \int_0^{\infty} \mathrm{D}_{\xi}\, \widehat{\boldsymbol{H}}(\tau,\boldsymbol{\xi}(t))\mathrm{d}\tau. \tag{7.82}$$

Thus, for two tensor functions $\boldsymbol{\xi}$ and $\boldsymbol{\zeta} \in C_b^1(\mathbb{R},\mathbb{R}^{3\times3})$, if they have the same value at time t, i.e.,

$$\boldsymbol{\xi}(t) = \boldsymbol{\zeta}(t), \tag{7.83}$$

then we have

$$\mathrm{D}_\xi\, \boldsymbol{P}(\boldsymbol{\xi}^t) = \mathrm{D}_\xi\, \boldsymbol{P}(\boldsymbol{\zeta}^t). \tag{7.84}$$

That is, the instantaneous change of the single-integral form with kernel of form (7.80) is independent of the history of $\boldsymbol{\xi}^t$ before time t. This is a very important case in applications.

Theorem 7.3. *The single-integral form* (7.30) *with kernel of form* (7.80) *is dissipative if and only if*

$$\mathrm{D}_\xi\, \boldsymbol{G}(\boldsymbol{\xi}), \mathrm{D}_\xi\, \widehat{\boldsymbol{H}}(\tau,\boldsymbol{\xi}) \text{ is symmetric}, \quad \forall \tau > 0,\ \boldsymbol{\xi} \in \mathbb{R}^{3\times3} \tag{7.85}$$

and

$$\partial_\tau \widehat{m}(\tau,\boldsymbol{\xi}) \text{ is concave (upward convex) with respect to } \boldsymbol{\xi}, \tag{7.86}$$

where

$$\widehat{m}(\tau,\boldsymbol{\xi}) = \int_{\boldsymbol{I}}^{\boldsymbol{\xi}} \widehat{\boldsymbol{H}}(\tau,\boldsymbol{\zeta}) \cdot \mathrm{d}\boldsymbol{\zeta}. \tag{7.87}$$

Proof. By Theorem 7.2, we need only illustrate (7.86). For the kernel $\boldsymbol{H}(\tau,\boldsymbol{\xi},\boldsymbol{\eta})$ of form (7.80), formula (7.53) in Theorem 7.2 is reduced to

$$\int_{\boldsymbol{\eta}}^{\boldsymbol{\xi}} \partial_\tau \widehat{\boldsymbol{H}}(\tau,\boldsymbol{\zeta}) \cdot \mathrm{d}\boldsymbol{\zeta} \leqslant \partial_\tau \widehat{\boldsymbol{H}}(\tau,\boldsymbol{\eta}) \cdot (\boldsymbol{\xi} - \boldsymbol{\eta}). \tag{7.88}$$

From (7.87), this can also be written as

$$\partial_\tau \widehat{m}(\tau,\boldsymbol{\xi}) - \partial_\tau \widehat{m}(\tau,\boldsymbol{\eta}) \leqslant \mathrm{D}_\xi\, \partial_\tau \widehat{m}(\tau,\boldsymbol{\eta}) \cdot (\boldsymbol{\xi} - \boldsymbol{\eta}), \qquad \forall \tau > 0,\ \boldsymbol{\xi},\boldsymbol{\eta} \in \mathbb{R}^{3\times3}, \tag{7.89}$$

which is equivalent to the concavity of $\partial_\tau \widehat{m}(\tau,\boldsymbol{\xi})$. □

Remark 7.2. It is not hard to see from Theorems 7.2 and 7.3 that, under conditions (7.85) and (7.86), the kernel of the dissipation potential $\omega(\boldsymbol{\xi}^t)$ of $\boldsymbol{P}(\boldsymbol{\xi}^t)$ is

$$m(\tau,\boldsymbol{\xi},\boldsymbol{\eta}) = \widehat{m}(\tau,\boldsymbol{\xi}) - \widehat{m}(\tau,\boldsymbol{\eta}) - \widehat{\boldsymbol{H}}(\tau,\boldsymbol{\eta}) \cdot (\boldsymbol{\xi} - \boldsymbol{\eta}), \tag{7.90}$$

and noting that $\widehat{m}(\tau,\boldsymbol{\xi}) \to 0$ as $\tau \to +\infty$, from (7.89) we get

$$\widehat{m}(\tau,\boldsymbol{\xi}) \text{ is convex (downward convex) with respect to } \boldsymbol{\xi}. \tag{7.91}$$

2. The kernel of the single-integral form (7.30) takes the following form:

$$\boldsymbol{H}(\tau,\boldsymbol{\xi},\boldsymbol{\eta}) = \widehat{\boldsymbol{H}}(\tau,\boldsymbol{\xi} - \boldsymbol{\eta}). \tag{7.92}$$

Then from Theorem 7.2 we immediately have the following result.

Theorem 7.4. *The single-integral form* (7.30) *with kernel of form* (7.92) *is dissipative if and only if*

$$\mathrm{D}_\xi\, \boldsymbol{G}(\boldsymbol{\xi}), \mathrm{D}_\xi\, \widehat{\boldsymbol{H}}(\tau,\boldsymbol{\xi}) \text{ are symmetric}, \quad \forall \tau > 0,\ \boldsymbol{\xi} \in \mathbb{R}^{3\times 3} \tag{7.93}$$

and

$$\partial_\tau \widehat{m}(\tau,\boldsymbol{\xi}) \leqslant 0, \quad \forall \tau > 0,\ \boldsymbol{\xi} \in \mathbb{R}^{3\times 3}, \tag{7.94}$$

where

$$\widehat{m}(\tau,\boldsymbol{\xi}) = \int_0^{\boldsymbol{\xi}} \widehat{\boldsymbol{H}}(\tau,\boldsymbol{\zeta}) \cdot \mathrm{d}\boldsymbol{\zeta}. \tag{7.95}$$

Remark 7.3. Under conditions (7.93) and (7.94), the kernel of the dissipation potential $\omega(\boldsymbol{\xi}^t)$ of the single-integral form $\boldsymbol{P}(\boldsymbol{\xi}^t)$ is

$$m(\tau,\boldsymbol{\xi},\boldsymbol{\eta}) = \widehat{m}(\tau,\boldsymbol{\xi}-\boldsymbol{\eta}), \tag{7.96}$$

and from (7.94) we get

$$\widehat{m}(\tau,\boldsymbol{\xi}) \geqslant 0, \quad \forall \tau > 0,\ \boldsymbol{\xi} \in \mathbb{R}^{3\times 3}. \tag{7.97}$$

Remark 7.4. The single-integral form with the above two special kinds of kernels, if regarded as the constitutive relation, does not satisfy the objectivity hypothesis in general.

7.2.3 Constitutive Equations of the Linear Viscoelastic Body

Suppose that the reference configuration Ω is under the natural state, i.e., the stress tensor is assumed to be zero. Suppose that there is a small deformation around the reference configuration, i.e., it is assumed that

$$\|\boldsymbol{F}^t(\cdot) - \boldsymbol{I}\|_{L^\infty(0,\infty)} \ll 1; \tag{7.98}$$

here the $L^\infty(0,\infty)$ norm of a second-order tensor means the maximum of the $L^\infty(0,\infty)$ norms of all its components.

Set

$$\nabla \boldsymbol{u}^t(\tau) = \boldsymbol{F}^t(\tau) - \boldsymbol{I}, \tag{7.99}$$

where $\boldsymbol{u} = \boldsymbol{y} - \boldsymbol{x}$ is the displacement vector. Then the constitutive equation (7.16) can be written as

$$\boldsymbol{P}(t) = \boldsymbol{G}(\boldsymbol{I} + \nabla \boldsymbol{u}(t)) + \int_0^\infty \boldsymbol{H}(\tau, \boldsymbol{I} + \nabla \boldsymbol{u}(t), \boldsymbol{I} + \nabla \boldsymbol{u}(t-\tau))\mathrm{d}\tau. \tag{7.100}$$

Here, for simplicity, we use $\boldsymbol{P}(t)$ to represent $\boldsymbol{P}(\boldsymbol{F}^t)$. Noting that $\boldsymbol{G}(\boldsymbol{I}) = 0$ and $\boldsymbol{H}(\tau,\boldsymbol{I},\boldsymbol{I}) = 0$ (see (7.31) and (7.19)), the above formula can be rewritten as

$$\begin{aligned}\boldsymbol{P}(t) = &\left(\mathrm{D}_\xi\, \boldsymbol{G}(\boldsymbol{I}) + \int_0^\infty \mathrm{D}_\xi\, \boldsymbol{H}(\tau,\boldsymbol{I},\boldsymbol{I})\mathrm{d}\tau\right) \nabla \boldsymbol{u}(t) \\ &+ \int_0^\infty \mathrm{D}_\eta\, \boldsymbol{H}(\tau,\boldsymbol{I},\boldsymbol{I}) \nabla \boldsymbol{u}(t-\tau)\mathrm{d}\tau \\ &+ o(\|\nabla \boldsymbol{u}^t(\cdot)\|_{L^\infty(0,\infty)}),\end{aligned} \tag{7.101}$$

where the product $\mathrm{D}_\xi \boldsymbol{G}\nabla\boldsymbol{u}$ of the fourth-order tensor and the second-order tensor is defined by the second-order tensor $\left(\sum_{i,j,k,l=1}^{3} \frac{\partial g_{ij}}{\partial \xi_{kl}} \frac{\partial u_k}{\partial x_l}\right)$; the others are similar.

From

$$\boldsymbol{A}'(\tau) = \mathrm{D}_\eta \boldsymbol{H}(\tau, \boldsymbol{I}, \boldsymbol{I}) \tag{7.102}$$

and

$$\boldsymbol{A}(0) = \mathrm{D}_\xi \boldsymbol{G}(\boldsymbol{I}) + \int_0^\infty \mathrm{D}_\xi \boldsymbol{H}(\tau, \boldsymbol{I}, \boldsymbol{I})\mathrm{d}\tau, \tag{7.103}$$

we can uniquely determine on $[0, +\infty)$ a function $\boldsymbol{A}(\tau) = (a_{ijkl}(\tau))$ assuming values in fourth-order tensors, called the *relaxation tensor* of the single-integral form (7.30) corresponding to the natural state. Thus, the constitutive equation (7.101) can be written as

$$\begin{aligned} \boldsymbol{P}(t) = {} & \boldsymbol{A}(0)\nabla\boldsymbol{u}(t) + \int_0^\infty \boldsymbol{A}'(\tau)\nabla\boldsymbol{u}(t-\tau)\mathrm{d}\tau \\ & + o(\|\nabla\boldsymbol{u}^t(\cdot)\|_{L^\infty(0,\infty)}). \end{aligned} \tag{7.104}$$

Theorem 7.5. *Suppose that $\boldsymbol{A}(\tau)$ is the relaxation tensor of the dissipative single-integral form* (7.30) *corresponding to the natural state; then*

(1) *for any given $\tau \geqslant 0$, $\boldsymbol{A}(\tau)$ is symmetric, i.e., $a_{ijkl}(\tau) = a_{klij}(\tau)$;*

(2) *for $\tau > 0$, $\boldsymbol{A}'(\tau)$ is negative semidefinite: $\boldsymbol{A}'(\tau) \leqslant 0$, i.e.,*

$$\sum_{i,j,k,l=1}^{3} a'_{ijkl}(\tau)\xi_{ij}\xi_{kl} \leqslant 0, \quad \forall \boldsymbol{\xi} = (\xi_{ij}) \in \mathbb{R}^{3\times 3};$$

(3) *for $\tau > 0$, $\boldsymbol{A}''(\tau)$ is positive semidefinite: $\boldsymbol{A}''(\tau) \geqslant 0$, i.e.,*

$$\sum_{i,j,k,l=1}^{3} a''_{ijkl}(\tau)\xi_{ij}\xi_{kl} \geqslant 0, \quad \forall \boldsymbol{\xi} = (\xi_{ij}) \in \mathbb{R}^{3\times 3}.$$

Proof. Since $\boldsymbol{H}(\tau, \boldsymbol{\xi}, \boldsymbol{\eta})$ satisfies hypothesis (A_1), it is easy to see that

$$\mathrm{D}_\xi \boldsymbol{H}(\tau, \boldsymbol{\xi}, \boldsymbol{\xi}) = -\mathrm{D}_\eta \boldsymbol{H}(\tau, \boldsymbol{\xi}, \boldsymbol{\xi}), \quad \forall \boldsymbol{\xi} \in \mathbb{R}^{3\times 3}. \tag{7.105}$$

Then, from Theorem 7.2, $\mathrm{D}_\eta \boldsymbol{H}(\tau, \boldsymbol{I}, \boldsymbol{I})$ is symmetric. Using (7.102), the symmetry of $\boldsymbol{A}'(\tau)$ for $\tau > 0$ is obtained. From Theorem 7.2 and (7.103), $\boldsymbol{A}(0)$ is also symmetric. This proves (1) .

Assume that $m(\tau, \boldsymbol{\xi}, \boldsymbol{\eta})$ is the kernel of the dissipation potential of the single-integral form (7.30); from (7.75) and (7.74) in the proof of Theorem 7.2, we have

$$m(\tau, \boldsymbol{\xi}, \boldsymbol{\eta}) \geqslant 0, \partial_\tau m(\tau, \boldsymbol{\xi}, \boldsymbol{\eta}) \leqslant 0, \quad \forall \tau > 0,\ \boldsymbol{\xi}, \boldsymbol{\eta} \in \mathbb{R}^{3\times 3}. \tag{7.106}$$

From this and noting $m(\tau,\boldsymbol{\xi},\boldsymbol{\xi})=\partial_\tau m(\tau,\boldsymbol{\xi},\boldsymbol{\xi})=0$, we know that $m(\tau,\boldsymbol{\xi},\boldsymbol{\eta})$ and $\partial_\tau m(\tau,\boldsymbol{\xi},\boldsymbol{\eta})$ achieve the minimum and the maximum at $\boldsymbol{\eta}=\boldsymbol{\xi}$, respectively. Then

$$\mathrm{D}_\eta m(\tau,\boldsymbol{\xi},\boldsymbol{\xi})=0, \tag{7.107}$$

$$\mathrm{D}_\eta^2 m(\tau,\boldsymbol{\xi},\boldsymbol{\xi})\geqslant 0, \tag{7.108}$$

and

$$\mathrm{D}_\eta^2\,\partial_\tau m(\tau,\boldsymbol{\xi},\boldsymbol{\xi})\leqslant 0 \tag{7.109}$$

for any given $\tau>0$ and $\boldsymbol{\xi}\in\mathbb{R}^{3\times 3}$. Similarly to (7.105), from (7.107) we get

$$\mathrm{D}_\xi\,\mathrm{D}_\eta m(\tau,\boldsymbol{\xi},\boldsymbol{\xi})=-\mathrm{D}_\eta^2 m(\tau,\boldsymbol{\xi},\boldsymbol{\xi}). \tag{7.110}$$

Moreover, from Note 7.2 after Theorem 7.2, $\boldsymbol{H}(\tau,\boldsymbol{\xi},\boldsymbol{\eta})=\mathrm{D}_\xi m(\tau,\boldsymbol{\xi},\boldsymbol{\eta})$. Then

$$\mathrm{D}_\eta \boldsymbol{H}(\tau,\boldsymbol{I},\boldsymbol{I})=\mathrm{D}_\xi\,\mathrm{D}_\eta m(\tau,\boldsymbol{I},\boldsymbol{I})=-\mathrm{D}_\eta^2 m(\tau,\boldsymbol{I},\boldsymbol{I}). \tag{7.111}$$

Thus, using (7.108)–(7.109) and (7.111), the conclusions of (2) and (3) follow immediately from the expression (7.102) of $\boldsymbol{A}'(\tau)$. Theorem 7.5 is proved. □

Linearizing (7.104) with higher-order infinitesimals neglected, we get

$$\boldsymbol{P}(t)=\boldsymbol{A}(0)\nabla\boldsymbol{u}(t)+\int_0^\infty \boldsymbol{A}'(\tau)\nabla\boldsymbol{u}(t-\tau)\mathrm{d}\tau, \tag{7.112}$$

which can be written in component form as

$$\begin{aligned}p_{ij}(t)=&\sum_{k,l=1}^{3}a_{ijkl}(0)\frac{\partial u_k}{\partial x_l}\\&+\sum_{k,l=1}^{3}\int_0^\infty a'_{ijkl}(\tau)\frac{\partial u_k}{\partial x_l}(t-\tau)\mathrm{d}\tau\quad(i,j=1,2,3).\end{aligned} \tag{7.113}$$

This is exactly the constitutive equation of the *linear viscoelastic body*.

After linearization with higher-order infinitesimals neglected, the Piola stress tensor $\boldsymbol{P}$ and the Cauchy stress tensor $\boldsymbol{T}$ are the same. Then, from the symmetry of the Cauchy stress tensor (see Theorem 5.2 in Chapter 5 of Volume I) and (7.113), we know that the relaxation tensor $\boldsymbol{A}(\tau)=(a_{ijkl}(\tau))$ also has the following symmetry:

$$a_{ijkl}(\tau)=a_{jikl}(\tau)\quad(i,j,k,l=1,2,3). \tag{7.114}$$

Noting also conclusion (1) in Theorem 7.5 (where the symmetry of $\boldsymbol{A}(\tau)$ means $a_{ijkl}(\tau)=a_{klij}(\tau)$), we get the following symmetry of $\boldsymbol{A}(\tau)$:

$$a_{ijkl}(\tau)=a_{klij}(\tau)=a_{jikl}(\tau)=a_{ijlk}(\tau). \tag{7.115}$$

Considering the above symmetry of $\boldsymbol{A}(\tau)$, the constitutive equation (7.113) of the linear viscoelastic body can be rewritten as

$$\begin{aligned}p_{ij}(t) &= \sum_{k,l=1}^{3} a_{ijkl}(0)e_{kl}(t)\\ &+ \sum_{k,l=1}^{3}\int_0^{\infty} a'_{ijkl}(\tau)e_{kl}(t-\tau)\mathrm{d}\tau \quad (i,j=1,2,3),\end{aligned} \tag{7.116}$$

where

$$e_{kl} = \frac{1}{2}\left(\frac{\partial u_k}{\partial x_l} + \frac{\partial u_l}{\partial x_k}\right) \tag{7.117}$$

are components of the infinitesimal strain tensor $\boldsymbol{E}$. Equation (7.116) can also be simplified as

$$\boldsymbol{P}(t) = \boldsymbol{A}(0)\boldsymbol{E}(t) + \int_0^{\infty} \boldsymbol{A}'(\tau)\boldsymbol{E}(t-\tau)\mathrm{d}\tau. \tag{7.118}$$

Equation (7.116) or (7.118) is the constitutive equation commonly used in linear viscoelasticity, where $\boldsymbol{A}(0)$ is also called the *instantaneous elastic tensor*. When $\boldsymbol{E}(\tau)$ does not depend on time τ, from (7.118) we have $\boldsymbol{P}(t) = \boldsymbol{A}(+\infty)\boldsymbol{E}$, where $\boldsymbol{A}(+\infty) = \lim_{\tau\to\infty}\boldsymbol{A}(\tau)$; then $\boldsymbol{A}(+\infty)$ is called the *equilibrium elastic tensor*.

It is easy to see that the constitutive equation (7.118) can also be written in the following form:

$$\boldsymbol{P}(t) = \boldsymbol{A}(0)\boldsymbol{E}(t) + \int_{-\infty}^{t} \boldsymbol{A}'(t-\tau)\boldsymbol{E}(\tau)\mathrm{d}\tau. \tag{7.119}$$

If we regard $\boldsymbol{E}(t)$ as the unknown function, the above formula is an integral system of Volterra type with convolution form for $\boldsymbol{E}(t)$. Under certain conditions, from (7.119) we can get that

$$\boldsymbol{E}(t) = \boldsymbol{B}(0)\boldsymbol{P}(t) + \int_{-\infty}^{t} \boldsymbol{B}'(t-\tau)\boldsymbol{P}(\tau)\mathrm{d}\tau, \tag{7.120}$$

where $\boldsymbol{B}(t)$ is a fourth-order tensor function called the *creep tensor*. Physical explanations of the relaxation tensor $\boldsymbol{A}(t)$ and the creep tensor $\boldsymbol{B}(t)$ will be given later.

Now we discuss isotropic viscoelastic materials. For isotropic materials, the fourth-order tensor $\boldsymbol{A}(\tau) = (a_{ijkl}(\tau))$ is an isotropic tensor. From the general form of fourth-order isotropic tensors (see Appendix A in Volume I), we have

$$a_{ijkl}(\tau) = \lambda(\tau)\delta_{ij}\delta_{kl} + \alpha(\tau)\delta_{ik}\delta_{jl} + \beta(\tau)\delta_{il}\delta_{jk}, \tag{7.121}$$

where $\lambda(\tau)$, $\alpha(\tau)$, and $\beta(\tau)$ are all scalar functions. Denote

$$\mu(\tau) = \frac{1}{2}(\alpha(\tau) + \beta(\tau)); \tag{7.122}$$

from (7.121) and noting the symmetry of the second-order tensor $\boldsymbol{E}(t)$, it is easy to see that the constitutive equation (7.116) can be written in the form

$$\begin{aligned} p_{ij}(t) &= \lambda(0)(e_{11}(t)+e_{22}(t)+e_{33}(t))\delta_{ij}+2\mu(0)e_{ij}(t) \\ &+\delta_{ij}\int_0^\infty \lambda'(\tau)(e_{11}(t-\tau)+e_{22}(t-\tau)+e_{33}(t-\tau))\mathrm{d}\tau \\ &+\int_0^\infty 2\mu'(\tau)e_{ij}(t-\tau)\mathrm{d}\tau, \end{aligned} \tag{7.123}$$

where $\lambda(\tau)$ and $\mu(\tau)$ are called the *Lamé relaxation modulus*. This is just the constitutive equation of the isotropic linear viscoelastic material. The above formula gives the stress–strain relation. Correspondingly, the constitutive equation of strain–stress form can also be given.

From (7.123) we immediately get

$$p_{ij}(t) = a_1(0)e_{ij}(t)+\int_0^\infty a_1'(\tau)e_{ij}(t-\tau)\mathrm{d}\tau \quad (i,j=1,2,3,i\neq j), \tag{7.124}$$

where

$$a_1(\tau) = 2\mu(\tau). \tag{7.125}$$

In addition, taking $i=j$ in (7.123) and summing over i from 1 to 3, we obtain

$$\begin{aligned} &p_{11}(t)+p_{22}(t)+p_{33}(t) \\ &= a_2(0)(e_{11}(t)+e_{22}(t)+e_{33}(t)) \\ &+\int_0^\infty a_2'(\tau)(e_{11}(t-\tau)+e_{22}(t-\tau)+e_{33}(t-\tau))\mathrm{d}\tau, \end{aligned} \tag{7.126}$$

where

$$a_2(\tau) = 3\lambda(\tau)+2\mu(\tau). \tag{7.127}$$

Similarly, we can also give the strain–stress relation corresponding to (7.124) and (7.126):

$$e_{ij}(t) = b_1(0)p_{ij}(t)+\int_0^\infty b_1'(\tau)p_{ij}(t-\tau)\mathrm{d}\tau \quad (i,j=1,2,3,i\neq j) \tag{7.128}$$

and

$$\begin{aligned} &e_{11}(t)+e_{22}(t)+e_{33}(t) \\ &= b_2(0)(p_{11}(t)+p_{22}(t)+p_{33}(t)) \\ &+\int_0^\infty b_2'(\tau)(p_{11}(t-\tau)+p_{22}(t-\tau)+p_{33}(t-\tau))\mathrm{d}\tau. \end{aligned} \tag{7.129}$$

Functions a_1, a_2 and b_1, b_2 appearing in the above formulas are called the *relaxation functions* and the *creep functions*, respectively. Just as μ represents the sheer modulus and $\lambda+\frac{2}{3}\mu$ represents the volume elastic modulus in elasticity, a_1, a_2 and b_1, b_2 correspond to the relaxation functions and the creep functions of the sheer deformation and the volume expansion, respectively.

Now we further discuss the significance and property of relaxation functions and creep functions. To this end, we unify (7.124) and (7.126) in the following form:

$$\sigma(t)=a(0)\varepsilon(t)+\int_0^\infty a'(\tau)\varepsilon(t-\tau)\mathrm{d}\tau, \tag{7.130}$$

where σ and ε are the corresponding stress and strain, respectively, and the relaxation function $a(\tau)$ can be taken as $a_1(\tau)$ or $a_2(\tau)$ according to the situation under consideration. Here we point out that the constitutive equation also takes the form (7.130) for one-dimensional problems (see section 7.3) and $a(\tau)=a_1(\tau)$ for pure sheer deformation. This indicates the significance of discussing the constitutive equation of the from (7.130).

Assume that a constant strain $\varepsilon(t)\equiv\varepsilon_0$ is produced starting from $t=0$, i.e.,

$$\varepsilon(t)=\varepsilon_0 H(t), \tag{7.131}$$

where ε_0 is a constant, and

$$H(t)=\begin{cases} 1, & t\geqslant 0, \\ 0, & t<0, \end{cases}$$

is the *Heaviside function*. By (7.130), corresponding to this strain, we have

$$\sigma(t)=\varepsilon_0 a(t). \tag{7.132}$$

From the mechanical point of view, it is obvious that we have

$$a(t)\geqslant 0. \tag{7.133}$$

Moreover, for such a constant strain, the stress of the viscoelastic body should decrease gradually with the increase of time, i.e., the state relaxes gradually (see Figure 7.7). This is why $a(t)$ is called the relaxation function. Therefore, the relaxation function $a(t)$ should be a monotonically decreasing function of t, i.e.,

$$a'(t)\leqslant 0. \tag{7.134}$$

The constitutive equation (7.130) shows that the stress $\sigma(t)$ at time t depends on the whole history $\varepsilon^t(\tau)=\varepsilon(t-\tau)$, $\tau\in[0,+\infty)$, before this moment. From the intuitive point of view, the dependence of the stress at current time t on the values of the strain at different times in the history is certainly different, and the longer the time is, the weaker the dependence is. This is usually called the *fading memory hypothesis*. Based on this hypothesis, in the dependence relation of the stress on the strain history, we should not equally treat the values of the strain at different times in the history, which can be reflected by introducing a weight function in the dependence relation. As a matter of fact, $a'(\tau)$ on the right-hand side of the constitutive equation (7.130) can work as a weight function. Thus, according to the fading memory hypothesis, the relaxation function $a(\tau)$ should further satisfy

$$|a'(t_1)|\geqslant|a'(t_2)|, \quad \forall\, 0<t_1<t_2. \tag{7.135}$$

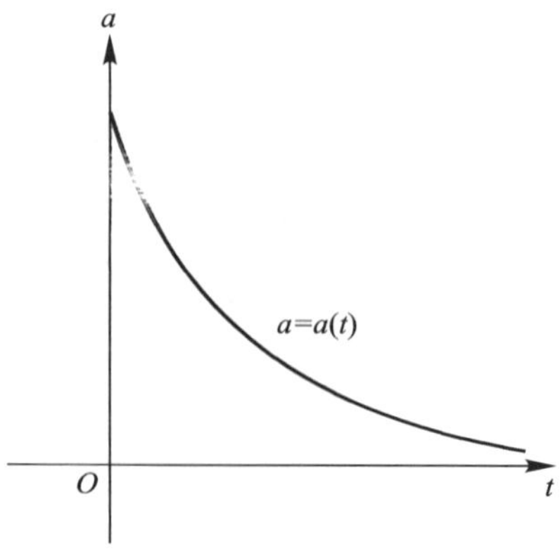

Figure 7.7

Noting (7.134), the above formula can be written as

$$a'(t_1) \leqslant a'(t_2), \quad \forall\, 0 < t_1 < t_2,$$

i.e., $a'(t)$ is a monotonically increasing function of t:

$$a''(t) \geqslant 0. \tag{7.136}$$

Formulas (7.134) and (7.136) are consistent with the results given in Theorem 7.5.

The strain–stress relation corresponding to the stress–strain relation (7.130) is

$$\varepsilon(t) = b(0)\sigma(t) + \int_0^{\infty} b'(\tau)\sigma(t-\tau)\mathrm{d}\tau, \tag{7.137}$$

where the creep function $b(\tau)$ is taken as $b_1(\tau)$ or $b_2(\tau)$ according to the situation under consideration. Suppose that a constant stress $\sigma(t) \equiv \sigma_0$ is imposed starting from $t = 0$, i.e.,

$$\sigma(t) = \sigma_0 H(t), \tag{7.138}$$

where σ_0 is a constant. Corresponding to this stress, from (7.137) we have

$$\varepsilon(t) = \sigma_0 b(t). \tag{7.139}$$

Obviously,

$$b(t) \geqslant 0. \tag{7.140}$$

Moreover, under the action of a constant stress, the strain should increase with the increase of time; therefore, $b(t)$ should be a monotonically increasing function of t, i.e.,

$$b'(t) \geqslant 0. \tag{7.141}$$

For the creep function $b(t)$, the fading memory hypothesis leads to

$$|b'(t_1)| \geqslant |b'(t_2)|, \quad \forall 0 < t_1 < t_2.$$

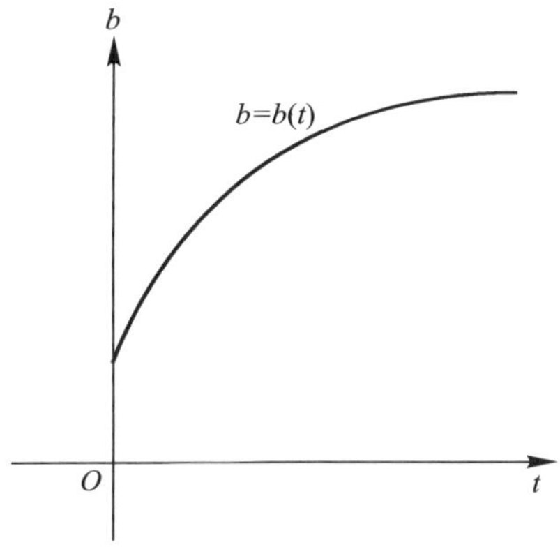

Figure 7.8

Noticing (7.141), the above formula can be written as

$$b'(t_1) \geqslant b'(t_2), \quad \forall 0 < t_1 < t_2,$$

i.e., $b'(t)$ is a monotonically decreasing function of t:

$$b''(t) \leqslant 0. \tag{7.142}$$

The above arguments show that the function $b(t)$ increases with the increase of time, but its speed of increase is getting slower and slower (see Figure 7.8). This situation is called the *creep* and explains why $b(t)$ is called the creep function.

Now we roughly analyze how we can tell whether the material under discussion is a viscoelastic solid or a viscoelastic fluid according to the relaxation function and the creep function. Just as stated in section 7.1.1, for an elastic body that is a solid deforming under loads, the deformation disappears immediately when it is unloaded, while for a viscous medium that is a fluid deforming under loads, the deformation does not disappear and a permanent deformation happens when it is unloaded. Viscoelastic material lies between the above two cases; it is difficult to classify it as a solid or a fluid. However, different materials have different characteristics, such as plastics and asphalt, which are both viscoelastic materials: the former has the characteristics of a solid, while the latter is closer to a fluid. Now we roughly (not strictly) distinguish between what kind of viscoelastic material is a solid and what is a fluid according to the above analysis.

By (7.141) and (7.142), $b'(t)$ is a nonnegative decreasing function. If

$$b'(+\infty) = \lim_{t \to +\infty} b'(t) = 0, \tag{7.143}$$

then the corresponding material is called a *viscoelastic solid*, while if

$$b'(+\infty) > 0, \tag{7.144}$$

then the corresponding material is called a *viscoelastic fluid*.

Now we explain the reasonableness of the above classification. Suppose that a sudden constant stress σ_0 is imposed at $t = t_1$ and this stress is suddenly removed at $t = t_2$. From (7.137) we have, as $t \to +\infty$,

$$\varepsilon(t) = (b(t-t_1) - b(t-t_2))\sigma_0 \to b'(+\infty)(t_2 - t_1)\sigma_0. \tag{7.145}$$

Therefore, under condition (7.143), the deformation eventually disappears with the increase of time t, and the material has the characteristics of a solid, while, under condition (7.144), the deformation will not disappear no matter how much time passes, and a permanent deformation happens as $t \to +\infty$; this meets the characteristics of a fluid.

For the Maxwell model given in Example 7.3 of section 7.1, it is easy to see from (7.7) that $a(t)$ and $b(t)$ in (7.130) and (7.137) are, respectively,

$$a(t) = k\,\mathrm{e}^{-\frac{k}{\mu}t}, \quad b(t) = \frac{1}{k} + \frac{1}{\mu}t. \tag{7.146}$$

Obviously we have $b'(+\infty) = b'(t) = \frac{1}{\mu} > 0$, so this model has the characteristics of a viscoelastic fluid. However, for the Kelvin model given in Example 7.4 of section 7.1, it is easy to derive from (7.12), correspondingly, that

$$b(t) = \frac{1}{k}(1 - \mathrm{e}^{-\frac{k}{\mu}t}). \tag{7.147}$$

On this occasion, $b'(+\infty) = 0$, so the Kelvin model has the characteristics of a viscoelastic solid.

In viscoelasticity, the relaxation function of a linear viscoelastic solid usually takes the form

$$a(t) = a_0 + a_1\,\mathrm{e}^{-\frac{t}{\mu}}, \tag{7.148}$$

where a_0, a_1, and μ are all positive constants. More generally, for some macromolecule polymers, the relaxation function takes the form

$$a(t) = a_0 + \sum_{i=1}^{N} a_i\,\mathrm{e}^{-\frac{t}{\mu_i}}, \tag{7.149}$$

where a_0, a_i, and μ_i $(i = 1, \ldots, N)$ are all positive constants.

Now we show that a viscoelastic material with relaxation function given by (7.149) must be a solid. For this, we prove a more general conclusion: if $a(t)$ is a sufficiently smooth function satisfying conditions (7.133)–(7.134) and (7.136), and

$$a(+\infty) \neq 0, \tag{7.150}$$

then (7.143) holds. For simplicity, let $\sigma \equiv \varepsilon \equiv 0$ as $t \leqslant 0$. Taking the Laplace transform to (7.130) and (7.137), respectively, and noticing the properties of the Laplace transform (see [4]), we easily get

$$\widehat{\sigma}(s) = s\widehat{a}(s)\widehat{\varepsilon}(s), \tag{7.151}$$
$$\widehat{\varepsilon}(s) = s\widehat{b}(s)\widehat{\sigma}(s), \tag{7.152}$$

where $\widehat{\sigma}$ stands for the Laplace transform of σ, and so forth. From the above two formulas we have

$$s^2\widehat{a}(s)\widehat{b}(s) = 1. \tag{7.153}$$

Noting furthermore the initial value theorem of the Laplace transform, we have

$$a(+0) = \lim_{\mathrm{Re}\, s\to+\infty} s\widehat{a}(s), \quad b(+0) = \lim_{\mathrm{Re}\, s\to+\infty} s\widehat{b}(s).$$

Then (7.153) gives

$$a(0)b(0) = 1. \tag{7.154}$$

Using (7.153)–(7.154) and the properties of the Laplace transform, it is easy to verify that

$$a(0)\widehat{b'}(s) + b(0)\widehat{a'}(s) + \widehat{a'}(s)\widehat{b'}(s) = 0. \tag{7.155}$$

Taking the Laplace inverse transform to (7.155), we get

$$a(0)b'(t) + b(0)a'(t) + \int_0^t a'(t-\tau)b'(\tau)\mathrm{d}\tau = 0. \tag{7.156}$$

Integrating this with respect to t from 0 to t, exchanging the order of integration, and using (7.154), we obtain

$$b(0)a(t) + \int_0^t a(\tau)b'(t-\tau)\mathrm{d}\tau = 1. \tag{7.157}$$

Noting that $a(+\infty) > 0$, if $b'(+\infty) > 0$, then as $t \to +\infty$, we have

$$\int_0^t a(\tau)b'(t-\tau)\mathrm{d}\tau \to +\infty. \tag{7.158}$$

But both (7.158) and (7.157) cannot be true simultaneously. Therefore, (7.143) must hold.

7.3 System of Viscoelastodynamics and Its Well-Posed Problems

7.3.1 System of Linear Viscoelastodynamics

As stated in the introduction of this chapter, the conservation equations of momentum given in Chapter 5 of Volume I (see (5.74) there) are still valid for a viscoelastic material. Substituting the constitutive equation (7.113) of the linear viscoelastic material into these conservation equations of momentum, we get

$$\begin{aligned}\rho_0\frac{\partial^2 u_i}{\partial t^2}(t,\boldsymbol{x}) = &\sum_{j,k,l=1}^{3} a_{ijkl}(0)\frac{\partial^2 u_k}{\partial x_j\partial x_l}(t,\boldsymbol{x})\\ &+\sum_{j,k,l=1}^{3}\int_0^\infty a'_{ijkl}(\tau)\frac{\partial^2 u_k}{\partial x_j\partial x_l}(t-\tau,\boldsymbol{x})\mathrm{d}\tau + \rho_0 b_i(t,\boldsymbol{x})\\ &(i=1,2,3).\end{aligned} \tag{7.159}$$

This is exactly the *system of linear viscoelastodynamics*. Setting $t-\tau$ as the new integration variable, the above system can also be written as

$$\rho_0 \frac{\partial^2 u_i}{\partial t^2}(t,\boldsymbol{x}) = \sum_{j,k,l=1}^{3} a_{ijkl}(0)\frac{\partial^2 u_k}{\partial x_j \partial x_l}(t,\boldsymbol{x}) + \sum_{j,k,l=1}^{3} \int_{-\infty}^{t} a'_{ijkl}(t-\tau)\frac{\partial^2 u_k}{\partial x_j \partial x_l}(\tau,\boldsymbol{x})\mathrm{d}\tau + \rho_0 b_i(t,\boldsymbol{x}) \quad (i=1,2,3). \tag{7.159$'$}$$

This is a *system of second-order linear integral–partial differential equations of Volterra type*.

Before discussing what kind of well-posed problems can be prescribed for this system, we first inspect the type of system. Since system (7.159) is not a pure system of partial differential equations, we cannot simply determine its type according to the conventional method of classifying second-order partial differential equations. There is no doubt that system (7.159) of linear viscoelastodynamics is a system of evolutionary type. Although both are evolutionary equations, there are essential differences between hyperbolic equations and parabolic equations, which are mainly reflected in the following two closely related aspects. The first one is involved with the propagation speed of the perturbation. For linear hyperbolic equations, the perturbation has finite propagation speed: if there is a perturbation of the solution around a certain point at the initial time, then this perturbation will propagate outward at a finite speed when the time increases. However, for linear parabolic equations, the propagation speed of the perturbation is infinite: the perturbation around a certain point at the initial time can propagate to anywhere instantaneously. The second aspect is involved with the smoothness of solutions. The solutions of linear hyperbolic equations (with sufficiently smooth coefficients, say, constant coefficients) will preserve the same smoothness as the initial data. That is, if the initial data have a certain discontinuity, then this discontinuity will be preserved at a later time. However, for linear parabolic equations, even if the solution has a certain discontinuity at the initial time $t=0$, this discontinuity will disappear instantaneously when $t>0$, and the solution becomes very smooth. Now we try to determine the type of the system (7.159) of integral–partial differential equations mainly according to these two characteristics. As is well known, pure elasticity results in the hyperbolicity of the system of dynamics, while pure viscosity leads to the parabolicity of the system of dynamics. Since the characteristics of viscoelastic materials lie between those of elasticity and viscosity, system (7.159) may lie between hyperbolic type and parabolic type. Determining which type it belongs to, hyperbolic or parabolic, depends on which factor dominates, elasticity or viscosity.

We still assume that the instantaneous elastic tensor $\boldsymbol{A}(0) = (a_{ijkl}(0))$ satisfies the *strong ellipticity condition*; i.e., there exists a positive constant α such that

$$\sum_{i,j,k,l=1}^{3} a_{ijkl}(0)\xi_i\xi_k\eta_j\eta_l \geqslant \alpha|\boldsymbol{\xi}|^2|\boldsymbol{\eta}|^2, \quad \forall \boldsymbol{\xi},\boldsymbol{\eta} \in \mathbb{R}^3. \tag{7.160}$$

Definition 7.1. *Suppose that the instantaneous elastic tensor $\boldsymbol{A}(0) = (a_{ijkl}(0))$ satisfies the strong ellipticity condition* (7.160), *and $\boldsymbol{A}'(\tau) = (a'_{ijkl}(\tau))$ has integrable singularity at $\tau = 0$; then system* (7.159) *is called the* system of linear hyperbolic integral–partial differential equations of Volterra type, *or simply* hyperbolic *for short.*

According to Definition 7.1, if the integral kernel $\boldsymbol{A}'(\tau)=(a'_{ijkl}(\tau))$ in system (7.159) is integrable around $\tau=0$, then whether this system is hyperbolic or not is determined only by the instantaneous elastic tensor $\boldsymbol{A}(0)=(a_{ijkl}(0))$. In the case that the kernel $\boldsymbol{A}'(\tau)$ is regular, i.e., it has no singularity at $\tau=0$, the above definition is reasonable from the viewpoint of either the finiteness of the propagation speed of perturbation or the smoothness of the solutions. However, when $\boldsymbol{A}'(\tau)$ has an integrable singularity at $\tau=0$, although the solution of the system has finite speed of propagation, the discontinuity of the solution at the initial time will not be preserved at later times in general. On this occasion, the system is actually in between hyperbolic type and parabolic type (see the discussion about the propagation of linear waves in the next section), but we still call it hyperbolic.

The typical well-posed problems for system (7.159) of second-order linear hyperbolic integral–partial differential equations are still the Cauchy problem and the mixed initial-boundary value problem. But different from pure partial differential equations, system (7.159) involves the integral of all the historical values of the unknown function $\boldsymbol{u}$ from $-\infty$ to the current time t. Therefore, when discussing the evolution of the solution to a well-posed problem of this kind of system starting from the initial time $t=0$, we have to know in advance all the historical values of $\boldsymbol{u}$ before time $t=0$, i.e., to assume that

$$u_i(t,\boldsymbol{x})=\phi_i(t,\boldsymbol{x}),\quad -\infty<t<0\quad (i=1,2,3), \tag{7.161}$$

where $\phi_i(t,\boldsymbol{x})$ $(i=1,2,3)$ are known functions defined in $(-\infty,0)\times\mathbb{R}^3$ or $(-\infty,0)\times\Omega$ (regarding the Cauchy problem or the mixed initial-boundary value problem). Hence, if we denote

$$f_i(t,\boldsymbol{x})=\rho_0 b_i(t,\boldsymbol{x})+\sum_{j,k,l=1}^{3}\int_{-\infty}^{0}a'_{ijkl}(t-\tau)\frac{\partial^2\phi_i(\tau,\boldsymbol{x})}{\partial x_j\partial x_l}\mathrm{d}\tau\quad (i=1,2,3), \tag{7.162}$$

then system (7.159) or (7.159)′ can be rewritten as

$$\begin{aligned}\rho_0\frac{\partial^2 u_i}{\partial t^2}(t,\boldsymbol{x})=&\sum_{j,k,l=1}^{3}a_{ijkl}(0)\frac{\partial^2 u_k}{\partial x_j\partial x_l}(t,\boldsymbol{x})\\&+\sum_{j,k,l=1}^{3}\int_0^t a'_{ijkl}(t-\tau)\frac{\partial^2 u_k(\tau,\boldsymbol{x})}{\partial x_j\partial x_l}\mathrm{d}\tau\\&+f_i(t,\boldsymbol{x})\quad (i=1,2,3),\end{aligned} \tag{7.163}$$

where $t>0$.

Cauchy problem. For any given positive number $T>0$, find a solution $\boldsymbol{u}(t,\boldsymbol{x})$ to system (7.163) of integral–partial differential equations in $(0,T)\times\mathbb{R}^3$, such that it satisfies the following initial conditions at $t=0$:

$$\begin{aligned}&u_i(0,\boldsymbol{x})=u_i^0(\boldsymbol{x}),\\&\frac{\partial u_i}{\partial t}(0,\boldsymbol{x})=u_i^1(\boldsymbol{x})\quad (i=1,2,3),\end{aligned} \tag{7.164}$$

where $\boldsymbol{x}\in\mathbb{R}^3$ and $u_i^0(\boldsymbol{x})$ and $u_i^1(\boldsymbol{x})$ $(i=1,2,3)$ are known functions.

Mixed initial-boundary value problem. Setting $\Omega \subset \mathbb{R}^3$, find a solution $\boldsymbol{u}(t,\boldsymbol{x})$ to system (7.163) of integral–partial differential equations in $(0,T)\times\Omega$ such that it satisfies initial conditions like (7.164) (where $\boldsymbol{x}\in\Omega$) at $t=0$ and the following boundary conditions on the boundary $\partial\Omega$ of Ω:

$$u_i|_{\partial\Omega} = d_i(t,\boldsymbol{x}) \quad (i=1,2,3) \tag{7.165}$$

or

$$\sum_{j,k,l=1}^{3}\left(a_{ijkl}(0)\frac{\partial u_k}{\partial x_l}+\int_0^t a'_{ijkl}(t-\tau)\frac{\partial u_k(\tau,\boldsymbol{x})}{\partial x_l}\mathrm{d}\tau\right)n_j\Big|_{\partial\Omega} = \sigma_i(t,\boldsymbol{x}) \quad (i=1,2,3), \tag{7.166}$$

where $T>0$ is any given positive number, d_i and σ_i $(i=1,2,3)$ are known functions defined on $(0,T)\times\partial\Omega$, and n_j $(j=1,2,3)$ are components of the unit outward normal vector $\boldsymbol{n}$ on $\partial\Omega$. The boundary conditions can also be prescribed using the following mixed form: suppose that the boundary $\partial\Omega$ of Ω is composed of two parts Γ_1 and Γ_2, and boundary conditions like (7.165) and (7.166) are imposed on Γ_1 and Γ_2, respectively.

The boundary conditions (7.165) and (7.166) imply that the displacement and the stress are given on the boundary $\partial\Omega$ of the viscoelastic body, respectively. For any given point $(t,\boldsymbol{x})$ $(t>0,\ \boldsymbol{x}\in\partial\Omega)$ on the lateral boundary $(0,T)\times\partial\Omega$ of $(0,T)\times\Omega$, a boundary condition like (7.166) depends not only on the values of the partial derivatives of the unknown function $\boldsymbol{u}$ at this point, but also on the whole range of values of the partial derivatives of the unknown function $\boldsymbol{u}$ at this point from 0 to t. This shows that (7.166) is a boundary condition of *nonlocal type*.

When discussing the well-posed problem of system (7.163) of integral–partial differential equations and the property of the solutions, we should sufficiently utilize the results provided by the dissipation inequality in addition to the requirement that the instantaneous elastic tensor $\boldsymbol{A}(0)=(a_{ijkl}(0))$ satisfy the strong ellipticity condition. From Theorem 7.5 we know that the relaxation tensor $\boldsymbol{A}(t)$ should be further required to satisfy

$$\boldsymbol{A}'(t) = (a'_{ijkl}(t)) \leqslant 0, \quad \forall t>0 \tag{7.167}$$

and

$$\boldsymbol{A}''(t) = (a''_{ijkl}(t)) \geqslant 0, \quad \forall t>0. \tag{7.168}$$

The above well-posed problem for a system of linear integral–partial differential equations can be handled by the method for the corresponding evolutionary partial differential equations in principle, but the particularity of its form should also be taken into consideration when specifically dealing with it. We take the following well-posed problem as an example to provide an approach to handle the well-posed problem of this kind of integral–partial differential equation using the semigroup operator method:

$$\begin{aligned}\frac{\partial^2 u_i}{\partial t^2}(t,\boldsymbol{x}) &= \sum_{j,k,l=1}^{3} a_{ijkl}(0)\frac{\partial^2 u_k}{\partial x_j \partial x_l}(t,\boldsymbol{x}) \\ &+ \sum_{j,k,l=1}^{3}\int_0^\infty a'_{ijkl}(\tau)\frac{\partial^2 u_k}{\partial x_j \partial x_l}(t-\tau,\boldsymbol{x})\mathrm{d}\tau, \\ &\qquad t>0, \boldsymbol{x}\in\Omega \quad (i=1,2,3),\end{aligned} \tag{7.169}$$

$$u_i|_{(0,\infty)\times\partial\Omega} = 0 \quad (i=1,2,3), \tag{7.170}$$

$$u_i(t,\boldsymbol{x}) = \phi_i(t,\boldsymbol{x}), \quad t\leqslant 0, \boldsymbol{x}\in\Omega \quad (i=1,2,3), \tag{7.171}$$

where $\phi_i(t,\boldsymbol{x})$ $(i=1,2,3)$ are known functions defined on $(-\infty,0]\times\Omega$.

Set

$$v_i(t,\boldsymbol{x}) = \frac{\partial u_i}{\partial t}(t,\boldsymbol{x}) \quad (i=1,2,3), \tag{7.172}$$

$$\begin{aligned} w_i(t,\boldsymbol{x},\tau) &= u_i^t(\tau,\boldsymbol{x}) - u_i(t,\boldsymbol{x}) \\ &= u_i(t-\tau,\boldsymbol{x}) - u_i(t,\boldsymbol{x}) \quad (i=1,2,3). \end{aligned} \tag{7.173}$$

It is easy to see that system (7.169) can be rewritten as

$$\frac{\partial u_i}{\partial t}(t,\boldsymbol{x}) = v_i(t,\boldsymbol{x}), \tag{7.174}$$

$$\begin{aligned}\frac{\partial v_i}{\partial t}(t,\boldsymbol{x}) &= \sum_{j,k,l=1}^{3} a_{ijkl}(+\infty)\frac{\partial^2 u_k}{\partial x_j \partial x_l}(t,\boldsymbol{x}) \\ &\quad + \sum_{j,k,l=1}^{3}\int_0^\infty a'_{ijkl}(\tau)\frac{\partial^2 w_k}{\partial x_j \partial x_l}(t,\boldsymbol{x},\tau)\mathrm{d}\tau,\end{aligned} \tag{7.175}$$

$$\frac{\partial w_i}{\partial t}(t,\boldsymbol{x},\tau) = -\frac{\partial w_i}{\partial \tau}(t,\boldsymbol{x},\tau) - v_i(t,\boldsymbol{x}), \tag{7.176}$$

where $i=1,2,3$. From condition (7.171) we get that, at $t=0$,

$$u_i(0,\boldsymbol{x}) = \phi_i(0,\boldsymbol{x}), \tag{7.177}$$

$$v_i(0,\boldsymbol{x}) = \frac{\partial \phi_i}{\partial t}(0,\boldsymbol{x}), \tag{7.178}$$

$$w_i(0,\boldsymbol{x},\tau) = \phi_i(-\tau,\boldsymbol{x}) - \phi_i(0,\boldsymbol{x}), \tag{7.179}$$

where $i=1,2,3$. In addition, unknown functions $w_i(t,\boldsymbol{x},\tau)$ satisfy the following conditions:

$$\text{at } \tau=0, \quad w_i(t,\boldsymbol{x},\tau) = 0 \quad (i=1,2,3). \tag{7.180}$$

Regarding system (7.174)–(7.176) as the evolutionary equations with time variable t, where variable τ $(0 \leqslant \tau < \infty)$ is in a similar position as the spatial variable $\boldsymbol{x}$, (7.180) is actually a set of boundary conditions at $\tau = 0$.

Denote

$$U(t,\boldsymbol{x},\tau) = (\boldsymbol{u}(t,\boldsymbol{x}), \boldsymbol{v}(t,\boldsymbol{x}), \boldsymbol{w}(t,\boldsymbol{x},\tau)). \tag{7.181}$$

System (7.174)–(7.176) can be written as

$$\frac{\partial U}{\partial t} = AU, \tag{7.182}$$

where A is a proper linear operator, and the initial conditions (7.177)–(7.179) can be written as

$$U(0,\boldsymbol{x},\tau) = U^0(\boldsymbol{x},\tau), \tag{7.183}$$

where $U^0(\boldsymbol{x},\tau)$ is a known vector function. Thus, the well-posed problem of integral–partial differential equations with given historical values of the unknown function before $t = 0$ is reduced to a standard form that can be handled by the semigroup operator method.

Now we specifically consider the case of isotropic materials. Substituting the constitutive equation (7.123) of the isotropic linear viscoelastic material into the conservation equations of momentum (see (5.74) in Chapter 5 of Volume I), we can obtain the following corresponding system of dynamical equations:

$$\begin{aligned}
&\rho_0 \frac{\partial^2 u_i}{\partial t^2}(t,\boldsymbol{x}) \\
&= \mu(0)\Delta u_i(t,\boldsymbol{x}) + (\lambda(0)+\mu(0)) \sum_{j=1}^{3} \frac{\partial^2 u_j}{\partial x_i \partial x_j}(t,\boldsymbol{x}) \\
&\quad + \int_0^\infty \mu'(\tau)\Delta u_i(t-\tau,\boldsymbol{x})\mathrm{d}\tau \\
&\quad + \int_0^\infty (\lambda'(\tau)+\mu'(\tau)) \sum_{j=1}^{3} \frac{\partial^2 u_j}{\partial x_i \partial x_j}(t-\tau,\boldsymbol{x})\mathrm{d}\tau \\
&\quad + \rho_0 b_i(t,\boldsymbol{x}) \quad (i = 1,2,3).
\end{aligned} \tag{7.184}$$

From (5.150) in Chapter 5 of Volume I, now the instantaneous elastic tensor should satisfy the following form of the strong ellipticity condition:

$$\mu(0) > 0, \quad \lambda(0) + 2\mu(0) > 0. \tag{7.185}$$

As to the formulation of various well-posed problems, it can be given correspondingly as before according to the general situation, which will not be covered here (see Exercise 10).

For the steady-state problem, i.e., the case that $\boldsymbol{u}$ does not depend on t, $\boldsymbol{b}$ should be independent of t, and it is easy to see that system (7.159) is reduced to

$$\sum_{j,k,l=1}^{3} a_{ijkl}(+\infty) \frac{\partial^2 u_k}{\partial x_j \partial x_l} + \rho_0 b_i = 0 \quad (i = 1,2,3), \tag{7.186}$$

where $\boldsymbol{A}(+\infty) = (a_{ijkl}(+\infty))$ is called the *equilibrium elastic tensor*. Under the assumption that the equilibrium elastic tensor $\boldsymbol{A}(+\infty)$ satisfies the strong ellipticity condition, system (7.186) is a *system of second-order linear elliptic partial differential equations*, whose well-posed problems have been discussed in section 5.6.1 in Chapter 5 of Volume I.

Different from the case of pure elasticity, there is also a kind of problem called the *quasi-steady state problem* in the case of viscoelasticity. On this occasion, although the displacement $\boldsymbol{u}$ depends on time t, the inertia term $\rho_0 \frac{\partial^2 \boldsymbol{u}}{\partial t^2}$ is very small and can be neglected, so system (7.159) can be reduced to

$$\begin{aligned}&\sum_{j,k,l=1}^{3} a_{ijkl}(0)\frac{\partial^2 u_k}{\partial x_j \partial x_l}(t,\boldsymbol{x})\\&+\sum_{j,k,l=1}^{3}\int_0^{\infty} a'_{ijkl}(\tau)\frac{\partial^2 u_k}{\partial x_j \partial x_l}(t-\tau,\boldsymbol{x})\mathrm{d}\tau+\rho_0 b_i(t,\boldsymbol{x})=0\\&\qquad (i=1,2,3).\end{aligned} \tag{7.187}$$

To solve this system, in addition to the boundary conditions subjected to the second-order elliptic partial differential equations, we need further to assume that the historical values of the displacement $\boldsymbol{u}$ before a certain time (say, $t=0$) are given. Similarly to the arguments about the dynamics problems (see (7.163)), system (7.187) can be reduced to

$$\begin{aligned}&\sum_{j,k,l=1}^{3} a_{ijkl}(0)\frac{\partial^2 u_k}{\partial x_j \partial x_l}(t,\boldsymbol{x})\\&+\sum_{j,k,l=1}^{3}\int_0^{t} a'_{ijkl}(t-\tau)\frac{\partial^2 u_k}{\partial x_j \partial x_l}(\tau,\boldsymbol{x})\mathrm{d}\tau+f_i(t,\boldsymbol{x})=0\\&\qquad (i=1,2,3).\end{aligned} \tag{7.188}$$

Similarly to the second-order linear elliptic partial differential equations, this system can be solved by using the boundary conditions given on the boundary $\partial\Omega$ of domain Ω.

7.3.2 System of Nonlinear Viscoelastodynamics

Plugging the constitutive relation (7.16) of single-integral form into the conservation equations of momentum (see (5.73) in Chapter 5 of Volume I), we obtain

$$\begin{aligned}\rho_0\frac{\partial^2 \boldsymbol{y}}{\partial t^2} &= \operatorname{div}\boldsymbol{G}(\nabla\boldsymbol{y}(t,\boldsymbol{x}))\\&+\int_0^{\infty}\operatorname{div}\boldsymbol{H}(\tau,\nabla\boldsymbol{y}(t,\boldsymbol{x}),\nabla\boldsymbol{y}(t-\tau,\boldsymbol{x}))\mathrm{d}\tau+\rho_0\boldsymbol{b}(t,\boldsymbol{x}),\end{aligned} \tag{7.189}$$

where div and ∇ represent, respectively, the divergence and the gradient with respect to the variable $\boldsymbol{x}=(x_1,x_2,x_3)$, and the divergence $\operatorname{div}\boldsymbol{G}$ of a second-order tensor function $\boldsymbol{G}=(g_{ij})$ is defined as the first-order tensor $\left(\sum_{j=1}^{3}\frac{\partial g_{ij}}{\partial x_j}\right)$, and so forth (see Appendix A in

Volume I). Writing the above formula in component form, we have

$$\begin{aligned}
&\rho_0 \frac{\partial^2 y_i}{\partial t^2}(t,\boldsymbol{x}) \\
&= \sum_{j=1}^{3} \frac{\partial}{\partial x_j} g_{ij}(\nabla \boldsymbol{y}(t,\boldsymbol{x})) \\
&+ \sum_{j=1}^{3} \int_0^\infty \frac{\partial}{\partial x_j} h_{ij}(\tau, \nabla \boldsymbol{y}(t,\boldsymbol{x}), \nabla \boldsymbol{y}(t-\tau,\boldsymbol{x}))\mathrm{d}\tau \\
&+ \rho_0 b_i(t,\boldsymbol{x}) \quad (i=1,2,3).
\end{aligned} \tag{7.190}$$

System (7.190) is a *system of second-order quasi-linear integral–partial differential equations of Volterra type*. Writing out the partial derivatives on the right-hand side of (7.190), this system of integral–partial differential equations can be rewritten as

$$\begin{aligned}
\rho_0 \frac{\partial^2 y_i}{\partial t^2}(t,\boldsymbol{x}) = &\sum_{j,k,l=1}^{3} \left(\frac{\partial g_{ij}}{\partial \xi_{kl}}(\nabla \boldsymbol{y}(t,\boldsymbol{x})) \right. \\
&+ \left. \int_0^\infty \frac{\partial h_{ij}}{\partial \xi_{kl}}(\tau, \nabla \boldsymbol{y}(t,\boldsymbol{x}), \nabla \boldsymbol{y}(t-\tau,\boldsymbol{x}))\mathrm{d}\tau \right) \times \frac{\partial^2 y_k}{\partial x_j \partial x_l}(t,\boldsymbol{x}) \\
&+ \sum_{j,k,l=1}^{3} \int_0^\infty \frac{\partial h_{ij}}{\partial \eta_{kl}}(\tau, \nabla \boldsymbol{y}(t,\boldsymbol{x}), \nabla \boldsymbol{y}(t-\tau,\boldsymbol{x})) \times \frac{\partial^2 y_k}{\partial x_j \partial x_l}(t-\tau,\boldsymbol{x})\mathrm{d}\tau \\
&+ \rho_0 b_i(t,\boldsymbol{x}) \quad (i=1,2,3).
\end{aligned} \tag{7.191}$$

Definition 7.2. *Suppose that* $\boldsymbol{H}(\tau,\boldsymbol{\xi},\boldsymbol{\eta})$ *satisfies hypotheses* (A_1)–(A_3) *in section* 7.2 *and* $\boldsymbol{G}(\boldsymbol{\xi})$ *satisfies* (7.31). *For the constitutive equation given by the single-integral form* (7.30), *if the instantaneous derivative* $D_{\xi}\boldsymbol{P}(\boldsymbol{\xi}^t)$ *satisfies the strong elliptic condition, i.e., for any given* $\boldsymbol{y} = \boldsymbol{y}(t,\boldsymbol{x})$, *it holds that*

$$\begin{aligned}
&\sum_{i,j,k,l=1}^{3} \left(\frac{\partial g_{ij}}{\partial \xi_{kl}}(\nabla \boldsymbol{y}(t,\boldsymbol{x})) \right. \\
&+ \left. \int_0^\infty \frac{\partial h_{ij}}{\partial \xi_{kl}}(\tau, \nabla \boldsymbol{y}(t,\boldsymbol{x}), \nabla \boldsymbol{y}(t-\tau,\boldsymbol{x}))\mathrm{d}\tau \right) \lambda_i \lambda_k \mu_j \mu_l > 0, \\
&\qquad \forall \boldsymbol{\lambda}, \boldsymbol{\mu} \in \mathbb{R}^3 \backslash \{0\},
\end{aligned} \tag{7.192}$$

then we say that system (7.191) *is a* system of second-order quasi-linear hyperbolic integral–partial differential equations, *or simply* hyperbolic *for short.*

In what follows we always assume that system (7.191) is hyperbolic.

The typical well-posed problems for system (7.191) of integral–partial differential equations are still the Cauchy problem and the mixed initial-boundary value problem.

When discussing these well-posed problems, as in the linear case, all the historical values of the unknown function $\boldsymbol{y} = \boldsymbol{y}(t,\boldsymbol{x})$ before a certain time (say, $t = 0$) have to be given:

$$y_i(t,\boldsymbol{x}) = \phi_i(t,\boldsymbol{x}) \quad (i = 1,2,3), \quad \forall t < 0, \tag{7.193}$$

where ϕ_i $(i = 1,2,3)$ are given functions.

The formulation of the Cauchy problem of system (7.191) is formally similar to that of linear viscoelastodynamics. But just as stated above, we usually take the displacement vector $\boldsymbol{u} = (u_1,u_2,u_3)$ as the unknown function for the linear case, while for the nonlinear case, we usually take the spatial coordinate $\boldsymbol{y} = (y_1,y_2,y_3)$ of particles after deformation as the unknown function.

The mixed initial-boundary value problem of system (7.191) is formulated like this: Let $T > 0$, $\Omega \subset \mathbb{R}^3$. Find a solution $\boldsymbol{y} = \boldsymbol{y}(t,\boldsymbol{x})$ of system (7.191) in $(0,T)\times\Omega$ such that it satisfies at $t = 0$ the initial conditions

$$y_i(0,\boldsymbol{x}) = y_i^0(\boldsymbol{x}), \quad \frac{\partial y_i}{\partial t}(0,\boldsymbol{x}) = y_i^1(\boldsymbol{x}) \quad (i = 1,2,3), \tag{7.194}$$

where $\boldsymbol{x} \in \Omega$ and $y_i^0(\boldsymbol{x})$, $y_i^1(\boldsymbol{x})$ $(i = 1,2,3)$ are given functions, and one of the following boundary conditions is also satisfied on the boundary $\partial\Omega$ of Ω:

$$y_i|_{\partial\Omega} = z_i(t,\boldsymbol{x}) \quad (i = 1,2,3) \tag{7.195}$$

or

$$\sum_{j=1}^{3}\Big(g_{ij}(\nabla\boldsymbol{y}(t,\boldsymbol{x})) + \int_{-\infty}^{t} h_{ij}(t-\tau,\nabla\boldsymbol{y}(t,\boldsymbol{x}),\nabla\boldsymbol{y}(\tau,\boldsymbol{x}))\mathrm{d}\tau\Big)n_j\Big|_{\partial\Omega}$$
$$= \sigma_i(t,\boldsymbol{x}) \quad (i = 1,2,3), \tag{7.196}$$

where $z_i(t,\boldsymbol{x})$ and $\sigma_i(t,\boldsymbol{x})$ $(i = 1,2,3)$ are known functions given in $(0,T)\times\partial\Omega$ and $\boldsymbol{n} = (n_1,n_2,n_3)$ is the unit outward normal vector on $\partial\Omega$.

Boundary condition (7.196) corresponds to imposing stress on the boundary $\partial\Omega$. It is a nonlinear boundary condition of nonlocal type with considerable complexity. On the other hand, it often happens in practical problems that the stress tensor $\boldsymbol{T}\boldsymbol{\nu}|_{\partial\Omega_t}$ is given on the boundary $\partial\Omega_t$ of the viscoelastic body Ω_t after deformation, where $\boldsymbol{T}$ is the Cauchy stress tensor and $\boldsymbol{\nu}$ is the unit outward normal vector on $\partial\Omega_t$. When reducing this boundary condition to the form of (7.196), the right-hand side may involve unknown functions and even the tangential partial derivatives of the unknown functions (see Example 5.1 in Chapter 5 of Volume I).

There are some results about the well-posedness problem of system (7.191). Roughly speaking, both the existence of global weak solutions and the existence of global classical solutions with small initial data can be proved under certain assumptions; the latter relies on a dissipation inequality stronger than what was stated in section 7.2. However, there is little work on the corresponding well-posedness problem with boundary conditions of the form (7.196).

7.3.3 Equation of One-Dimensional Nonlinear Viscoelastodynamics

The pure axial deformation of a viscoelastic rod is given by

$$y_1 = x_1 + u_1(t,x_1), \quad y_2 = x_2, \quad y_3 = x_3. \tag{7.197}$$

On this occasion, we can see from the constitutive equation (7.16) that the Piola stress tensor $\boldsymbol{P}$ is a function of only t and x_1; assume further that $\boldsymbol{b}$ is independent of x_2 and x_3. The first equation in the viscoelastodynamic system (7.190) is then reduced to

$$\begin{aligned}\rho_0 \frac{\partial^2 u_1}{\partial t^2} = & \frac{\partial}{\partial x_1} g\left(\frac{\partial u_1}{\partial x_1}\right) \\ & + \int_0^\infty \frac{\partial}{\partial x_1} h\left(\tau, \frac{\partial u_1}{\partial x_1}(t, x_1), \frac{\partial u_1}{\partial x_1}(t-\tau, x_1)\right) \mathrm{d}\tau \\ & + \rho_0 b_1(t, x_1),\end{aligned} \tag{7.198}$$

where we denote for short $g(\frac{\partial u_1}{\partial x_1}) = g_{11}(\boldsymbol{I} + \nabla \boldsymbol{u})$, $\nabla \boldsymbol{u} = \operatorname{diag}\{\frac{\partial u_1}{\partial x_1}, 0, 0\}$, and so forth.

The pure sheer deformation of the viscoelastic material is given by

$$y_1 = x_1 + u_1(t, x_2), \quad y_2 = x_2, \quad y_3 = x_3. \tag{7.199}$$

Now we need to assume that $\boldsymbol{b}$ is independent of x_1 and x_3. For this kind of deformation, the first equation in the viscoelastodynamic system (7.190) is reduced to

$$\begin{aligned}\rho_0 \frac{\partial^2 u_1}{\partial t^2} = & \frac{\partial}{\partial x_2} g\left(\frac{\partial u_1}{\partial x_2}\right) \\ & + \int_0^\infty \frac{\partial}{\partial x_2} h\left(\tau, \frac{\partial u_1}{\partial x_2}(t, x_2), \frac{\partial u_1}{\partial x_2}(t-\tau, x_2)\right) \mathrm{d}\tau \\ & + \rho b_1(t, x_2),\end{aligned} \tag{7.200}$$

where $g(\frac{\partial u_1}{\partial x_2}) = g_{12}(\boldsymbol{I} + \nabla \boldsymbol{u})$, $\boldsymbol{u} = (u_1(t, x_2), 0, 0)$, and so on.

Equations (7.198) and (7.200) can be unified as

$$\rho_0 u_{tt} = (g(u_x))_x + \int_0^\infty (h(\tau, u_x(t, x), u_x(t-\tau, x)))_x \mathrm{d}\tau + \rho_0 b(t, x), \tag{7.201}$$

where u, g, and h are all scalars. Inspecting the strong ellipticity condition for the pure axial deformation (7.197) and the pure sheer deformation (7.199), respectively, we obtain that for the unified form (7.201) of equations (7.198) and (7.200), the strong ellipticity condition means that

$$g'(u_x) + \int_0^\infty h_\xi(\tau, u_x(t, x), u_x(t-\tau, x)) \mathrm{d}\tau > 0, \tag{7.202}$$

where $h_\xi(\tau, u_x(t, x),\ u_x(t-\tau, x))$ stands for the derivative of $h(\tau, u_x(t, x),\ u_x(t-\tau, x))$ with respect to the variable $u_x(t, x)$.

Now we investigate the expression of the dissipation inequality in the one-dimensional case and its consequences. For the one-dimensional viscoelastodynamic equation (7.201), the dissipation inequality corresponding to (7.53) in Theorem 7.2 reduces to

$$\int_\eta^\xi \partial_\tau h(\tau, \zeta, \eta) \mathrm{d}\zeta \leqslant 0, \quad \forall \xi, \eta \in \mathbb{R}, \tau > 0. \tag{7.203}$$

Now we discuss further a special case that is very important in both theory and application. Consider the equation of the form

$$\rho_0 u_{tt} = (\phi(u_x))_x + \int_0^\infty a'(\tau)(\psi(u_x(t-\tau,x)))_x \mathrm{d}\tau + b(t,x). \tag{7.204}$$

Assume that $a(+\infty)=0$; it is easy to see that, for this equation, g and h in (7.201) are, respectively,

$$g(\xi) = \phi(\xi) - a(0)\psi(\xi), \quad h(\tau,\xi,\eta) = a'(\tau)(\psi(\eta)-\psi(\xi)). \tag{7.205}$$

Then the strong ellipticity condition (7.202) is reduced to

$$\phi'(\xi) > 0, \quad \forall \xi \in \mathbb{R}, \tag{7.206}$$

while the dissipation inequality (7.203) is reduced to

$$a''(\tau)\int_\eta^\xi (\psi(\eta)-\psi(\zeta))\mathrm{d}\zeta \leqslant 0, \quad \forall \xi,\eta \in \mathbb{R}, \tau > 0. \tag{7.207}$$

Formula (7.207) implies that $a''(\tau)$ should have a definite sign. Without loss of generality, we assume that

$$a''(\tau) \geqslant 0, \quad \forall \tau > 0. \tag{7.208}$$

We assume furthermore that $a'(+\infty)=0$. Integrating the above formula from τ to $+\infty$, we immediately get

$$a'(\tau) \leqslant 0, \quad \forall \tau > 0. \tag{7.209}$$

Together with $a(+\infty)=0$, we can similarly get

$$a(\tau) \geqslant 0, \quad \forall \tau > 0. \tag{7.210}$$

Moreover, from (7.207) and (7.208) we also have

$$\int_\eta^\xi (\psi(\zeta)-\psi(\eta))\mathrm{d}\zeta \geqslant 0, \quad \forall \xi,\eta \in \mathbb{R}. \tag{7.211}$$

Then it is easy to see that $\psi(\xi)$ is a monotonically increasing function, i.e.,

$$\psi'(\xi) \geqslant 0, \quad \forall \xi \in \mathbb{R} \tag{7.212}$$

(see Exercise 11).

Definition 7.3. *Suppose that $a \in L^1_{\mathrm{loc}}[0,\infty)$. If for any given $w \in C[0,\infty)$ and any given $T > 0$, we have*

$$\int_0^T w(t) \int_0^t a(t-\tau) w(\tau) \mathrm{d}\tau \mathrm{d}t \geqslant 0, \tag{7.213}$$

then function $a(t)$ is called positive definite. *Moreover, if there exists $\varepsilon > 0$ such that function $a(t) - \varepsilon \mathrm{e}^{-t}$ is positive definite, then function a is called* strongly positive definite.

It can be proved that if $a \in C^2[0,\infty)$, $a' \not\equiv 0$, and

$$(-1)^j a^{(j)}(t) \geqslant 0, \quad \forall t \geqslant 0 \quad (j = 0,1,2), \tag{7.214}$$

then a is strongly positive definite (see [8]).

From (7.208)–(7.210) we can see that, for integral–partial differential equations like (7.204), under some conditions, such as $a \in C^2[0,\infty)$ and so on, the dissipation inequality actually requires that $a(t)$ be strongly positive definite. Most of the existing research results about equation (7.204) were obtained under the assumption that $a(t)$ is strongly positive definite (see [12] and the references cited therein).

7.4 Singularity of the Kernel and the Propagation of Linear Waves

In this section we will explore the influence of the singularity of the kernel on the behaviors of solutions for the system of linear viscoelastodynamics. For simplicity, we will take the system of one-dimensional linear viscoelastodynamics as an example to guide our discussion.

The system of linear viscoelastodynamics is, in the one-dimensional case (see the discussion in section 7.3.3), reduced to

$$\rho_0 u_{tt}(t,x) = a(0) u_{xx}(t,x) + \int_{-\infty}^t a'(t-\tau) u_{xx}(\tau,x) \mathrm{d}\tau + \rho_0 b(t,x). \tag{7.215}$$

For simplicity, let $\rho_0 \equiv 1$ and $b \equiv 0$. If $a(\tau)$ satisfies condition (7.214), then $a(\tau)$ is a monotonically decreasing nonnegative function, while $a'(\tau)$ is a monotonically increasing nonpositive function. Assume that

$$a(\tau) \to c \text{ as } \tau \to +\infty, \tag{7.216}$$

where $c \geqslant 0$; then we have

$$\int_{-\infty}^t a'(t-\tau) \mathrm{d}\tau = -a(0) + c. \tag{7.217}$$

Thus, equation (7.215) can be written as

$$u_{tt}(t,x) = c u_{xx}(t,x) + \int_{-\infty}^t m(t-\tau)(u_{xx}(t,x) - u_{xx}(\tau,x)) \mathrm{d}\tau, \tag{7.218}$$

where $m(\tau) = -a'(\tau)$.

We now start directly from the integral–partial differential equation (7.218) to study the influence of the singularity at $\tau = 0$ of the kernel $m(\tau)$ (possibly nonintegrable) on the solution. Here, we assume that $c \geqslant 0$ is a constant, $m(\tau) \not\equiv 0$ is a monotonically decreasing nonnegative function that is integrable at infinity, and $\tau m(\tau)$ is integrable at $\tau = 0$.

We consider the *Rayleigh problem* of equation (7.218), that is, to find the solution $u(t,x)$ of equation (7.218) in the first quadrant $\{(t,x) \mid t > 0, x > 0\}$ such that it satisfies the following conditions:

$$u(t,x) = 0 \quad \text{as } t \leqslant 0, x > 0, \tag{7.219}$$

$$u(t,0) = 1 \quad \text{as } t > 0. \tag{7.220}$$

Now we use the Laplace transform method to solve the above Rayleigh problem. Denote by $\widehat{u}(s,x)$ and $\widehat{m}(s)$ the Laplace transforms of functions $u(t,x)$ and $m(t)$ with respect to t, separately, that is,

$$\widehat{u}(s,x) = \int_0^\infty \mathrm{e}^{-st}\, u(t,x)\mathrm{d}t, \tag{7.221}$$

$$\widehat{m}(s) = \int_0^\infty \mathrm{e}^{-st}\, m(t)\mathrm{d}t. \tag{7.222}$$

Taking the Laplace transform of equation (7.218) with respect to variable t, and noticing that

$$\begin{aligned}\widehat{u}_{tt}(s,x) &= s^2\widehat{u}(s,x) - su(0,x) - u_t(0,x)\\ &= s^2\widehat{u}(s,x)\end{aligned} \tag{7.223}$$

(here we used (7.219)) and

$$\widehat{\left(\int_0^t m(t-\tau)u(\tau,x)\mathrm{d}\tau\right)}(s,x) = \widehat{m}(s)\widehat{u}(s,x), \tag{7.224}$$

we obtain

$$s^2\widehat{u}(s,x) = (c + \widehat{m}(0) - \widehat{m}(s))\widehat{u}_{xx}(s,x), \tag{7.225}$$

where $c \geqslant 0$ is the coefficient in equation (7.218) and

$$\widehat{m}(0) = \int_0^\infty m(t)\mathrm{d}t.$$

Moreover, the boundary condition (7.220), under the Laplace transform, reads as

$$\widehat{u}(s,0) = \frac{1}{s}. \tag{7.226}$$

If $m(\tau)$ is not integrable at $\tau = 0$, both $\widehat{m}(0)$ and $\widehat{m}(s)$ do not exist; however, noticing that $\tau m(\tau)$ is integrable at $\tau = 0$, $\widehat{m}(0) - \widehat{m}(s)$ can still be defined by

$$\widehat{m}(0) - \widehat{m}(s) = \int_0^\infty (1 - \mathrm{e}^{-st})m(t)\mathrm{d}t, \quad \forall \mathrm{Re}\, s \geqslant 0. \tag{7.227}$$

Therefore, even if $m(\tau)$ is not integrable at $\tau = 0$, (7.225) is reasonable in the above sense. Considering that $m(t)$ is nonnegative and not identically equal to zero and using (7.227), we can easily verify that

$$\mathrm{Re}(\widehat{m}(0) - \widehat{m}(s)) > 0, \quad \forall \mathrm{Re}\, s \geqslant 0,\ s \neq 0. \tag{7.228}$$

In addition, noticing that $m(t)$ is monotonically decreasing, it is easy to verify that

$$\mathrm{sign}\left(\int_0^\infty \mathrm{e}^{-s_1 t} m(t) \sin s_2 t \,\mathrm{d}t\right) = \mathrm{sign}\, s_2, \quad \forall s_1 \geqslant 0,\ s_2 \in \mathbb{R} \tag{7.229}$$

(see Exercise 5). Then, taking the imaginary part of (7.227), we easily get

$$\mathrm{sign}\,\mathrm{Im}(\widehat{m}(0) - \widehat{m}(s)) = \mathrm{sign}\,\mathrm{Im}\, s. \tag{7.230}$$

In the sense of (7.227), $\widehat{m}(0) - \widehat{m}(s)$ is always reasonable under the assumption of $m(t)$ at the beginning. So for the ordinary differential equation (7.225) with respect to variable x, its solution, which satisfies condition (7.226) and is bounded with respect to $x \geqslant 0$ for any given $s \neq 0$ with $\mathrm{Re}\, s \geqslant 0$, can be given by

$$\widehat{u}(s,x) = \frac{1}{s} \mathrm{e}^{-s(c + \widehat{m}(0) - \widehat{m}(s))^{-\frac{1}{2}} x}. \tag{7.231}$$

Then by the Laplace inverse transform formula we get

$$\begin{aligned} u(t,x) &= \frac{1}{2\pi\,\mathrm{i}} \int_{\gamma - \mathrm{i}\infty}^{\gamma + \mathrm{i}\infty} \mathrm{e}^{st}\, \widehat{u}(s,x) \mathrm{d}s \\ &= \frac{1}{2\pi\,\mathrm{i}} \int_{\gamma - \mathrm{i}\infty}^{\gamma + \mathrm{i}\infty} \frac{1}{s} \mathrm{e}^{-s((c + \widehat{m}(0) - \widehat{m}(s))^{-\frac{1}{2}} x - t)} \mathrm{d}s, \end{aligned} \tag{7.232}$$

where γ is any given real positive constant.

We first show that the integral on the right-hand side of (7.232) makes sense under the previous assumptions. Using (7.228) and (7.230), it is easy to verify that

$$\mathrm{Re}(c + \widehat{m}(0) - \widehat{m}(s))^{-\frac{1}{2}} > 0, \quad \forall \mathrm{Re}\, s \geqslant 0,\ s \neq 0, \tag{7.233}$$

and

$$\mathrm{signIm}(c + \widehat{m}(0) - \widehat{m}(s))^{-\frac{1}{2}} = -\,\mathrm{signIm}\, s, \quad \forall \mathrm{Re}\, s \geqslant 0,\ s \neq 0. \tag{7.234}$$

Hence,

$$\begin{aligned} & |\mathrm{e}^{-s(c + \widehat{m}(0) - \widehat{m}(s))^{-\frac{1}{2}} x}| \\ = {} & \mathrm{e}^{-\mathrm{Re}\, s \cdot \mathrm{Re}(c + \widehat{m}(0) - \widehat{m}(s))^{-\frac{1}{2}} x + \mathrm{Im}\, s \cdot \mathrm{Im}(c + \widehat{m}(0) - \widehat{m}(s))^{-\frac{1}{2}} x} \\ \leqslant {} & 1, \quad \forall x \geqslant 0, \end{aligned}$$

so from the expression (7.231) of $\widehat{u}(s,x)$ we get

$$|\widehat{u}(s,x)| \leqslant \frac{1}{|s|}, \quad \forall \mathrm{Re}\, s \geqslant 0,\ s \neq 0,\ x \geqslant 0. \tag{7.235}$$

Denote $s=\gamma+\mathrm{i}\beta$. From the above formula we have

$$|\widehat{u}(s,t)|^2 \leqslant \frac{1}{\gamma^2+\beta^2}.$$

This implies that, for any fixed $\gamma>0$, $\widehat{u}(s,x)$, regarded as a function of the imaginary part β of s, belongs to $L^2(\mathbb{R})$. Noticing that the integral in (7.232) can be rewritten as

$$\frac{1}{2\pi\,\mathrm{i}}\int_{\gamma-\mathrm{i}\infty}^{\gamma+\mathrm{i}\infty} \mathrm{e}^{st}\,\widehat{u}(s,x)\mathrm{d}s=\frac{1}{2\pi}\mathrm{e}^{\gamma t}\int_{-\infty}^{\infty}\mathrm{e}^{\mathrm{i}\beta t}\,\widehat{u}(\gamma+\mathrm{i}\beta,x)\mathrm{d}\beta,$$

from the Fourier transform theory of L^2 functions, (7.232) is meaningful in the following sense:

$$u(t,x)=\frac{1}{2\pi}\mathrm{e}^{\gamma t}\lim_{N\to\infty}\int_{-N}^{N}\mathrm{e}^{\mathrm{i}\beta t}\,\widehat{u}(\gamma+\mathrm{i}\beta,x)\mathrm{d}\beta, \tag{7.236}$$

where the limit should be understood as the limit in the sense of $L^2(\mathbb{R})$.

Next we will study properties of $u(t,x)$ given by (7.232) according to different situations of the singularity of $m(t)$ at $t=0$.

We first assume that $m\in L^1(0,\infty)$. Then from the Riemann–Lebesgue theorem we know that

$$\lim_{s\to\infty}\widehat{m}(s)=0, \qquad \text{for } \mathrm{Re}\,s\geqslant 0, \tag{7.237}$$

so when $\mathrm{Re}\,s\geqslant 0$, we have

$$\lim_{s\to\infty}((c+\widehat{m}(0)-\widehat{m}(s))^{-\frac{1}{2}}x-t)=\frac{x}{\sqrt{c+\widehat{m}(0)}}-t. \tag{7.238}$$

In addition, since the integrand on the right-hand side of (7.232) has no singular point when $\mathrm{Re}\,s>0$, by the Cauchy theorem we have

$$\frac{1}{2\pi\,\mathrm{i}}\left(\int_{\gamma-\mathrm{i}\rho}^{\gamma+\mathrm{i}\rho}+\int_{\Gamma}\right)\frac{1}{s}\mathrm{e}^{-s((c+\widehat{m}(0)-\widehat{m}(s))^{-\frac{1}{2}}x-t)}\mathrm{d}s=0, \tag{7.239}$$

where Γ is the part of the circle $|s-\gamma|=\rho$ lying on the right-hand side of the straight line $s=\gamma$ on the complex plane (see Figure 7.9).

Therefore, when

$$x>at, \tag{7.240}$$

where

$$a=\sqrt{c+\widehat{m}(0)}, \tag{7.241}$$

we get

$$\lim_{\rho\to\infty}\int_{\Gamma}\frac{1}{s}\mathrm{e}^{-s((c+\widehat{m}(0)-\widehat{m}(s))^{-\frac{1}{2}}x-t)}\mathrm{d}s=0; \tag{7.242}$$

then from (7.239) we have

$$\lim_{\rho\to\infty}\int_{\gamma-\mathrm{i}\rho}^{\gamma+\mathrm{i}\rho}\frac{1}{s}\mathrm{e}^{-s((c+\widehat{m}(0)-\widehat{m}(s))^{-\frac{1}{2}}x-t)}\mathrm{d}s=0. \tag{7.243}$$

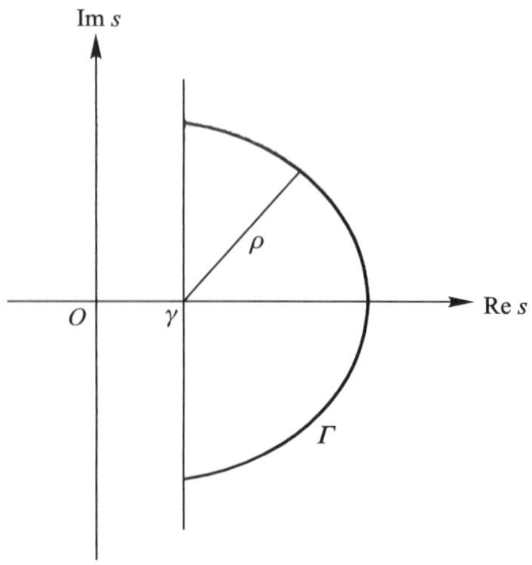

Figure 7.9

Noticing (7.232) and (7.236), we obtain

$$u(t,x) \equiv 0 \quad \text{if } x > at. \tag{7.244}$$

This implies that the linear wave determined by the integral–partial differential equation (7.218) propagates at the finite speed $a = \sqrt{c + \int_0^\infty m(t)\mathrm{d}t}$ when the kernel $m(t)$ is integrable, which reflects a hyperbolic characteristic. However, when $m(t)$ is not integrable at $t = 0$, we cannot expect a finite propagation speed of the wave.

First, we consider the case when $m(t)$ is integrable and study the discontinuity of the solution of problem (7.218)–(7.220) along the straight lines $x = at$. We first assume that m is a regular kernel, i.e., $m \in C^\infty[0,\infty)$, and its arbitrary-order derivatives are integrable on $[0,\infty)$. Then from (7.222) and integrating by parts, we can easily see that $\widehat{m}(s)$ can be expressed by the following form as $s \to \infty$:

$$\widehat{m}(s) = \frac{m(0)}{s} + \frac{m'(0)}{s^2} + \cdots, \quad \mathrm{Re}\, s \geqslant 0, s \neq 0. \tag{7.245}$$

Thus, we easily get

$$\begin{aligned} &\mathrm{e}^{-s(c+\widehat{m}(0)-\widehat{m}(s))^{-\frac{1}{2}}x} \\ &= \mathrm{e}^{-(c+\widehat{m}(0))^{-\frac{1}{2}}sx - \frac{1}{2}m(0)(c+\widehat{m}(0))^{-\frac{3}{2}}x} + O\left(\frac{1}{s}\right). \end{aligned} \tag{7.246}$$

Substituting this into the right-hand side of (7.232), the contribution of the first term on the right-hand side of the above formula is

$$\begin{aligned} &v(t,x) \\ &= \mathrm{e}^{-\frac{1}{2}m(0)(c+\widehat{m}(0))^{-\frac{3}{2}}x} \cdot \frac{1}{2\pi\,\mathrm{i}} \int_{\gamma-\mathrm{i}\infty}^{\gamma+\mathrm{i}\infty} \frac{1}{s} \mathrm{e}^{-s((c+\widehat{m}(0))^{-\frac{1}{2}}x-t)}\,\mathrm{d}s \\ &= \mathrm{e}^{-\frac{1}{2}m(0)(c+\widehat{m}(0))^{-\frac{3}{2}}x} H(t-(c+\widehat{m}(0))^{-\frac{1}{2}}x), \end{aligned} \tag{7.247}$$

where $H(\cdot)$ is the Heaviside function. Here we used the fact that the Laplace transform of $H(t)$ is $\frac{1}{s}$. In addition, it is easy to see that (7.232) yields a continuous function of t and x by substituting the second term on the right-hand side of (7.246) into the integral on the right-hand side of (7.232). Therefore, the solution $u(t,x)$ of problem (7.218)–(7.220) has a discontinuity across the straight line $x = at$, and the jump of this discontinuity is

$$A(t) = \exp\left(-\frac{m(0)t}{2(c+\widehat{m}(0))}\right). \tag{7.248}$$

So under the above assumptions on $m(t)$, this jump decays to zero exponentially with the increase of t. In this case, the solution behaves similarly to that of the hyperbolic partial differential equations with dissipation.

Now we assume that $m(t)$ has an integrable singularity at $t=0$. On this occasion, $\widehat{m}(0)$ is still finite although $m(0)=\infty$; thus $a < +\infty$, and the solution of problem (7.218)–(7.220) still has a finite propagation speed. But (7.248) enlightens us that $u(t,x)$ should remain continuous across $x = at$ in this case. That is, the integral–partial differential equation (7.218) with this kind of kernel has the parabolic characteristic that the initial discontinuity is mollified instantaneously while still preserving the hyperbolic characteristic that the wave propagates at a finite speed. Now we illustrate this point by an example with the function

$$m(t) = \sum_{n=1}^{\infty} \mathrm{e}^{-n^{\alpha}t} \tag{7.249}$$

given by (7.21). From (7.25), the behavior of $m(t)$ near $t=0$ is

$$m(t) \sim t^{-\frac{1}{\alpha}}. \tag{7.250}$$

Then, when $\alpha > 1$, $m(t)$ has an integrable singularity at $t=0$, and its Laplace transform is

$$\widehat{m}(s) = \sum_{n=1}^{\infty} \frac{1}{n^{\alpha}+s}, \quad \forall \mathrm{Re}\, s \geqslant 0. \tag{7.251}$$

Denote $f(\lambda) = \frac{1}{\lambda^{\alpha}+s}$, $s = \gamma + \mathrm{i}\beta$. It is easy to verify that

$$\left|\int_0^{\infty} f(\lambda)\mathrm{d}\lambda - \sum_{n=1}^{\infty} f(n)\right| \leqslant \int_0^{\infty} |f'(\lambda)|\mathrm{d}\lambda \tag{7.252}$$

and

$$\int_0^{\infty} |f'(\lambda)|\mathrm{d}\lambda = \frac{1}{|\beta|}\left(\frac{\pi}{2} - \arctan\frac{\gamma}{|\beta|}\right). \tag{7.253}$$

Moreover, we also have

$$\int_0^{\infty} f(\lambda)\mathrm{d}\lambda = \frac{\pi}{\alpha} s^{\frac{1}{\alpha}-1} \csc\frac{\pi}{\alpha} \tag{7.254}$$

(see [7]). From (7.251)–(7.254) we get

$$\left|\frac{\pi}{\alpha} s^{\frac{1}{\alpha}-1} \csc\frac{\pi}{\alpha} - \widehat{m}(s)\right| \leqslant \frac{1}{|\beta|}\left(\frac{\pi}{2} - \arctan\frac{\gamma}{|\beta|}\right).$$

It is easy to see that, when s tends to infinity in any angular domain that does not contain the negative real axis, it always holds that

$$\frac{1}{|\beta|}\left(\frac{\pi}{2}-\arctan\frac{\gamma}{|\beta|}\right)=O\left(\frac{1}{|s|}\right).$$

Then we have, correspondingly,

$$\widehat{m}(s)=\frac{\pi}{\alpha}s^{\frac{1}{\alpha}-1}\csc\frac{\pi}{\alpha}+O\left(\frac{1}{|s|}\right). \tag{7.255}$$

Since $|\mathrm{e}^{-st}|$ is bounded for t and γ varying in a bounded domain, and noticing that, when $|\beta|$ is sufficiently large,

$$\begin{aligned}&s(c+\widehat{m}(0)-\widehat{m}(s))^{-\frac{1}{2}}\\&=(c+\widehat{m}(0))^{-\frac{1}{2}}\left(s+\frac{\pi\csc\frac{\pi}{\alpha}}{2\alpha(c+\widehat{m}(0))}s^{\frac{1}{\alpha}}+o(1)s^{\frac{1}{\alpha}}\right)\end{aligned}$$

and

$$\mathrm{Re}\,s\geqslant|s|^{\frac{1}{\alpha}}\cos\frac{\pi}{2\alpha},$$

substituting the above formula of $\widehat{m}(s)$ into the integrand on the right-hand side of (7.232) gives

$$\left|\frac{1}{s}\mathrm{e}^{-s((c+\widehat{m}(0)-\widehat{m}(s))^{-\frac{1}{2}}x-t)}\right|\leqslant M\,\mathrm{e}^{-kx|s|^{\frac{1}{\alpha}}}, \tag{7.256}$$

where M and k are positive constants. This implies that the rate of decay to zero of the integrand on the right-hand side of (7.232) as $s\to\infty$ is faster than that of any power of $|s|^{-1}$, and then the solution $u(t,x)$ of problem (7.218)–(7.220) is infinitely differentiable for $t>0$, $x>0$. This shows that the discontinuity of the solution at the initial time is mollified instantaneously for kernels like $m(t)$ that have an integrable singularity at $t=0$. Further discussion can show that if $m(t)$ has a weaker singularity (e.g., a logarithmic singularity) at $t=0$, then the smoothness of solution $u(t,x)$ at $x=at$ will decrease accordingly.

Finally we study the case that $m(t)$ has a nonintegrable singularity at $t=0$. We still conduct our discussion by taking the same function $m(t)$ given by (7.251) as an example. When $\frac{1}{2}<\alpha\leqslant 1$, $m(t)$ has a nonintegrable singularity at $t=0$. At this time, we cannot define $\widehat{m}(0)$ and $\widehat{m}(s)$ directly, but from (7.227) we have

$$\widehat{m}(0)-\widehat{m}(s)=\sum_{n=1}^{\infty}\frac{s}{n^{\alpha}(n^{\alpha}+s)}. \tag{7.257}$$

For simplicity, we only discuss the case $\alpha=1$; then

$$\widehat{m}(0)-\widehat{m}(s)=\sum_{n=1}^{\infty}\left(\frac{1}{n}-\frac{1}{n+s}\right). \tag{7.258}$$

When N is very large, the behavior of $\sum_{n=1}^{N} \frac{1}{n}$ is like $\ln N + C$, that is,

$$\lim_{N\to\infty} \left(\sum_{n=1}^{N} \frac{1}{n} - \ln N \right) = C, \tag{7.259}$$

where C is the Euler constant. Set $f(\lambda) = \frac{1}{\lambda+s}$, where $\mathrm{Re}\, s \geqslant 0$, $s \neq 0$. It is not difficult to verify that

$$\lim_{N\to\infty} \left| \int_0^N f(\lambda)\mathrm{d}\lambda - \sum_{n=1}^{N} f(n) \right| \leqslant \int_0^{\infty} |f'(\lambda)| \mathrm{d}\lambda. \tag{7.260}$$

In addition, it is easy to directly verify (7.253) for $\alpha = 1$. From (7.260) and (7.253), we have

$$\lim_{N\to\infty} \left(\ln(N+s) - \sum_{n=1}^{N} \frac{1}{n+s} \right) = \ln s + O\left(\frac{1}{|s|}\right). \tag{7.261}$$

Noting (7.259), we get

$$\lim_{N\to\infty} \sum_{n=1}^{N} \left(\frac{1}{n} - \frac{1}{n+s} \right) = \ln s + C + O\left(\frac{1}{|s|}\right), \tag{7.262}$$

that is,

$$\widehat{m}(0) - \widehat{m}(s) = \ln s + C + O\left(\frac{1}{|s|}\right). \tag{7.263}$$

This implies that the integrand on the right-hand side of (7.232) decays faster than any power of $|s|^{-1}$ as $|s| \to \infty$ along the integration path; therefore, the solution $u(t,x)$ of problem (7.218)–(7.220) is still infinitely differentiable for $t > 0$, $x > 0$. For the case $0 < \alpha < \frac{1}{2}$, we can get similar results. Further discussions show that when $\frac{1}{2} < \alpha \leqslant 1$, the solution $u(t,x)$ is actually analytic with respect to t and x for $t > 0$, $x > 0$. This indicates that, for the Rayleigh problem (7.218)–(7.220) in the case that the kernel $m(t)$ has a nonintegrable singularity at $t = 0$, not only is the discontinuity of the solution $u(t,x)$ at the initial time mollified instantaneously, and the solution becomes very smooth and even analytic, but also, since $u(t,x)$ is analytic, it cannot be equivalently zero in any region in $t > 0$, $x > 0$ (unless $u(t,x)$ is always zero); the wave therefore propagates at an infinite speed. Thus, equation (7.218) is more like a parabolic partial differential equation.

The above discussions can be summarized as follows:

1. If m is smooth on $[0,\infty)$, the singularity of solution $u(t,x)$ propagates at the finite speed

$$a = \sqrt{c + \int_0^{\infty} m(t)\mathrm{d}t} \tag{7.264}$$

and the jump of the solution along $x = at$ decays exponentially to zero.

2. If m is integrable at $t=0$, but $m(0)=\infty$, the propagation speed is still finite, but the initial singularity of the solution can be removed instantaneously, and the smoothness of the solution depends on the level of the singularity of m at $t=0$.

3. If m has a nonintegrable singularity at $t=0$, the propagation speed is infinite; the solution has higher smoothness; it is even analytic.

The discussion of the linear case in this section tells us that, for an equation (equations) of nonlinear viscoelastodynamics, the solution to the well-posed problem has better properties, such as higher smoothness, if the kernel has a singularity at $t=0$ than the purely elastic (hyperbolic) case. As for the weak solution, the global existence theorem in the multidimensional case has been established (see [6], [11], etc.) accordingly. However, the global existence of classical solutions for any given initial data (maybe not sufficiently small) is still open even for the one-dimensional case.

7.5 Propagation of Acceleration Waves

Similarly to the system of thermoelastodynamics in the last chapter, we study the propagation of the acceleration wave for the system of viscoelastodynamics in this section. We take the following equation of one-dimensional viscoelastodynamics as an example:

$$u_{tt}=(g(u_x))_x+\int_{-\infty}^{t} m(t-\tau)(h(u_x(t,x),u_x(\tau,x)))_x\mathrm{d}\tau, \tag{7.265}$$

and we assume that m is a regular kernel.

Let Γ be a curve on the (t,x) plane:

$$x=x(t). \tag{7.266}$$

If u is suitably smooth and satisfies equation (7.265) except on the curve Γ, and u and its first-order partial derivatives u_t and u_x are continuous across the curve Γ, while the second-order partial derivatives have discontinuities of the first type on Γ, then the curve Γ is called an *acceleration wave* curve.

Suppose that the viscoelastic body is under a state of uniform deformation before the incoming propagation of the acceleration wave, that is,

$$u\equiv u_0 \quad \text{as } x>x(t), \tag{7.267}$$

where u_0 is a constant (see Figure 7.10), and assume that the acceleration wave propagates from left to right, i.e., $x'(t)>0$. Denote

$$\frac{\mathrm{d}}{\mathrm{d}_\Gamma t}=\frac{\partial}{\partial t}+x'(t)\frac{\partial}{\partial x}. \tag{7.268}$$

By the continuity of u_t and u_x across Γ, we have

$$\left[\frac{\mathrm{d}u_t}{\mathrm{d}_\Gamma t}\right]=\left[\frac{\mathrm{d}u_x}{\mathrm{d}_\Gamma t}\right]=0, \tag{7.269}$$

where $[\cdot]$ stands for the jump of the current function across Γ. From (7.269) we get, respectively,

$$[u_{tt}]=-x'(t)[u_{xt}] \tag{7.270}$$

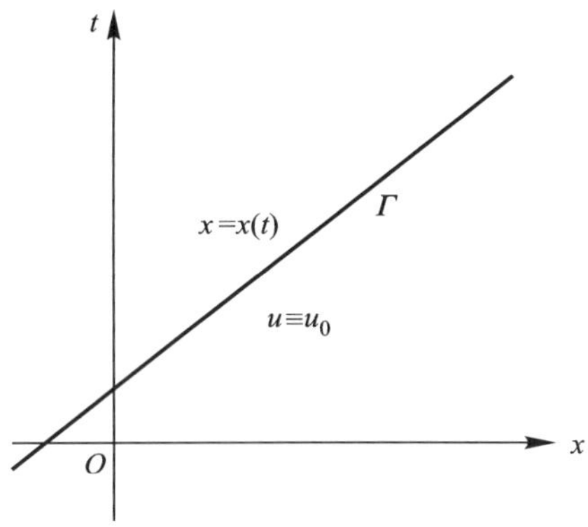

Figure 7.10

and

$$[u_{xt}] = -x'(t)[u_{xx}]. \tag{7.271}$$

Thus, we have

$$[u_{tt}] = (x'(t))^2[u_{xx}]. \tag{7.272}$$

Using (7.267) and the continuity of u_x across Γ, from equation (7.265) we have

$$[u_{tt}] = \left(g'(0) + h_1(0,0) \int_0^\infty m(s)\mathrm{d}s \right) [u_{xx}], \tag{7.273}$$

where $h_1(\cdot,\cdot)$ represents the partial derivative of h with respect to its first argument. Comparing (7.272) and (7.273), we obtain

$$g'(0) + h_1(0,0) \int_0^\infty m(s)\mathrm{d}s = (x'(t))^2. \tag{7.274}$$

From the strong ellipticity condition, the left-hand side of the above formula remains positive. Then, for the acceleration wave propagating toward the right, we know that

$$x'(t) = \sqrt{g'(0) + h_1(0,0) \int_0^\infty m(s)\mathrm{d}s} \stackrel{\mathrm{d}}{=} V \tag{7.275}$$

is a positive constant.

Differentiating equation (7.215) once with respect to t, we get

$$\begin{aligned} u_{ttt} = {} & g'(u_x)u_{xxt} + g''(u_x)u_{xt}u_{xx} \\ & + m(0)(h_1(u_x,u_x) + h_2(u_x,u_x))u_{xx} \\ & + \int_{-\infty}^t m'(t-\tau)(h(u_x(t,x),u_x(\tau,x)))_x \mathrm{d}\tau \\ & + \int_{-\infty}^t m(t-\tau)(h(u_x(t,x),u_x(\tau,x)))_{xt} \mathrm{d}\tau, \end{aligned} \tag{7.276}$$

where $h_2(\cdot,\cdot)$ represents the partial derivative of h with respect to its second argument. Using assumption (7.267) and noting (7.275), and assuming further

$$m(+\infty) = 0, \tag{7.277}$$

from the above equation we easily get that, on the curve $x = x(t)$,

$$\begin{aligned}[u_{ttt}] = V^2[u_{xxt}] \\ &+ \left(g''(0) + h_{11}(0,0)\int_0^\infty m(s)\mathrm{d}s\right)[u_{xx}][u_{xt}] \\ &+ m(0)h_2(0,0)[u_{xx}],\end{aligned} \tag{7.278}$$

where $h_{11}(0,0)$ represents the second-order partial derivative of h with respect to its first argument. It is easy to see that

$$\frac{\mathrm{d}}{\mathrm{d}t}[u_{tt}] = [u_{ttt}] + V[u_{xtt}], \tag{7.279}$$

$$\frac{\mathrm{d}}{\mathrm{d}t}[u_{xt}] = [u_{xtt}] + V[u_{xxt}]. \tag{7.280}$$

Noting (7.270), we get

$$[u_{ttt}] = 2\frac{\mathrm{d}}{\mathrm{d}t}[u_{tt}] + V^2[u_{xxt}]. \tag{7.281}$$

Plugging (7.281), (7.270), and (7.272) into (7.278), we obtain that the jump

$$a(t) = [u_{tt}] \tag{7.282}$$

of u_{tt} across Γ satisfies the differential equation

$$a'(t) + \alpha a^2(t) + \beta a(t) = 0, \tag{7.283}$$

where

$$\alpha = \frac{1}{2V^3}\left(g''(0) + h_{11}(0,0)\int_0^\infty m(s)\mathrm{d}s\right), \tag{7.284}$$

$$\beta = -\frac{1}{2V^2}m(0)h_2(0,0). \tag{7.285}$$

Equation (7.283) is exactly the same as (6.177) in Chapter 6 in form; hence, similar to Theorem 6.1 in Chapter 6, we have the following theorem.

Theorem 7.6. *For system* (7.265) *of one-dimensional viscoelastodynamics, the amplitude of the acceleration wave given by* (7.267) *and propagating from left to right into the state of uniform deformation is*

$$a(t) = \frac{\beta}{\left(\frac{\beta}{a(0)} + \alpha\right)\mathrm{e}^{\beta t} - \alpha}, \tag{7.286}$$

where α and β are given by (7.284) *and* (7.285), *respectively. Moreover, we have the following.*

1. *Assume that*

$$m(0)h_2(0,0) < 0. \tag{7.287}$$

If

$$|a(0)| < \left|\frac{\beta}{\alpha}\right| \tag{7.288}$$

or

$$\operatorname{sign} a(0) = \operatorname{sign}\left(g''(0) + h_{11}(0,0)\int_0^{\infty} m(s)\mathrm{d}s\right), \tag{7.289}$$

then $a(t)$ *decays exponentially to zero as* $t \to +\infty$ *with the same decay rate as* $\mathrm{e}^{-\beta t}$.

2. *Assume that*

$$g''(0) + h_{11}(0,0)\int_0^{\infty} m(s)\mathrm{d}s \neq 0. \tag{7.290}$$

If

$$|a(0)| > \left|\frac{\beta}{\alpha}\right| \tag{7.291}$$

and

$$\operatorname{sign} a(0) = -\operatorname{sign}\left(g''(0) + h_{11}(0,0)\int_0^{\infty} m(s)\mathrm{d}s\right), \tag{7.292}$$

then $a(t) \to \infty$ *as* $t \to t_\infty$, *where*

$$t_\infty = -\frac{1}{\beta}\ln\left(1 + \frac{\beta}{\alpha a(0)}\right). \tag{7.293}$$

Now we give a simple explanation of the conditions and conclusions of this theorem. From Theorem 7.2, the constitutive relation corresponding to equation (7.265) is dissipative if and only if

$$m'(\tau)\int_\eta^\xi h(\zeta,\eta)\mathrm{d}\zeta \leqslant 0, \quad \forall \xi,\eta \in \mathbb{R},\ \tau \geqslant 0. \tag{7.294}$$

From this we see that $m'(\tau)$ should preserve a definite sign. Without loss of generality, suppose that

$$m'(\tau) \geqslant 0, \quad \forall \tau \geqslant 0. \tag{7.295}$$

Assume that $m(+\infty) = 0$ (see (7.277)); integrating (7.295) from τ to $+\infty$, we get

$$m(\tau) \leqslant 0, \quad \forall \tau \geqslant 0. \tag{7.296}$$

In addition, from (7.294) and (7.295) we can see that

$$\int_\eta^\xi h(\zeta,\eta)\mathrm{d}\zeta \leqslant 0, \quad \forall \xi,\eta \in \mathbb{R}. \tag{7.297}$$

Noticing $h(\eta,\eta) = 0$, we immediately obtain that $h(\xi,\eta)$ is monotonically decreasing with respect to ξ at least near η, i.e.,

$$h_1(\xi,\eta) \leqslant 0, \quad \forall \xi \in \mathbb{R} \text{ and near } \eta. \tag{7.298}$$

Hence, noting that $h(\eta,\eta)=0$ yields $h_2(0,0)=-h_1(0,0)$, from (7.296) and (7.298) we can see that (7.287) actually reflects the requirement of the dissipation inequality. Then conclusion 1 of Theorem 7.6 implies that, under some suitable dissipation assumptions, the jump of the discontinuity of the solution will decay exponentially to zero with increasing time if the initial jump is sufficiently small. This conclusion suggests that the system of viscoelastodynamics may have global classical solutions for sufficiently small initial data under some suitable dissipation assumptions. On the other hand, condition (7.290) ensures that equation (7.265) is *genuinely nonlinear* when $u_t=u_x=0$. Conclusion 2 of Theorem 7.6 suggests that the classical solutions of equation (7.265) may blow up in finite time for suitably large initial data under the condition of genuine nonlinearity. The above results on the propagation of the acceleration wave for the viscoelastodynamic equation were obtained by Coleman and Gurtin in the mid-1960s (see [3]). Part of the above conjectures has been confirmed by subsequent studies. We refer the reader to [12] and the references cited therein for the one-dimensional case. There has also been some progress on the study of the multidimensional case in recent years.

Exercises

1. For the viscoelastic model made up of N Maxwell models and one single spring in parallel as shown in Figure 7.11, prove that its stress–strain relation is

$$\sigma(t)=k_s\varepsilon(t)+\sum_{i=1}^{N}k_i\int_{-\infty}^{t}\mathrm{e}^{-\lambda_i(t-\tau)}\frac{\mathrm{d}\varepsilon(\tau)}{\mathrm{d}\tau}\mathrm{d}\tau,$$

where $\lambda_i=k_i/\mu_i$ $(i=1,2,\ldots,N)$.

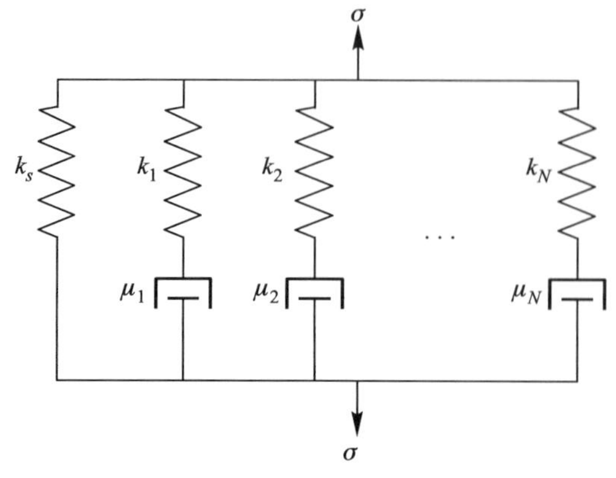

Figure 7.11

2. Use the constitutive equations of the Maxwell model and the Kelvin model to determine the constitutive equation of the Maxwell–Kelvin combined model as shown in Figure 7.12.

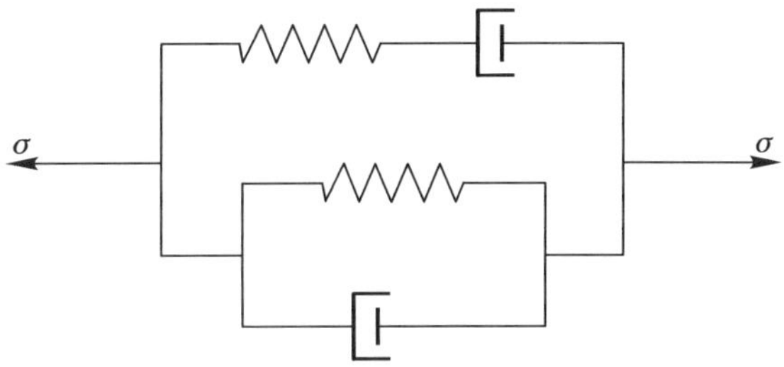

Figure 7.12

3. Prove that any given single-integral form

$$P(\xi^t) = G(\xi(t)) + \int_0^\infty H(\tau, \xi(t), \xi(t-\tau))\mathrm{d}\tau$$

can always be rewritten so that its kernel $H(\tau, \xi, \eta)$ satisfies hypothesis (A_1).

4. Prove Lemma 7.3.

5. Suppose that $m(t)$ is a monotonically decreasing nonnegative function in $(0, +\infty)$ and $m \in L^1(\rho, \infty)$ for any given $\rho > 0$. Suppose furthermore that $tm(t)$ is integrable at $t = 0$. Prove that

$$\int_0^\infty m(t) \sin st \mathrm{d}t \geqslant 0, \quad \forall s \in \mathbb{R},\ s \geqslant 0.$$

6. Let $a \in L^1(0, \infty)$. Prove that a is strongly positive definite if and only if there exists $\varepsilon > 0$ such that

$$\mathrm{Re}\widehat{a}(\mathrm{i}\xi) \geqslant \frac{\varepsilon}{1+\xi^2}, \quad \forall \xi \in \mathbb{R},$$

where $\widehat{a}$ stands for the Laplace transform of a.

7. Suppose that $a, a', a'' \in L^1(0, \infty)$ and a is strongly positive definite. Prove that

$$a(0) > 0, \quad a'(0) < 0.$$

8. Consider the following initial-boundary problem with boundary conditions of non-local type:

$$\begin{cases} u_{tt} = (\phi(u_x))_x + \displaystyle\int_0^t a'(t-\tau)(\psi(u_x(\tau,x)))_x \mathrm{d}\tau, 0 < x < l, \\ \left(\phi(u_x) + \displaystyle\int_0^t a'(t-\tau)\psi(u_x(\tau,x))\mathrm{d}\tau\right)\Bigg|_{x=0,l} = 0, \\ u|_{t=0} = u^0(x), \quad u_t|_{t=0} = u^1(x), \end{cases}$$

where u^0 and u^1 are given functions. Suppose that a is a regular and positive definite kernel, $\phi(0) = \psi(0) = 0$, and $\psi'(\xi) > 0$, $\phi'(\xi) > a(0)\psi'(\xi)$, $\forall \xi \in \mathbb{R}$. Prove that the boundary conditions of this problem can be reduced to a local type, namely, are equivalent to

$$u_x|_{x=0} = u_x|_{x=l} = 0.$$

9. Suppose that $\Omega \subset \mathbb{R}^3$ is a bounded domain with smooth boundary $\partial\Omega$, and that $\boldsymbol{y} = \boldsymbol{y}(t, \boldsymbol{x})$ is the solution of system (7.190) of nonlinear viscoelastodynamics with the constitutive equation of the dissipative single-integral form (7.30), satisfying on $\partial\Omega$ the boundary condition

$$\boldsymbol{u} = \boldsymbol{y} - \boldsymbol{x} = 0.$$

Suppose furthermore that the volume force $\boldsymbol{b} \equiv 0$. Prove that

$$E(t) = \frac{1}{2}\int_\Omega |\boldsymbol{y}_t(t,\boldsymbol{x})|^2 \mathrm{d}x + \int_\Omega l(\nabla \boldsymbol{y}(t,\boldsymbol{x}))\mathrm{d}x + \int_\Omega \int_0^\infty m(\tau, \nabla \boldsymbol{y}(t,\boldsymbol{x}), \nabla \boldsymbol{y}(t-\tau,\boldsymbol{x}))\mathrm{d}\tau \mathrm{d}x$$

is the Lyapunov functional of this problem, i.e.,

$$\frac{\mathrm{d}E(t)}{\mathrm{d}t} \leqslant 0, \quad \forall t \in \mathbb{R},$$

where

$$l(\boldsymbol{\xi}) = \int_{\boldsymbol{I}}^{\boldsymbol{\xi}} \boldsymbol{G}(\boldsymbol{\zeta}) \cdot \mathrm{d}\boldsymbol{\zeta}, \quad m(\tau, \boldsymbol{\xi}, \boldsymbol{\eta}) = \int_{\boldsymbol{\eta}}^{\boldsymbol{\xi}} \boldsymbol{H}(\tau, \boldsymbol{\zeta}, \boldsymbol{\eta}) \cdot \mathrm{d}\boldsymbol{\zeta}.$$

10. Formulate corresponding well-posed problems for system (7.184) of isotropic linear viscoelastodynamics.
11. Suppose that $\psi(\xi)$ is a continuously differentiable function and

$$\int_\eta^\xi (\psi(\zeta) - \psi(\eta))\mathrm{d}\zeta \geqslant 0, \quad \forall \xi, \eta \in \mathbb{R}.$$

Prove that

$$\psi'(\xi) \geqslant 0, \quad \forall \xi \in \mathbb{R}.$$

12. Verify (7.252).

Bibliography

[1] Christensen R. M. *Theory of Viscoelasticity: An Introduction*. New York: Academic Press, 1982.

[2] Coleman B. D. *Thermodynamics of materials with memory*. Arch. Rat. Mech. Anal., 17 (1964), 1–46.

[3] Coleman B. D., Gurtin M. E. *Waves in materials with memory* III, *Thermodynamic influences on growth and decay of acceleration waves*. Arch. Rat. Mech. Anal., 19 (1965), 266–298.

[4] Doetsch G. *Introduction to the Theory and Application of the Laplace Transformation*. New York: Springer–Verlag, 1974.

[5] Doi M., Edwards S. F. *Dynamics of concentrated polymer systems*, Part 1. J. Chem. Soc. Faraday II, 74 (1978), 1789–1801; Part 2. 1802–1817; Part 3. 1818–1832; Part 4. 75 (1979), 38–44.

[6] Engler H. *Weak solution of a class of quasilinear hyperbolic integro–differential equations describing viscoelastic materials*. Arch. Rat. Mech. Anal., 113 (1991), 1–38.

[7] Gradshteyn I. S., Ryzhik L. M. *Table of Integrals, Series and Products*. New York: Academic Press, 1980.

[8] Gripenberg G., Londen S.-O., Staffans O. *Volterra Integral and Functional Equations*. Cambridge: Cambridge University Press, 1990.

[9] Gurtin M. E., Hrusa W. J. *On energies for nonlinear viscoelastic materials of single-integral type*. Quart. Appl. Math., 46 (1988), 381–392.

[10] Gurtin M. E., Hrusa W. J. *On the thermodynamics of viscoelastic materials of single-integral type*. Quart. Appl. Math., 49 (1991), 67–85.

[11] Qin T. H. *Global existence of weak solutions to the boundary value problem for a three-dimensional viscoelastic dynamic system*. J. Elasticity, 42 (1996), 99–114.

[12] Renardy M., Hrusa W. J., Nohel J. A. *Mathematical Problems in Viscoelasticity*. Burnt Mill, UK: Longman Scientific & Technical, 1987.

Chapter 8

Kinetic Theory of Gases

8.1 Introduction

In Chapters 2 to 4 of Volume I, we studied fluid dynamics from a macroscopic point of view. There, the object we studied is not the microscopic property of individual molecules, but the macroscopic property of fluid micelles made up of a large number of molecules. However, since the macroscopic state of motion of a fluid should be determined by the microscopic state of motion of fluid molecules in nature, we can further understand the essence of a problem only when it is brought into the framework of the microscopic level.

In this chapter, we will start from the microscopic phenomenon of molecules to clarify the transport process of gases. Not only is this an important basis for research on rarefied gas dynamics, but also the macroscopic phenomenon of gas dynamics can be regarded on this basis as a special limiting case to deal with in order to figure out the effective range of various approximations. This naturally suggests the approach of how to determine the viscosity, the heat conductivity, and the equation of state.

Now we give some general descriptions.

It is known that a macroscopic gas sample contains a large number of molecules under normal temperature and pressure. For example, under standard conditions, i.e., when the pressure $p = 1$ atmosphere (about 1.013×10^5 Pa) and the temperature is $T = 273$ K (i.e. 0°C), the number of molecules in one cubic centimeter of air is

$$n \approx 2.69 \times 10^{19}.$$

This huge number tells us that the detailed kinetic description of individual molecules is not only impossible but also unnecessary. From a practical point of view, it should be reasonable to average the motion of a large number of individual molecules and to take advantage of the statistical description.

Under standard conditions, the average distance between molecules in the air is

$$r_0 \approx \frac{1}{\sqrt[3]{n}} \approx 3 \times 10^{-7} \text{ (cm)}.$$

This is much bigger than the size of a molecule (which is approximately equal to 0.53×10^{-8} cm). Therefore, the interaction between gas molecules is negligible except in collisions. Moreover, since the possibility of collisions between molecules is relatively small, we can neglect the chance of simultaneous collisions of three or more molecules and consider the effect of molecular collisions only from the viewpoint of molecular interaction in pairs.

Generally speaking, quantum mechanics (see Chapter 10) should be employed in the study of the motions and collisions of microscopic particles, but, as a matter of fact, it is sufficient to use classical mechanics for many application problems in gas dynamics. In this chapter, by means of classical mechanics, we will determine the molecular orbit of each collision, to establish the transport equation of molecules, and thus to derive some fundamental relations of thermodynamics and fluid dynamics.

8.2 Boltzmann Equation

8.2.1 Distribution Function

Suppose that only one kind of molecule is involved. For any given time t, the state of a molecule can be described by its position $\boldsymbol{x} = (x_1, x_2, x_3)$ and velocity $\boldsymbol{v} = (v_1, v_2, v_3)$. In order to describe the distribution of molecules, we introduce the *distribution function*

$$f = f(t, \boldsymbol{x}, \boldsymbol{v}),$$

which means that the average number of molecules at time t located in an infinitesimal volume element $\mathrm{d}x$ around $\boldsymbol{x}$ with velocity $\boldsymbol{v}$ in an infinitesimal volume element $\mathrm{d}v$ around $\boldsymbol{v}$ is

$$\mathrm{d}N = f(t, \boldsymbol{x}, \boldsymbol{v})\mathrm{d}x\mathrm{d}v, \tag{8.1}$$

where $\mathrm{d}x = \mathrm{d}x_1\mathrm{d}x_2\mathrm{d}x_3$ and $\mathrm{d}v = \mathrm{d}v_1\mathrm{d}v_2\mathrm{d}v_3$. Then f is the average number of molecules at time t in unit volume ranging in unit velocity at $(\boldsymbol{x}, \boldsymbol{v})$. This is a distribution function of density. Here, "average" means that function f is to be given by taking an average over the results from many similar experiments measuring the distribution of molecules.

If the distribution function $f(t, \boldsymbol{x}, \boldsymbol{v})$ is given, it can be used to determine many macroscopic quantities. For instance, the total number of molecules in any given volume V is

$$N = \int_V \mathrm{d}x \int_{\mathbb{R}^3} f(t, \boldsymbol{x}, \boldsymbol{v})\mathrm{d}v. \tag{8.2}$$

From the definition of f, the total number of molecules at time t in unit volume at $\boldsymbol{x}$ is

$$n(t, \boldsymbol{x}) = \int_{\mathbb{R}^3} f(t, \boldsymbol{x}, \boldsymbol{v})\mathrm{d}v. \tag{8.3}$$

Then the density of the gas under consideration at time t and position $\boldsymbol{x}$ is

$$\rho(t, \boldsymbol{x}) = nM = M \int_{\mathbb{R}^3} f(t, \boldsymbol{x}, \boldsymbol{v})\mathrm{d}v, \tag{8.4}$$

where M is the mass of a single molecule while the average velocity $\boldsymbol{V}(t,\boldsymbol{x}) = (V_1, V_2, V_3)$ at time t and position $\boldsymbol{x}$ is determined by

$$\boldsymbol{V}(t,\boldsymbol{x}) = \frac{1}{n}\int_{\mathbb{R}^3} \boldsymbol{v} f(t,\boldsymbol{x},\boldsymbol{v})\mathrm{d}\boldsymbol{v} = \frac{\displaystyle\int_{\mathbb{R}^3} \boldsymbol{v} f(t,\boldsymbol{x},\boldsymbol{v})\mathrm{d}\boldsymbol{v}}{\displaystyle\int_{\mathbb{R}^3} f(t,\boldsymbol{x},\boldsymbol{v})\mathrm{d}\boldsymbol{v}}. \tag{8.5}$$

In addition, in the kinetic theory of gases, the temperature $T(t,\boldsymbol{x})$ at time t and position $\boldsymbol{x}$ is determined by

$$\frac{3}{2}nkT(t,\boldsymbol{x}) = \int_{\mathbb{R}^3} \frac{1}{2}M|\boldsymbol{v}-\boldsymbol{V}|^2 f(t,\boldsymbol{x},\boldsymbol{v})\mathrm{d}\boldsymbol{v}, \tag{8.6}$$

where n and $\boldsymbol{V}$ are given by (8.3) and (8.5), respectively, M is the mass of a single molecule, and

$$k = 1.380 \times 10^{-16} \text{ J/K}$$

is the *Boltzmann constant*. We will see later that, for an ideal gas under thermodynamic equilibrium, (8.6) reduces to the normal definition of temperature, and even for a gas under nonequilibrium states, the definition of temperature can also be given by (8.6).

The right-hand side of (8.6) represents the heat energy per unit volume. In fact, $\boldsymbol{v}-\boldsymbol{V}$ is the velocity of the molecule with the macroscopic velocity being subtracted, $\frac{1}{2}M|\boldsymbol{v}-\boldsymbol{V}|^2$ is the kinetic energy of a single molecule, and $f\mathrm{d}\boldsymbol{v}$ is the number of molecules in unit volume with velocity in the volume element $\mathrm{d}\boldsymbol{v}$. Therefore, the right-hand side of (8.6) is the energy caused by the heat motion of molecules in unit volume and should be a function of temperature. This implies that the gas under consideration is an ideal gas.

The pressure tensor at time t and position $\boldsymbol{x}$ is defined by

$$p_{ij}(t,\boldsymbol{x}) = \int_{\mathbb{R}^3} M(v_i - V_i)(v_j - V_j) f(t,\boldsymbol{x},\boldsymbol{v})\mathrm{d}\boldsymbol{v} \qquad (i,j = 1,2,3). \tag{8.7}$$

We will see later that it differs from the stress tensor defined in section 2.2 in Chapter 2 of Volume I by only a minus sign.

In many cases that we are interested in, f is largely spherically symmetric with respect to variable $\boldsymbol{v}-\boldsymbol{V}$, since the coordinate orientation has almost no effect on f, and there will be no dominant direction after subtracting the average velocity $\boldsymbol{V}$ from $\boldsymbol{v}$. In particular, when f is a function of only $|\boldsymbol{v}-\boldsymbol{V}|^2$, noticing that

$$\int_{\mathbb{R}^3} u_i u_j f(u_1^2+u_2^2+u_3^2)\mathrm{d}u = 0 \quad \text{if } i \neq j,$$

$$\begin{aligned}\int_{\mathbb{R}^3} u_1^2 f(u_1^2+u_2^2+u_3^2)\mathrm{d}u &= \int_{\mathbb{R}^3} u_2^2 f(u_1^2+u_2^2+u_3^2)\mathrm{d}u \\ &= \int_{\mathbb{R}^3} u_3^2 f(u_1^2+u_2^2+u_3^2)\mathrm{d}u,\end{aligned}$$

we can easily see that

$$p_{ij}(t,\boldsymbol{x}) = \begin{cases} 0 & \text{if } i \neq j, \\ p(t,\boldsymbol{x}) & \text{if } i = j, \end{cases} \tag{8.8}$$

and, noting (8.6), we have

$$p = \frac{1}{3}\sum_{i=1}^{3} p_{ii} = \frac{1}{3}\int_{\mathbb{R}^3} M|\boldsymbol{v}-\boldsymbol{V}|^2 f(t,\boldsymbol{x},\boldsymbol{v})\mathrm{d}v = nkT(t,\boldsymbol{x}). \tag{8.9}$$

This just corresponds to the case of ideal gases. It also suggests that the above arguments can be applied to ideal gases.

Finally, the total energy at time t and position $\boldsymbol{x}$ is

$$\begin{aligned}
&\int_{\mathbb{R}^3} \frac{M}{2}|\boldsymbol{v}|^2 f(t,\boldsymbol{x},\boldsymbol{v})\mathrm{d}v \\
&= \int_{\mathbb{R}^3} \frac{M}{2}(|\boldsymbol{v}-\boldsymbol{V}|^2 + 2(\boldsymbol{v}-\boldsymbol{V})\cdot\boldsymbol{V} + |\boldsymbol{V}|^2) f\mathrm{d}v \\
&= \int_{\mathbb{R}^3} \frac{M}{2}|\boldsymbol{v}-\boldsymbol{V}|^2 f\mathrm{d}v + \int_{\mathbb{R}^3} M(\boldsymbol{v}\cdot\boldsymbol{V}) f\mathrm{d}v \\
&\quad -\frac{M}{2}|\boldsymbol{V}|^2\int_{\mathbb{R}^3} f\mathrm{d}v \\
&= \frac{3}{2}nkT + nM|\boldsymbol{V}|^2 - \frac{1}{2}nM|\boldsymbol{V}|^2 \\
&= \frac{3}{2}nkT + \frac{1}{2}\rho|\boldsymbol{V}|^2.
\end{aligned} \tag{8.10}$$

Here we used (8.6), (8.5), and (8.3)–(8.4). The total energy is the sum of the internal energy and the macroscopic kinetic energy. Noting $\frac{3}{2}nkT = \frac{3}{2}\frac{kT}{M}\rho$, the internal energy per unit mass is

$$e = \frac{3}{2}\frac{kT}{M}. \tag{8.11}$$

8.2.2 Boltzmann Equation

In order to determine the distribution function $f(t,\boldsymbol{x},\boldsymbol{v})$ and to discuss the general theory of nonequilibrium properties of gases, we need to establish the equation satisfied by f—the *Boltzmann equation*. This is the mathematical description of the conservation law of the number of molecules.

To obtain this equation, we consider the motion of a molecule at time t under state $(\boldsymbol{x},\boldsymbol{v})$. Suppose that the external force acting on the molecule is $\boldsymbol{F}(t,\boldsymbol{x},\boldsymbol{v})$ (it can often be assumed to be independent of $\boldsymbol{v}$, since external forces are usually regarded as macroscopic actions); then the force acting on a single molecule per unit mass is

$$\boldsymbol{g}(t,\boldsymbol{x},\boldsymbol{v}) = \frac{\boldsymbol{F}(t,\boldsymbol{x},\boldsymbol{v})}{M}. \tag{8.12}$$

In a time interval $\mathrm{d}t$, assume that molecules do not collide with each other; then the molecule that is under state $(\boldsymbol{x},\boldsymbol{v})$ at time t will be under state $(\boldsymbol{x}',\boldsymbol{v}')$ at time $t+\mathrm{d}t$, where

$$\begin{aligned}\boldsymbol{x}' &= \boldsymbol{x}+\boldsymbol{v}\mathrm{d}t,\\ \boldsymbol{v}' &= \boldsymbol{v}+\boldsymbol{g}\mathrm{d}t.\end{aligned} \tag{8.13}$$

Correspondingly, the molecules in the volume element $\mathrm{d}x\mathrm{d}v$ around state $(\boldsymbol{x},\boldsymbol{v})$ at time t will be in the volume element $\mathrm{d}x'\mathrm{d}v'$ around state $(\boldsymbol{x}',\boldsymbol{v}')$ at time $t+\mathrm{d}t$ due to movement. Noticing that

$$\mathrm{d}x'\mathrm{d}v' = \left|\det\frac{\partial(\boldsymbol{x}',\boldsymbol{v}')}{\partial(\boldsymbol{x},\boldsymbol{v})}\right|\mathrm{d}x\mathrm{d}v,$$

and that the Jacobi determinant $\det\frac{\partial(\boldsymbol{x}',\boldsymbol{v}')}{\partial(\boldsymbol{x},\boldsymbol{v})}$ of transformation (8.13) is

$$\begin{vmatrix} 1 & 0 & 0 & \mathrm{d}t & 0 & 0\\ 0 & 1 & 0 & 0 & \mathrm{d}t & 0\\ 0 & 0 & 1 & 0 & 0 & \mathrm{d}t\\ \frac{\partial g_1}{\partial x_1}\mathrm{d}t & \frac{\partial g_1}{\partial x_2}\mathrm{d}t & \frac{\partial g_1}{\partial x_3}\mathrm{d}t & 1 & 0 & 0\\ \frac{\partial g_2}{\partial x_1}\mathrm{d}t & \frac{\partial g_2}{\partial x_2}\mathrm{d}t & \frac{\partial g_2}{\partial x_3}\mathrm{d}t & 0 & 1 & 0\\ \frac{\partial g_3}{\partial x_1}\mathrm{d}t & \frac{\partial g_3}{\partial x_2}\mathrm{d}t & \frac{\partial g_3}{\partial x_3}\mathrm{d}t & 0 & 0 & 1 \end{vmatrix} = 1+O(\mathrm{d}t^2),$$

neglecting the higher-order infinitesimals, which will be inoperative in later discussions, we have

$$\mathrm{d}x'\mathrm{d}v' = \mathrm{d}x\mathrm{d}v. \tag{8.14}$$

Then the number of molecules in the volume element $\mathrm{d}x\mathrm{d}v$ around $(\boldsymbol{x},\boldsymbol{v})$ at time t is

$$\mathrm{d}N = f(t,\boldsymbol{x},\boldsymbol{v})\mathrm{d}x\mathrm{d}v, \tag{8.15}$$

while the number of molecules in the corresponding volume element around $(\boldsymbol{x}',\boldsymbol{v}')$ at time $t+\mathrm{d}t$ is

$$\mathrm{d}N' = f(t+\mathrm{d}t,\boldsymbol{x}',\boldsymbol{v}')\mathrm{d}x\mathrm{d}v. \tag{8.16}$$

If there is no interaction (collision) between molecules, all the molecules in the volume element around $(\boldsymbol{x},\boldsymbol{v})$ at time t will move into the corresponding volume element around $(\boldsymbol{x}',\boldsymbol{v}')$ at time $t+\mathrm{d}t$; thus we have

$$\mathrm{d}N = \mathrm{d}N'. \tag{8.17}$$

But actually, due to intermolecular interactions (collisions), on one hand, some molecules that are not in the volume element $\mathrm{d}x\mathrm{d}v$ around $(\boldsymbol{x},\boldsymbol{v})$ at time t will be scattered into the corresponding volume element around $(\boldsymbol{x}',\boldsymbol{v}')$ at time $t+\mathrm{d}t$; on the other hand, some molecules that are in the volume element $\mathrm{d}x\mathrm{d}v$ around $(\boldsymbol{x},\boldsymbol{v})$ at time t will be scattered out of the corresponding volume element around $(\boldsymbol{x}',\boldsymbol{v}')$ at $t+\mathrm{d}t$. Therefore, (8.17) is not true in general and should be replaced by

$$\mathrm{d}N' - \mathrm{d}N = J\mathrm{d}t\mathrm{d}x\mathrm{d}v. \tag{8.18}$$

The right-hand side of the above formula represents the net gain of particles in the volume element caused by the scattering; it should be proportional to $\mathrm{d}x$, $\mathrm{d}v$, and $\mathrm{d}t$, which is why we have the above expression, and J will be determined later.

From (8.15)–(8.16) and (8.13), neglecting higher-order infinitesimals, we have

$$\begin{aligned} & \mathrm{d}N' - \mathrm{d}N \\ &= f(t+\mathrm{d}t, \boldsymbol{x}', \boldsymbol{v}')\mathrm{d}x\mathrm{d}v - f(t,\boldsymbol{x},\boldsymbol{v})\mathrm{d}x\mathrm{d}v \\ &= (f(t+\mathrm{d}t, \boldsymbol{x}+\boldsymbol{v}\mathrm{d}t, \boldsymbol{v}+\boldsymbol{g}\mathrm{d}t) - f(t,\boldsymbol{x},\boldsymbol{v}))\mathrm{d}x\mathrm{d}v \\ &= \left(\frac{\partial f}{\partial t} + \boldsymbol{v}\cdot\nabla_x f + \boldsymbol{g}\cdot\nabla_v f\right)\mathrm{d}x\mathrm{d}v\mathrm{d}t, \end{aligned} \tag{8.19}$$

where we denote

$$\nabla_x = \left(\frac{\partial}{\partial x_1}, \frac{\partial}{\partial x_2}, \frac{\partial}{\partial x_3}\right)$$

and

$$\nabla_v = \left(\frac{\partial}{\partial v_1}, \frac{\partial}{\partial v_2}, \frac{\partial}{\partial v_3}\right).$$

From (8.18) and (8.19), we obtain the following partial differential equation satisfied by $f(t,\boldsymbol{x},\boldsymbol{v})$:

$$\frac{\partial f}{\partial t} + \boldsymbol{v}\cdot\nabla_x f + \boldsymbol{g}\cdot\nabla_v f = J, \tag{8.20}$$

which is called the *Boltzmann equation*.

The right-hand side J of the above equation can be written as

$$J = J_+ - J_-, \tag{8.21}$$

where J_+ and J_- have the following meanings: $J_-\mathrm{d}t\mathrm{d}x\mathrm{d}v$ represents the total number of molecules leaving the volume element $\mathrm{d}x\mathrm{d}v$ in time interval $[t, t+\mathrm{d}t]$ due to collisions, namely, the number of collisions with the following property: one of the two collided molecules is in the volume element $\mathrm{d}x\mathrm{d}v$ around $(\boldsymbol{x},\boldsymbol{v})$ before collision. $J_+\mathrm{d}t\mathrm{d}x\mathrm{d}v$ represents the total number of molecules entering the volume element $\mathrm{d}x\mathrm{d}v$ in time interval $[t, t+\mathrm{d}t]$ due to collisions, namely, the number of collisions with the following property: one of the two collided molecules is in the volume element $\mathrm{d}x\mathrm{d}v$ around $(\boldsymbol{x},\boldsymbol{v})$ after collision. Hence, J is exactly the net gain of particle numbers in the volume element $\mathrm{d}x\mathrm{d}v$ around $(\boldsymbol{x},\boldsymbol{v})$ in time interval $[t, t+\mathrm{d}t]$. J is called the *collision term*.

It demands detailed study on the properties of collisions to determine J_+ and J_-.

8.2.3 Two-Body Collision

We investigate the *elastic collision* in free space between two molecules with equal mass and without internal structure. The state of each molecule is described by its position and velocity.

Suppose that these two molecules have velocity $\boldsymbol{v}$ and $\boldsymbol{w}$ before collision and have velocity $\boldsymbol{v}'$ and $\boldsymbol{w}'$ after collision, respectively. By the conservation laws of momentum and energy, we have

$$\boldsymbol{v} + \boldsymbol{w} = \boldsymbol{v}' + \boldsymbol{w}' \tag{8.22}$$

and

$$|\boldsymbol{v}|^2+|\boldsymbol{w}|^2=|\boldsymbol{v}'|^2+|\boldsymbol{w}'|^2. \tag{8.23}$$

Set

$$\overline{\boldsymbol{v}}=\frac{1}{2}(\boldsymbol{v}+\boldsymbol{w}),\quad \overline{\boldsymbol{v}}'=\frac{1}{2}(\boldsymbol{v}'+\boldsymbol{w}')$$

and

$$\boldsymbol{u}=\boldsymbol{w}-\boldsymbol{v},\quad \boldsymbol{u}'=\boldsymbol{w}'-\boldsymbol{v}'.$$

Then (8.22)–(8.23) can be rewritten as

$$\overline{\boldsymbol{v}}=\overline{\boldsymbol{v}}', \tag{8.24}$$

$$|\boldsymbol{u}|=|\boldsymbol{u}'|. \tag{8.25}$$

In fact, (8.22) is equivalent to (8.24). From (8.22) we also have

$$|\boldsymbol{v}|^2+2\boldsymbol{v}\cdot\boldsymbol{w}+|\boldsymbol{w}|^2=|\boldsymbol{v}'|^2+2\boldsymbol{v}'\cdot\boldsymbol{w}'+|\boldsymbol{w}'|^2.$$

Noting (8.23), the above formula yields

$$\boldsymbol{v}\cdot\boldsymbol{w}=\boldsymbol{v}'\cdot\boldsymbol{w}'.$$

Thus,

$$|\boldsymbol{v}|^2-2\boldsymbol{v}\cdot\boldsymbol{w}+|\boldsymbol{w}|^2=|\boldsymbol{v}'|^2-2\boldsymbol{v}'\cdot\boldsymbol{w}'+|\boldsymbol{w}'|^2,$$

which is exactly (8.25). The derivation is obvious the other way around.

We can see from (8.24)–(8.25) that the elastic collision does not change the average velocity $\overline{\boldsymbol{v}}'=\overline{\boldsymbol{v}}$, and $\boldsymbol{u}'$ is just a rotation of $\boldsymbol{u}$ keeping the same size. Then the collision can be completely determined for given $\overline{\boldsymbol{v}}$, $\boldsymbol{u}$ and the rotation angle (θ,ϕ) of $\boldsymbol{u}'$ with respect to $\boldsymbol{u}$ (see Figure 8.1). The angles θ and ϕ are called the *scattering angles*.

If $\overline{\boldsymbol{v}}$ and $\boldsymbol{u}$ vary in their nearby volume elements $\mathrm{d}\overline{v}$ and $\mathrm{d}u$, respectively, and the scattering angles θ and ϕ remain unchanged, then the quantities $\overline{\boldsymbol{v}}'$ and $\boldsymbol{u}'$ after collision will vary in their corresponding volume elements $\mathrm{d}\overline{v}'$ and $\mathrm{d}u'$. By $\overline{\boldsymbol{v}}=\overline{\boldsymbol{v}}'$, we know that $\mathrm{d}\overline{v}=\mathrm{d}\overline{v}'$. From the invariance of θ and ϕ, we know that $|\mathrm{d}\boldsymbol{u}|=|\mathrm{d}\boldsymbol{u}'|$. Thus, we have

$$\mathrm{d}\overline{v}\mathrm{d}u=\mathrm{d}\overline{v}'\mathrm{d}u'. \tag{8.26}$$

In addition, from the definitions of $\overline{\boldsymbol{v}}$ and $\boldsymbol{u}$, we easily know by direct calculations that

$$\mathrm{d}\overline{v}\mathrm{d}u=\mathrm{d}v\mathrm{d}w. \tag{8.27}$$

Similarly, we have

$$\mathrm{d}\overline{v}'\mathrm{d}u'=\mathrm{d}v'\mathrm{d}w'. \tag{8.28}$$

Therefore, if we change $\boldsymbol{v}$ and $\boldsymbol{w}$ slightly while keeping the scattering angles unchanged, then $\boldsymbol{v}'$ and $\boldsymbol{w}'$ will also change slightly, and

$$\mathrm{d}v\mathrm{d}w=\mathrm{d}v'\mathrm{d}w'. \tag{8.29}$$

Here, the average velocity $\overline{\boldsymbol{v}}$ is not a quantity of particular importance. In fact, under a new coordinate system moving at a uniform velocity $\overline{\boldsymbol{v}}$, it is only necessary to take the

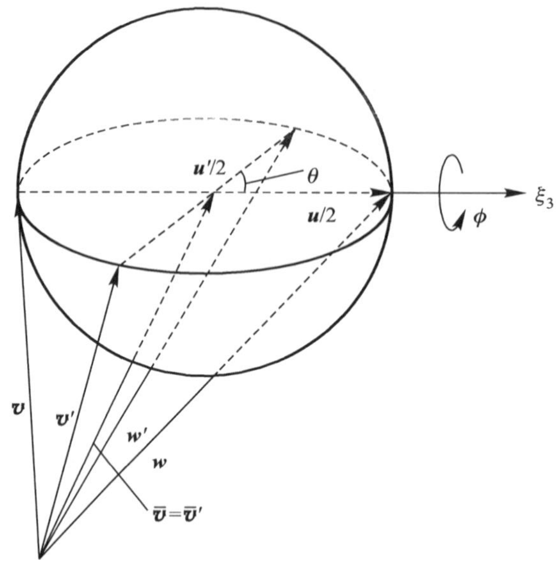

Figure 8.1

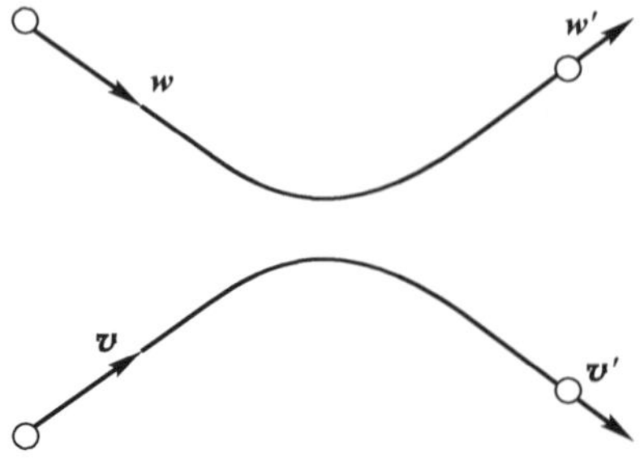

Figure 8.2

relative velocities $\boldsymbol{u}$ and $\boldsymbol{u}'$ into account. This coordinate system is called the *centroid coordinate system*.

The collision process under the laboratory coordinate system (i.e., the coordinate system of the observer) and the centroid coordinate system are depicted in Figure 8.2 and Figure 8.3, respectively.

Under the centroid coordinate system, we only need to focus on one single collided molecule, since the other molecule it collided with has a velocity in the opposite direction. Hence, the problem is equivalently reduced to a problem about the scattering of a molecule with respect to an imaginary fixed force center O. When a certain molecule approaches point O at velocity $\boldsymbol{u}$, the vertical distance b between the incident orbit and point O is called the *collision parameter*. We take O as the origin, and the ξ_3 axis parallel to the direction of $\boldsymbol{u}$. Since $|\boldsymbol{u}| = |\boldsymbol{u}'|$, the final state of this molecule is determined by two scattering angles θ and ϕ, where θ is the angle between $\boldsymbol{u}'$ and the ξ_3 axis, and

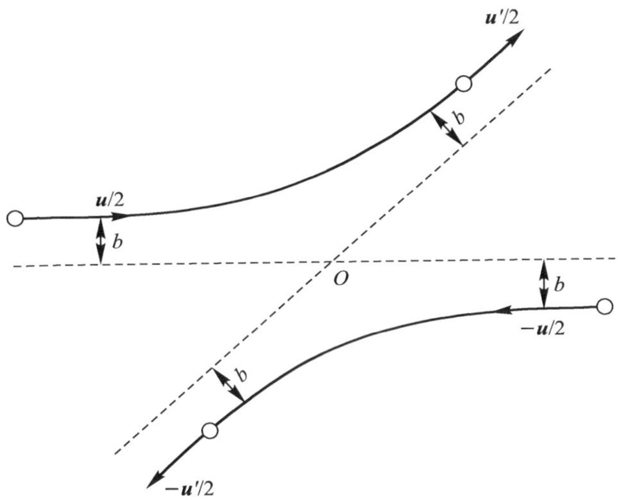

Figure 8.3

ϕ is the azimuth angle of $\boldsymbol{u}'$ with respect to the ξ_3 axis. Hereafter, we denote θ and ϕ together by Ω.

Now we introduce an important concept, called the *differential scattering cross section* $\sigma(\Omega)$.

We see that the initial velocities $\boldsymbol{v}$ and $\boldsymbol{w}$ cannot determine the collision uniquely, since they cannot determine the collision parameter b. For any given $\boldsymbol{v}$ and $\boldsymbol{w}$, since $\overline{\boldsymbol{v}} = \frac{1}{2}(\boldsymbol{v}+\boldsymbol{w})$ is given, equivalently, a kind of collision is given, and they have the same centroid coordinate system. This kind of collision is composed of collisions corresponding to all possible collision parameters (and thus all possible scattering angles). In order to describe this kind of collision, we can imagine that a molecular incident beam, which is distributed uniformly in space and with initial velocity $\boldsymbol{u}$, shoots onto the force center O; see Figure 8.4.

Let I be the number of molecules in the incident beam passing in unit time (second) across a unit area perpendicular to the beam, called the *incident flux*. Then the differential scattering cross section $\sigma(\Omega)$ is defined as follows:

$$I\sigma(\Omega)\mathrm{d}\Omega = \text{number of molecules deflected into the solid angle } \mathrm{d}\Omega \text{ around } \Omega \text{ due to scattering.} \tag{8.30}$$

The differential scattering cross section $\sigma(\Omega)$ has the dimension of area, and it has the following geometrical meaning:

$$\begin{aligned}&\text{number of molecules deflected into } [\Omega, \Omega+\mathrm{d}\Omega]\\&\text{in unit time due to scattering}\\=\;&\text{number of molecules in the incident beam passing}\\&\text{vertically across area } \sigma(\Omega)\mathrm{d}\Omega \text{ in unit time.}\end{aligned}$$

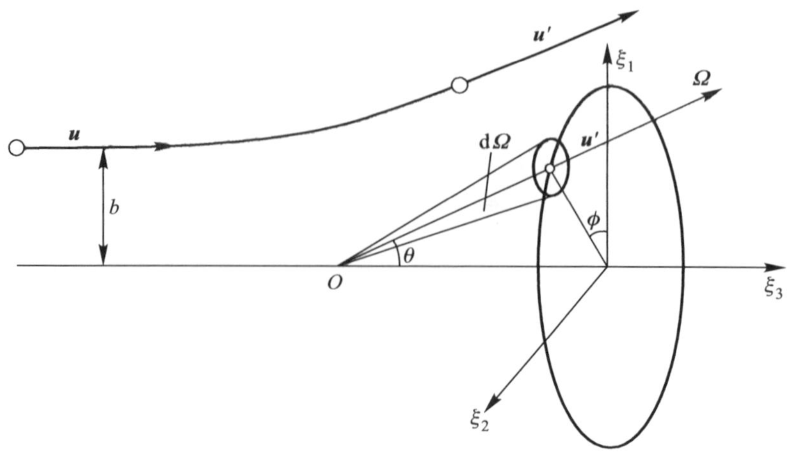

Figure 8.4

The integral of $\sigma(\Omega)$ with respect to all the solid-angle elements is denoted by

$$\sigma_{tot} = \int \sigma(\Omega)\mathrm{d}\Omega, \tag{8.31}$$

called the *total cross section*. Here, both $\sigma(\Omega)$ and σ_{tot} should be regarded as being related to $|\boldsymbol{u}|$.

The differential scattering cross section is a quantity that can be measured directly by experiments. If the internal molecular potential is known, $\sigma(\Omega)$ can also be obtained by computation, but it has to be treated by methods on the basis of quantum mechanics. This is because there will be overlaps between molecules in the process of collision so that classical mechanics is no longer applicable. Here, we assume that $\sigma(\Omega)$ is given for the gas under consideration.

Now we denote

$$\sigma(\Omega) = \sigma(\boldsymbol{v},\boldsymbol{w} \to \boldsymbol{v}',\boldsymbol{w}'),$$

where $\boldsymbol{v}$ and $\boldsymbol{w}$ are the velocities of two molecules before collision, $\boldsymbol{v}'$ and $\boldsymbol{w}'$ are the velocities after collision, and Ω are the angles between $\boldsymbol{w}-\boldsymbol{v}$ and $\boldsymbol{w}'-\boldsymbol{v}'$ and the azimuth angle.

The symmetric properties of $\sigma(\Omega)$ are as follows:

(a) Invariance of inversion with respect to time:

$$\sigma(\boldsymbol{v},\boldsymbol{w} \to \boldsymbol{v}',\boldsymbol{w}') = \sigma(-\boldsymbol{v}',-\boldsymbol{w}' \to -\boldsymbol{v},-\boldsymbol{w}). \tag{8.32}$$

This implies that every molecule will backtrack if the time is inverted.

(b) Invariance with respect to rotation and reflection:

$$\sigma(\boldsymbol{v},\boldsymbol{w} \to \boldsymbol{v}',\boldsymbol{w}') = \sigma(\boldsymbol{v}^*,\boldsymbol{w}^* \to \boldsymbol{v}'^*,\boldsymbol{w}'^*), \tag{8.33}$$

where $\boldsymbol{v}^*$, $\boldsymbol{w}^*$, and so on stand for the vectors obtained by $\boldsymbol{v}$, $\boldsymbol{w}$, and so on after the same rotation or specular reflection.

Given a collision, we can define its *anticollision*. It is composed of the interchange between the initial state and the final state of the original collision. As a consequence of the

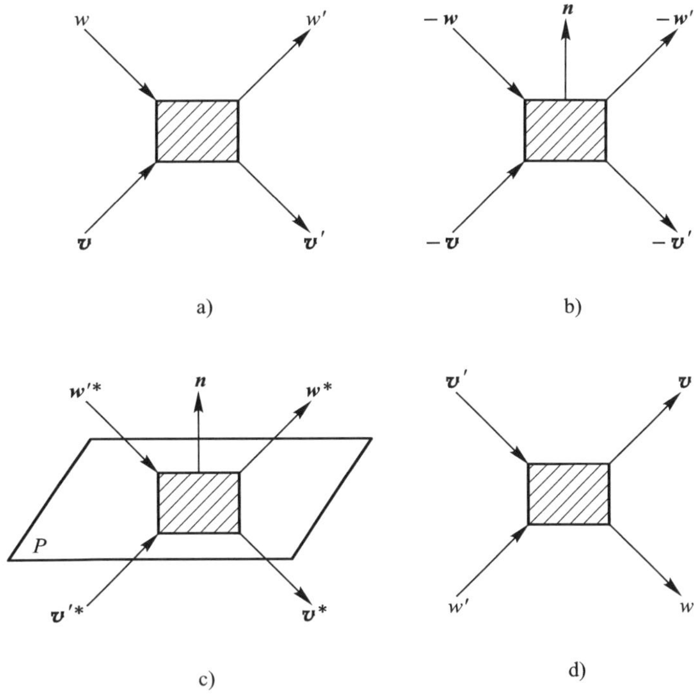

Figure 8.5

above symmetric properties, we can prove that the anticollision has the same differential scattering cross section as the original collision, i.e.,

$$\sigma(\boldsymbol{v},\boldsymbol{w} \to \boldsymbol{v}',\boldsymbol{w}') = \sigma(\boldsymbol{v}',\boldsymbol{w}' \to \boldsymbol{v},\boldsymbol{w}). \tag{8.34}$$

In fact, we consider a collision as shown in Figure 8.5a. From property (a), the differential scattering cross section is invariant when the time is inverted; then we have

$$\sigma(\boldsymbol{v},\boldsymbol{w} \to \boldsymbol{v}',\boldsymbol{w}') = \sigma(-\boldsymbol{v}',-\boldsymbol{w}' \to -\boldsymbol{v},-\boldsymbol{w});$$

see Figure 8.5b. Take a certain axial direction $\boldsymbol{n}$ perpendicular to $\overline{v} = \overline{v}'$, and then rotate the angle π around $\boldsymbol{n}$. From property (b), the differential scattering cross section is still invariant, i.e.,

$$\sigma(-\boldsymbol{v}',-\boldsymbol{w}' \to -\boldsymbol{v},-\boldsymbol{w}) = \sigma(\boldsymbol{v}'^{*},\boldsymbol{w}'^{*} \to \boldsymbol{v}^{*},\boldsymbol{w}^{*}).$$

Here $\boldsymbol{v}^*$, $\boldsymbol{w}^*$, and so on are vectors obtained by $-\boldsymbol{v}$, $-\boldsymbol{w}$, and so on after the above rotation; see Figure 8.5c. Now take a reflection of Figure 8.5c with respect to the plane P perpendicular to $\boldsymbol{n}$, and we easily know that $\boldsymbol{v}^*$, $\boldsymbol{v}'^*$, and so on are changed into $\boldsymbol{v}$, $\boldsymbol{v}'$, and so on, respectively; see Figure 8.5d. From property (b), the differential scattering cross section is still invariant, and we have

$$\sigma(\boldsymbol{v}'^{*},\boldsymbol{w}'^{*} \to \boldsymbol{v}^{*},\boldsymbol{w}^{*}) = \sigma(\boldsymbol{v}',\boldsymbol{w}' \to \boldsymbol{v},\boldsymbol{w}).$$

This proves (8.34).

The key point of the above derivation is the following: any given vector $\boldsymbol{v}$ turns into $-\boldsymbol{v}$ after rotating the angle π around $\boldsymbol{n}$ and then reflecting with respect to the plane vertical to $\boldsymbol{n}$. It goes back to $\boldsymbol{v}$ after a time inversion. Certainly, now the position of this vector changes (compare Figures 8.5a and 8.5d).

8.2.4 Determination of Collision Term J

Now we discuss how to determine the collision term J on the right-hand side of the Boltzmann equation (8.20). For this purpose, we make the following assumptions:

(a) Consider only collisions between two molecules, namely, assume that the gas is quite dilute.

(b) Do not take the container wall of the gas into account, since this influences only the boundary conditions.

(c) Neglect the effect of external forces on the collision cross section.

(d) The state of motion (velocity and position) of each molecule is not related to others when no collision occurs; that is, the probability distribution of each molecule is independent. This is called the *molecular chaos hypothesis*. It implies that the number of molecular pairs in the spatial volume element $\mathrm{d}x$ at position $\boldsymbol{x}$ and the spatial volume element $\mathrm{d}y$ at position $\boldsymbol{y}$, with velocity lying in the element $\mathrm{d}v$ around $\boldsymbol{v}$ and the element $\mathrm{d}w$ around $\boldsymbol{w}$, is

$$(f(t,\boldsymbol{x},\boldsymbol{v})\mathrm{d}x\mathrm{d}v)(f(t,\boldsymbol{y},\boldsymbol{w})\mathrm{d}y\mathrm{d}w). \tag{8.35}$$

Now we determine J_- in (8.21). As stated above, J_- represents the reduced number of molecules due to collision in unit time, in unit volume element, and in unit velocity element. We inspect molecules located in an infinitesimal element $\mathrm{d}x$ around $\boldsymbol{x}$ with velocity in an infinitesimal element $\mathrm{d}v$ around $\boldsymbol{v}$. In the same volume element $\mathrm{d}x$, molecules in the velocity element $\mathrm{d}w$ around any velocity $\boldsymbol{w}$ can be regarded as an incident molecular beam acting on the molecule with velocity $\boldsymbol{v}$. Noticing that the relative velocity of the incident molecular beam is $\boldsymbol{w}-\boldsymbol{v}$ when the target is fixed, we can easily see that the incident flux of this molecular beam is

$$\begin{aligned} I &= \text{the number of molecules with velocity in } \mathrm{d}w \text{ around } \boldsymbol{w} \\ &\quad\ \text{in the incident molecular beam passing across the unit} \\ &\quad\ \text{area vertical to } \boldsymbol{w}-\boldsymbol{v} \text{ in unit time} \\ &= f(t,\boldsymbol{x},\boldsymbol{w})\mathrm{d}w|\boldsymbol{w}-\boldsymbol{v}|. \end{aligned} \tag{8.36}$$

Then, by (8.30), the number of molecules scattered into the solid angle $\mathrm{d}\Omega$ around Ω in time interval $[t,t+\mathrm{d}t]$ by the collision $\{\boldsymbol{v},\boldsymbol{w}\to\boldsymbol{v}',\boldsymbol{w}'\}$ is

$$I\sigma(\Omega)\mathrm{d}\Omega\mathrm{d}t = f(t,\boldsymbol{x},\boldsymbol{w})\mathrm{d}w|\boldsymbol{w}-\boldsymbol{v}|\sigma(\Omega)\mathrm{d}\Omega\mathrm{d}t, \tag{8.37}$$

where $\sigma(\Omega)=\sigma(\boldsymbol{v},\boldsymbol{w}\to\boldsymbol{v}',\boldsymbol{w})$ is the differential scattering cross section under the centroid coordinate system, and Ω are the angles between $\boldsymbol{w}-\boldsymbol{v}$ and $\boldsymbol{w}'-\boldsymbol{v}'$ and the azimuth angle.

Integrating (8.37) with respect to $\boldsymbol{w}$ and Ω, we obtain that the number of molecules colliding with a molecule with velocity $\boldsymbol{v}$ in time interval $[t,t+\mathrm{d}t]$ is

$$\int_{\mathbb{R}^3}\mathrm{d}w\int\mathrm{d}\Omega\sigma(\Omega)|\boldsymbol{w}-\boldsymbol{v}|f(t,\boldsymbol{x},\boldsymbol{w})\mathrm{d}t. \tag{8.38}$$

Considering furthermore that the number of molecules in the infinitesimal element $\mathrm{d}x\mathrm{d}v$ around $(\boldsymbol{x},\boldsymbol{v})$ is

$$f(t,\boldsymbol{x},\boldsymbol{v})\mathrm{d}x\mathrm{d}v, \tag{8.39}$$

by the molecular chaos hypothesis (see (8.35)), we obtain that the number of molecules leaving the original element due to collision in the element $\mathrm{d}x\mathrm{d}v$ around $(\boldsymbol{x},\boldsymbol{v})$ in time interval $[t,t+\mathrm{d}t]$ is

$$\int_{\mathbb{R}^3}\mathrm{d}w\int \mathrm{d}\Omega\sigma(\Omega)|\boldsymbol{w}-\boldsymbol{v}|f(t,\boldsymbol{x},\boldsymbol{v})f(t,\boldsymbol{x},\boldsymbol{w})\mathrm{d}x\mathrm{d}v\mathrm{d}t. \tag{8.40}$$

Thus, we get

$$J_- = \int_{\mathbb{R}^3}\mathrm{d}w\int \mathrm{d}\Omega\sigma(\Omega)|\boldsymbol{w}-\boldsymbol{v}|f(t,\boldsymbol{x},\boldsymbol{v})f(t,\boldsymbol{x},\boldsymbol{w}), \tag{8.41}$$

where the integration is with respect to $\boldsymbol{w}$ and the scattering angles Ω.

Although the integral with respect to Ω on the right-hand side of the above formula can be directly written into the form of the total cross section,

$$\int \sigma(\Omega)\mathrm{d}\Omega = \sigma_{tot}, \tag{8.42}$$

for the convenience of the following discussion, we still adopt (8.41), which looks more complicated.

Now we determine J_+ in (8.21). We consider a collision like $\{\boldsymbol{v}',\boldsymbol{w}' \to \boldsymbol{v},\boldsymbol{w}\}$, where $\boldsymbol{v}$ is fixed as the velocity of the molecule under consideration. For a molecule with velocity $\boldsymbol{v}'$, after having been shot by an incident molecular beam with velocity $\boldsymbol{w}'$, its incident flux is

$$I = f(t,\boldsymbol{x},\boldsymbol{w}')\mathrm{d}w'|\boldsymbol{w}'-\boldsymbol{v}'| \tag{8.43}$$

(see (8.36)). Therefore, in time interval $[t,t+\mathrm{d}t]$, the number of molecules scattered into the solid angle $\mathrm{d}\Omega$ around Ω by this kind of collision is

$$f(t,\boldsymbol{x},\boldsymbol{w}')\mathrm{d}w'|\boldsymbol{w}'-\boldsymbol{v}'|\sigma'(\Omega)\mathrm{d}\Omega\mathrm{d}t, \tag{8.44}$$

where

$$\sigma'(\Omega) = \sigma(\boldsymbol{v}',\boldsymbol{w}' \to \boldsymbol{v},\boldsymbol{w}). \tag{8.45}$$

Integrating (8.44) with respect to $\boldsymbol{w}'$ and Ω, we know that the total number of molecules colliding with the molecules with velocity $\boldsymbol{v}'$ in time interval $[t,t+\mathrm{d}t]$ is

$$\int_{\mathbb{R}^3}\mathrm{d}w'\int \mathrm{d}\Omega\sigma'(\Omega)|\boldsymbol{w}'-\boldsymbol{v}'|f(t,\boldsymbol{x},\boldsymbol{w}')\mathrm{d}t. \tag{8.46}$$

Since the number of molecules in the element $\mathrm{d}x\mathrm{d}v'$ around $(\boldsymbol{x},\boldsymbol{v}')$ is

$$f(t,\boldsymbol{x},\boldsymbol{v}')\mathrm{d}x\mathrm{d}v', \tag{8.47}$$

the number of molecules entering the element $\mathrm{d}x\mathrm{d}v$ around $(\boldsymbol{x},\boldsymbol{v})$ due to collisions in time interval $[t,t+\mathrm{d}t]$ is

$$\int_{\mathbb{R}^3}\mathrm{d}w'\int \mathrm{d}\Omega\sigma'(\Omega)|\boldsymbol{w}'-\boldsymbol{v}'|f(t,\boldsymbol{x},\boldsymbol{v}')f(t,\boldsymbol{x},\boldsymbol{w}')\mathrm{d}x\mathrm{d}v'\mathrm{d}t, \tag{8.48}$$

where the integration is with respect to $\boldsymbol{w}'$ and Ω, and $\boldsymbol{v}'$ is the target of the incident flow.

Since the anticollision has the same differential scattering cross section as the original collision (see (8.34)), we have

$$\sigma'(\Omega) = \sigma(\Omega), \tag{8.49}$$

where $\sigma(\Omega) = \sigma(\boldsymbol{v}, \boldsymbol{w} \to \boldsymbol{v}', \boldsymbol{w}')$. In addition, from (8.25) and (8.29) we have

$$|\boldsymbol{w}' - \boldsymbol{v}'| = |\boldsymbol{w} - \boldsymbol{v}| \tag{8.50}$$

and

$$\mathrm{d}v'\mathrm{d}w' = \mathrm{d}v\mathrm{d}w. \tag{8.51}$$

Using (8.49)–(8.51), we easily obtain from (8.48) that

$$J_+ = \int_{\mathbb{R}^3} \mathrm{d}w \int \mathrm{d}\Omega\sigma(\Omega)|\boldsymbol{w} - \boldsymbol{v}| f(t, \boldsymbol{x}, \boldsymbol{v}') f(t, \boldsymbol{x}, \boldsymbol{w}'). \tag{8.52}$$

In the above formula, $\boldsymbol{v}'$ and $\boldsymbol{w}'$ are functions of $\boldsymbol{v}$, $\boldsymbol{w}$, and Ω for any given $\boldsymbol{v}$.

Substituting (8.41) and (8.52) into (8.21), we get the collision term

$$J = \int_{\mathbb{R}^3} \mathrm{d}w \int \mathrm{d}\Omega\sigma(\Omega)|\boldsymbol{w} - \boldsymbol{v}|(f(t, \boldsymbol{x}, \boldsymbol{v}') f(t, \boldsymbol{x}, \boldsymbol{w}') - f(t, \boldsymbol{x}, \boldsymbol{v}) f(t, \boldsymbol{x}, \boldsymbol{w})), \tag{8.53}$$

where $\sigma(\Omega) = \sigma(\boldsymbol{v}, \boldsymbol{w} \to \boldsymbol{v}', \boldsymbol{w}')$ is the differential scattering cross section.

Hence, the Boltzmann equation (8.20) can be written specifically as

$$\begin{aligned} &\left(\frac{\partial}{\partial t} + \boldsymbol{v} \cdot \nabla_x + \boldsymbol{g} \cdot \nabla_v\right) f(t, \boldsymbol{x}, \boldsymbol{v}) \\ &= \int_{\mathbb{R}^3} \mathrm{d}w \int \mathrm{d}\Omega\sigma(\Omega)|\boldsymbol{w} - \boldsymbol{v}|(f(t, \boldsymbol{x}, \boldsymbol{v}') f(t, \boldsymbol{x}, \boldsymbol{w}') - f(t, \boldsymbol{x}, \boldsymbol{v}) f(t, \boldsymbol{x}, \boldsymbol{w})). \end{aligned} \tag{8.54}$$

This is a *nonlinear integrodifferential equation* of distribution function $f(t, \boldsymbol{x}, \boldsymbol{v})$. The problem of studying the kinetic theory of gases is then reduced to solving the above equation.

8.3 Equilibrium State of Dilute Gases

8.3.1 Boltzmann H-Theorem

The solution independent of time t of the Boltzmann equation (8.54) is called the *equilibrium distribution function*. It should also be the limit state of the distribution function as $t \to +\infty$.

Assume that there is no external force ($\boldsymbol{g} \equiv 0$), and the corresponding distribution function is independent of $\boldsymbol{x}$: $f = f(t, \boldsymbol{v})$. At this point the equilibrium distribution function is denoted by $f_0(\boldsymbol{v})$. From the Boltzmann equation (8.54), $f_0(\boldsymbol{v})$ should satisfy

$$\int_{\mathbb{R}^3} \mathrm{d}w \int \mathrm{d}\Omega\sigma(\Omega)|\boldsymbol{w} - \boldsymbol{v}|(f_0(\boldsymbol{v}') f_0(\boldsymbol{w}') - f_0(\boldsymbol{v}) f_0(\boldsymbol{w})) = 0, \tag{8.55}$$

where $\boldsymbol{v}'$ and $\boldsymbol{w}'$ are functions of $\boldsymbol{v}$, $\boldsymbol{w}$, and Ω for any given $\boldsymbol{v}$.

Obviously, if $f_0(\boldsymbol{v})$ satisfies

$$f_0(\boldsymbol{v}')f_0(\boldsymbol{w}') - f_0(\boldsymbol{v})f_0(\boldsymbol{w}) = 0, \tag{8.56}$$

where $\{\boldsymbol{v},\boldsymbol{w} \to \boldsymbol{v}',\boldsymbol{w}'\}$ means all possible collisions (i.e., collisions with a nonzero scattering cross section), then $f_0(\boldsymbol{v})$ is a solution of (8.55) and then is also the equilibrium solution of the Boltzmann equation (8.54).

Now we show that if $f_0(\boldsymbol{v})$ is a solution of (8.55), then it must satisfy (8.56). So, under the above assumptions, $f_0(\boldsymbol{v})$ is an equilibrium solution of the Boltzmann equation (8.54) if and only if (8.56) is satisfied.

To prove this, according to Boltzmann, we define the functional

$$H(t) = \int_{\mathbb{R}^3} \mathrm{d}\boldsymbol{v} f(t,\boldsymbol{v}) \ln f(t,\boldsymbol{v}), \tag{8.57}$$

where $f(t,\boldsymbol{v})$ is the distribution function of the corresponding nonequilibrium state (depending on time t!). From the Boltzmann equation (8.54), $f(t,\boldsymbol{v})$ should satisfy

$$\frac{\partial f(t,\boldsymbol{v})}{\partial t} = \int_{\mathbb{R}^3} \mathrm{d}\boldsymbol{w} \int \mathrm{d}\Omega\sigma(\Omega)|\boldsymbol{w}-\boldsymbol{v}|(f(t,\boldsymbol{v}')f(t,\boldsymbol{w}') - f(t,\boldsymbol{v})f(t,\boldsymbol{w})). \tag{8.58}$$

Differentiating (8.57) with respect to t, we get

$$\frac{\mathrm{d}H(t)}{\mathrm{d}t} = \int_{\mathbb{R}^3} \mathrm{d}\boldsymbol{v} \frac{\partial f(t,\boldsymbol{v})}{\partial t}(1+\ln f(t,\boldsymbol{v})). \tag{8.59}$$

Then we obtain that if f is the equilibrium distribution function, namely, $\frac{\partial f}{\partial t} \equiv 0$, then

$$\frac{\mathrm{d}H(t)}{\mathrm{d}t} = 0. \tag{8.60}$$

Therefore, we need only prove the equivalence between (8.60) and (8.56). For this purpose, we first prove the following theorem.

Boltzmann H-Theorem. *If $f(t,\boldsymbol{v})$ satisfies the Boltzmann equation* (8.54) *(where $\boldsymbol{g} \equiv 0$), then we have*

$$\frac{\mathrm{d}H(t)}{\mathrm{d}t} \leqslant 0. \tag{8.61}$$

Proof. Substituting (8.58) into (8.59), we get

$$\frac{\mathrm{d}H(t)}{\mathrm{d}t} = \int_{\mathbb{R}^3} \mathrm{d}\boldsymbol{v} \int_{\mathbb{R}^3} \mathrm{d}\boldsymbol{w} \int \mathrm{d}\Omega\sigma(\Omega)|\boldsymbol{w}-\boldsymbol{v}|(f_1' f_2' - f_1 f_2)(1+\ln f_1), \tag{8.62}$$

where, for simplicity, we denote $f_1 = f(t,\boldsymbol{v})$, $f_1' = f(t,\boldsymbol{v}')$, and so on.

Noting that the interchange between $\boldsymbol{v}$ and $\boldsymbol{w}$ does not change the value of $\sigma(\Omega)$ (because the two colliding molecules do not change), it is easy to obtain from the above formula that

$$\frac{\mathrm{d}H(t)}{\mathrm{d}t} = \int_{\mathbb{R}^3} \mathrm{d}v \int_{\mathbb{R}^3} \mathrm{d}w \int \mathrm{d}\Omega \sigma(\Omega)|\boldsymbol{w}-\boldsymbol{v}|(f_1' f_2' - f_1 f_2)(1+\ln f_2). \tag{8.63}$$

Adding (8.62) and (8.63) together, we have

$$\frac{\mathrm{d}H(t)}{\mathrm{d}t} = \frac{1}{2}\int_{\mathbb{R}^3} \mathrm{d}v \int_{\mathbb{R}^3} \mathrm{d}w \int \mathrm{d}\Omega \sigma(\Omega)|\boldsymbol{w}-\boldsymbol{v}|(f_1' f_2' - f_1 f_2)(2+\ln(f_1 f_2)). \tag{8.64}$$

Exchanging $\{\boldsymbol{v},\boldsymbol{w}\}$ and $\{\boldsymbol{v}',\boldsymbol{w}'\}$ in the integral on the right-hand side of the above formula, we get

$$\frac{\mathrm{d}H(t)}{\mathrm{d}t} = \frac{1}{2}\int_{\mathbb{R}^3} \mathrm{d}v' \int_{\mathbb{R}^3} \mathrm{d}w' \int \mathrm{d}\Omega \sigma'(\Omega)|\boldsymbol{w}'-\boldsymbol{v}'|(f_1 f_2 - f_1' f_2')(2+\ln(f_1' f_2')). \tag{8.65}$$

Because any collision $\{\boldsymbol{v},\boldsymbol{w}\} \to \{\boldsymbol{v}',\boldsymbol{w}'\}$ has its inverse collision $\{\boldsymbol{v}',\boldsymbol{w}'\} \to \{\boldsymbol{v},\boldsymbol{w}\}$, and they share the same differential scattering cross section,

$$\sigma'(\Omega) = \sigma(\Omega) \tag{8.66}$$

(see (8.34)), and noticing (8.25) and (8.29), (8.65) can be rewritten as

$$\frac{\mathrm{d}H(t)}{\mathrm{d}t} = \frac{1}{2}\int_{\mathbb{R}^3} \mathrm{d}v \int_{\mathbb{R}^3} \mathrm{d}w \int \mathrm{d}\Omega \sigma(\Omega)|\boldsymbol{w}-\boldsymbol{v}|(f_1 f_2 - f_1' f_2')(2+\ln(f_1' f_2')). \tag{8.67}$$

Adding the above formula and (8.64) together, we obtain

$$\frac{\mathrm{d}H(t)}{\mathrm{d}t} = \frac{1}{4}\int_{\mathbb{R}^3} \mathrm{d}v \int_{\mathbb{R}^3} \mathrm{d}w \int \mathrm{d}\Omega \sigma(\Omega)|\boldsymbol{w}-\boldsymbol{v}|(f_1' f_2' - f_1 f_2)(\ln(f_1 f_2) - \ln(f_1' f_2')). \tag{8.68}$$

Since $\ln x$ is a strictly increasing function in its domain, we have

$$(b-a)(\ln b - \ln a) \geqslant 0, \quad \forall a,b > 0; \tag{8.69}$$

then the integrand on the right-hand side of (8.68) cannot be positive, and consequently (8.61) holds. The proof is finished. □

As a by-product of the above proof, it is not difficult to see that (8.60) holds if and only if the integrand on the right-hand side of (8.68) is identically equal to zero, i.e.,

$$f_1' f_2' - f_1 f_2 = 0. \tag{8.70}$$

It follows immediately that (8.60) is equivalent to (8.56).

We can guess from this that $f(t,\boldsymbol{v}) \to f_0(\boldsymbol{v})$ as $t \to +\infty$. However, for a specific well-posed problem, this remains to be proved rigorously.

By the way, the Boltzmann-H theorem implies the irreversibility of the Boltzmann equation. In fact, we can prove that for ideal gases, function H differs from the entropy by only a negative constant factor. For further discussion on the irreversibility of the Boltzmann equation, see [3] and [6].

8.3.2 Maxwell–Boltzmann Distribution

As stated above, the equilibrium distribution function $f_0(\boldsymbol{v})$ is the solution of equation (8.56). This distribution is called the *Maxwell–Boltzmann distribution*.

Now we determine the specific expression of this distribution.

Taking the logarithm of (8.56), we have

$$\ln f_0(\boldsymbol{v}) + \ln f_0(\boldsymbol{w}) = \ln f_0(\boldsymbol{v}') + \ln f_0(\boldsymbol{w}'). \tag{8.71}$$

Since $\{\boldsymbol{v},\boldsymbol{w}\}$ and $\{\boldsymbol{v}',\boldsymbol{w}'\}$ are the initial and final states of any possible collision, respectively, the above formula is in the form of conservation laws. Thus, if $\chi(\boldsymbol{v})$ is a quantity related to a molecule with velocity $\boldsymbol{v}$, such that $\chi(\boldsymbol{v}) + \chi(\boldsymbol{w})$ is conserved in the collision of two molecules,

$$\chi(\boldsymbol{v}) + \chi(\boldsymbol{w}) = \chi(\boldsymbol{v}') + \chi(\boldsymbol{w}'), \tag{8.72}$$

then (8.71) has a solution

$$\ln f_0(\boldsymbol{v}) = \chi(\boldsymbol{v}). \tag{8.73}$$

By the superposition principle, the general solution of (8.71) is

$$\ln f_0(\boldsymbol{v}) = \chi_1(\boldsymbol{v}) + \chi_2(\boldsymbol{v}) + \cdots, \tag{8.74}$$

where $\chi_1, \chi_2, \ldots$ exhaust all possible independent conserved quantities.

For a molecule assumed to be without self-spin (no internal structure), its mass (being a constant), momentum (or $\boldsymbol{v}$), and energy (or $|\boldsymbol{v}|^2$) are all conserved quantities. Therefore, $\ln f_0(\boldsymbol{v})$ should be a linear combination of $|\boldsymbol{v}|^2$, the three components of $\boldsymbol{v}$, and a constant, which can be written as

$$\ln f_0(\boldsymbol{v}) = -A|\boldsymbol{v} - \boldsymbol{v}_0|^2 + \ln C \tag{8.75}$$

or

$$f_0(\boldsymbol{v}) = C\,\mathrm{e}^{-A|\boldsymbol{v}-\boldsymbol{v}_0|^2}, \tag{8.76}$$

where C, A, and the three components of $\boldsymbol{v}_0$ are five constants that can be determined by observing the system.

By substituting (8.76) into (8.3), it is easy to see that the number of molecules per unit volume is

$$\begin{aligned} n &= C \int_{\mathbb{R}^3} \mathrm{e}^{-A|\boldsymbol{v}-\boldsymbol{v}_0|^2}\,\mathrm{d}\boldsymbol{v} \\ &= C \int_{\mathbb{R}^3} \mathrm{e}^{-A|\boldsymbol{v}|^2}\,\mathrm{d}\boldsymbol{v} \\ &= C\left(\frac{\pi}{A}\right)^{3/2}. \end{aligned} \tag{8.77}$$

From this we have $A > 0$, and

$$C = \left(\frac{A}{\pi}\right)^{3/2} n. \tag{8.78}$$

Substituting (8.76) into (8.5), we know that the average velocity of the gas molecules is

$$\begin{aligned} V &= \frac{C}{n} \int_{\mathbb{R}^3} \boldsymbol{v}\, \mathrm{e}^{-A|\boldsymbol{v}-\boldsymbol{v}_0|^2} \mathrm{d}\boldsymbol{v} \\ &= \frac{C}{n} \int_{\mathbb{R}^3} (\boldsymbol{v}+\boldsymbol{v}_0)\mathrm{e}^{-A|\boldsymbol{v}|^2} \mathrm{d}\boldsymbol{v} \\ &= \boldsymbol{v}_0. \end{aligned} \tag{8.79}$$

Here we used (8.77) and

$$\int_{\mathbb{R}^3} \boldsymbol{v}\, \mathrm{e}^{-A|\boldsymbol{v}|^2} \mathrm{d}\boldsymbol{v} = 0.$$

Thus, if it is assumed that there is no translation for the gas as a whole, we must take

$$\boldsymbol{v}_0 = 0. \tag{8.80}$$

Now we calculate the average heat energy of the molecules. For this, we must take $\boldsymbol{v}_0 = 0$ to exclude the macroscopic kinetic energy. Noting the physical meaning of the right-hand side of (8.6), it is easy to get the average molecular heat energy

$$\begin{aligned} \epsilon &= \frac{\displaystyle\int_{\mathbb{R}^3} \frac{M}{2}|\boldsymbol{v}|^2 f(t,\boldsymbol{v})\mathrm{d}\boldsymbol{v}}{n} \\ &= \frac{M}{2n}\int_{\mathbb{R}^3} |\boldsymbol{v}|^2 f_0(\boldsymbol{v})\mathrm{d}\boldsymbol{v} \\ &= \frac{MC}{2n}\int_{\mathbb{R}^3} |\boldsymbol{v}|^2 \mathrm{e}^{-A|\boldsymbol{v}|^2} \mathrm{d}\boldsymbol{v} \\ &= \frac{3}{4}\frac{M}{A}. \end{aligned} \tag{8.81}$$

Thus,

$$A = \frac{3}{4}\frac{M}{\epsilon}. \tag{8.82}$$

Together with (8.78), we have

$$C = \left(\frac{3M}{4\pi\epsilon}\right)^{3/2} n. \tag{8.83}$$

In order to express the average heat energy by a quantity that can be directly measured, we seek the equation of state corresponding to the equilibrium distribution function, that is, giving the corresponding expression of pressure. By definition, the pressure means the average force of the gas acting on a unit area of the fully reflecting plane. Suppose the disc in Figure 8.6 stands for this unit area, with the x_1 axis as the normal direction. Under the assumption that the average molecular velocity is zero, a molecule can reach the plane only when the component of its velocity $\boldsymbol{v}$ along the x_1 direction satisfies

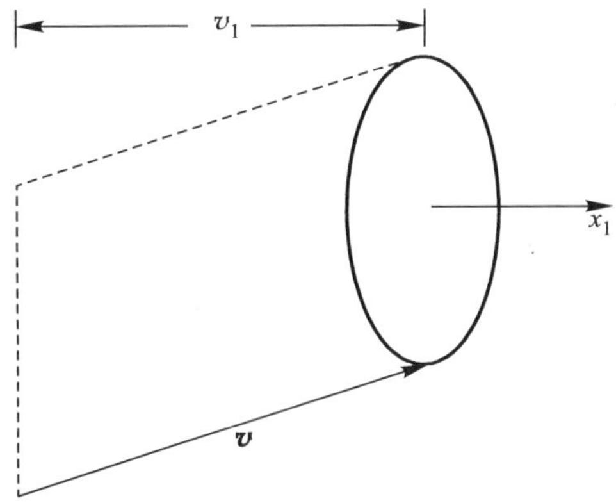

Figure 8.6

$v_1 > 0$. However, the reflection of a molecule on this plane only changes the sign of this velocity component, and then the molecule loses momentum $2Mv_1$. It is not hard to see that the number of molecules with velocity in an infinitesimal element $\mathrm{d}v$ around $\boldsymbol{v}$ ($v_1 > 0$) reflected in unit time by this disc is equal to $f(t,\boldsymbol{v})v_1\mathrm{d}v$, which is the number of molecules included in a cylinder with this disc being its bottom, its generating line parallel to $\boldsymbol{v}$, and v_1 its height.

Thus, the loss of momentum is

$$2Mv_1^2 f(t,\boldsymbol{v})\mathrm{d}v.$$

Therefore, for a gas with zero average molecular velocity, the pressure, namely, the total loss of momentum caused by the above in unit time, is

$$\begin{aligned} p &= \int_{v_1>0} 2Mv_1^2 f(t,\boldsymbol{v})\mathrm{d}v \\ &= \int_{v_1>0} 2Mv_1^2 f_0(\boldsymbol{v})\mathrm{d}v \\ &= MC\int_{\mathbb{R}^3} v_1^2 \mathrm{e}^{-A|\boldsymbol{v}|^2}\mathrm{d}v \\ &= \frac{MC}{3}\int_{\mathbb{R}^3} |\boldsymbol{v}|^2 \mathrm{e}^{-A|\boldsymbol{v}|^2}\mathrm{d}v. \end{aligned}$$

Together with (8.81), from the above formula we have

$$p = \frac{2}{3}n\epsilon. \tag{8.84}$$

This is exactly the equation of state.

Furthermore, from (8.9), when $f = f_0(\boldsymbol{v}) = C\mathrm{e}^{-A|\boldsymbol{v}|^2}$, we have

$$p = nkT,$$

where k is the Boltzmann constant. Thus, (8.84) can be rewritten as

$$\epsilon = \frac{3}{2}kT. \tag{8.85}$$

Hence, according to (8.82)–(8.83), we obtain

$$A = \frac{M}{2kT}, \quad C = n\left(\frac{M}{2\pi kT}\right)^{3/2}. \tag{8.86}$$

Thus, in terms of the temperature T, the average velocity $\boldsymbol{v}_0$, the molecular density n, and the molecular mass M, the equilibrium distribution function for a dilute gas in the absence of external forces can be written as

$$f_0(\boldsymbol{v}) = n\left(\frac{M}{2\pi kT}\right)^{3/2} \mathrm{e}^{-M|\boldsymbol{v}-\boldsymbol{v}_0|^2/2kT}. \tag{8.87}$$

This is just the Maxwell–Boltzmann distribution.

Now we discuss the situation in the presence of an external force field. In the case that the external force received by the particles is the conservative force given by

$$\boldsymbol{F} = -\nabla_x \phi(\boldsymbol{x}), \tag{8.88}$$

it is not difficult to prove that

$$f(\boldsymbol{x},\boldsymbol{v}) = f_0(\boldsymbol{v})\mathrm{e}^{-\phi(\boldsymbol{x})/kT} \tag{8.89}$$

is a solution independent of time t to the Boltzmann equation (8.54), that is, the steady-state solution, where $f_0(\boldsymbol{v})$ is the Maxwell–Boltzmann distribution given by (8.87) (there $\boldsymbol{v}_0 = 0$). In fact, it is obvious that $\frac{\partial f}{\partial t} = 0$, and from (8.53), by using (8.56), the collision term is

$$J = \mathrm{e}^{-2\phi(\boldsymbol{x})/kT} \int_{\mathbb{R}^3} \mathrm{d}w \int \mathrm{d}\Omega \sigma(\Omega)|\boldsymbol{w}-\boldsymbol{v}| \\ \cdot (f_0(\boldsymbol{v}')f_0(\boldsymbol{w}') - f_0(\boldsymbol{v})f_0(\boldsymbol{w})) = 0.$$

In addition, we can directly verify that

$$\left(\boldsymbol{v}\cdot\nabla_x + \frac{\boldsymbol{F}}{M}\cdot\nabla_v\right) f(\boldsymbol{x},\boldsymbol{v}) = 0. \tag{8.90}$$

Absorbing the factor $\mathrm{e}^{-\phi(\boldsymbol{x})/kT}$ into the molecular density n, we can rewrite the equilibrium distribution function (8.89) as

$$f(\boldsymbol{x},\boldsymbol{v}) = n(\boldsymbol{x})\left(\frac{M}{2\pi kT}\right)^{\frac{3}{2}} \mathrm{e}^{-M|\boldsymbol{v}|^2/2kT}, \tag{8.91}$$

where

$$n(\boldsymbol{x}) = \int_{\mathbb{R}^3} f_0(\boldsymbol{v})\mathrm{d}v\,\mathrm{e}^{-\phi(\boldsymbol{x})/kT} = \int_{\mathbb{R}^3} f(\boldsymbol{x},\boldsymbol{v})\mathrm{d}v. \tag{8.92}$$

When the external force received by the particles is a constant vector $\boldsymbol{F} = M\boldsymbol{g}$ (this is the case for the gravitational field), it is easy to verify directly that

$$f(t,\boldsymbol{v}) = n\left(\frac{M}{2\pi kT}\right)^{3/2} \mathrm{e}^{-M|\boldsymbol{v}-\boldsymbol{g}t|^2/2kT} \tag{8.93}$$

is a solution of the Boltzmann equation (8.54), i.e., it satisfies

$$\frac{\partial f}{\partial t} + \boldsymbol{g}\cdot\nabla_v f = 0, \tag{8.94}$$

while the collision term J is still zero. The distribution function given by (8.93) is not the steady-state solution of the Boltzmann equation, but it still describes the equilibrium state of gas, but only for a gas in macroscopic uniformly accelerated motion as a whole.

8.4 Conservation Laws

To study the nonequilibrium phenomenon of gases, we have to figure out the corresponding distribution function by solving the Boltzmann equation with given initial condition. Even if we do not solve the Boltzmann equation directly, we can still derive some accurate properties of solutions of the Boltzmann equation by starting from kinetic quantities that are strictly conserved in any molecular collision.

Suppose that $\chi(\boldsymbol{x},\boldsymbol{v})$ is a quantity related to the molecule at position $\boldsymbol{x}$ with velocity $\boldsymbol{v}$, such that for any collision $\{\boldsymbol{v},\boldsymbol{w}\to\boldsymbol{v}',\boldsymbol{w}'\}$ occurring at point $\boldsymbol{x}$, we always have

$$\chi_1 + \chi_2 = \chi_1' + \chi_2', \tag{8.95}$$

where $\chi_1 = \chi(\boldsymbol{x},\boldsymbol{v})$, $\chi_2 = \chi(\boldsymbol{x},\boldsymbol{w})$, and so on; then χ is called a *conserved quantity*.

Theorem 8.1. *It always holds that*

$$\int_{\mathbb{R}^3} \mathrm{d}v\,\chi(\boldsymbol{x},\boldsymbol{v})J(\boldsymbol{x},\boldsymbol{v}) = 0, \tag{8.96}$$

where $J(\boldsymbol{x},\boldsymbol{v})$ is the term on the right-hand side of the Boltzmann equation (8.54), *i.e.,*

$$\begin{aligned} J(\boldsymbol{x},\boldsymbol{v}) = \int_{\mathbb{R}^3} \mathrm{d}w \int \mathrm{d}\Omega\,\sigma(\Omega)|\boldsymbol{w}-\boldsymbol{v}| \\ \cdot\left(f(t,\boldsymbol{x},\boldsymbol{v}')f(t,\boldsymbol{x},\boldsymbol{w}') - f(t,\boldsymbol{x},\boldsymbol{v})f(t,\boldsymbol{x},\boldsymbol{w})\right), \end{aligned} \tag{8.97}$$

where $f(t,\boldsymbol{x},\boldsymbol{v})$ is not required to be a solution of the Boltzmann equation. Here, for simplicity, we do not mark out the explicit dependence of J on t.

Proof. From (8.97) we have

$$\begin{aligned} &\int_{\mathbb{R}^3} \mathrm{d}v\,\chi(\boldsymbol{x},\boldsymbol{v})J(\boldsymbol{x},\boldsymbol{v}) \\ &= \int_{\mathbb{R}^3} \mathrm{d}v \int_{\mathbb{R}^3} \mathrm{d}w \int \mathrm{d}\Omega\,\sigma(\Omega)|\boldsymbol{w}-\boldsymbol{v}|\chi(\boldsymbol{x},\boldsymbol{v}) \\ &\quad\cdot\left(f(t,\boldsymbol{x},\boldsymbol{v}')f(t,\boldsymbol{x},\boldsymbol{w}') - f(t,\boldsymbol{x},\boldsymbol{v})f(t,\boldsymbol{x},\boldsymbol{w})\right). \end{aligned} \tag{8.98}$$

Interchanging the integration variables $\boldsymbol{v}$ and $\boldsymbol{w}$ on the right-hand side of (8.98) while noticing that $\sigma(\Omega)$ remains unchanged, and adding the result to (8.98), we get

$$\begin{aligned}&\int_{\mathbb{R}^3} \mathrm{d}\boldsymbol{v}\chi(\boldsymbol{x},\boldsymbol{v})J(\boldsymbol{x},\boldsymbol{v})\\&=\frac{1}{2}\int_{\mathbb{R}^3}\mathrm{d}\boldsymbol{v}\int_{\mathbb{R}^3}\mathrm{d}\boldsymbol{w}\int \mathrm{d}\Omega\sigma(\Omega)|\boldsymbol{w}-\boldsymbol{v}|(f_1'f_2'-f_1f_2)(\chi_1+\chi_2),\end{aligned} \tag{8.99}$$

where $f_1=f_1(t,\boldsymbol{x},\boldsymbol{v})$, $f_2=f_2(t,\boldsymbol{x},\boldsymbol{w})$, and so forth.

Interchanging $\{\boldsymbol{v},\boldsymbol{w}\}$ and $\{\boldsymbol{v}',\boldsymbol{w}'\}$ on the right-hand side of (8.99), and noting formulas (8.25), (8.29), and (8.34), we can rewrite (8.99) as

$$\begin{aligned}&\int_{\mathbb{R}^3} \mathrm{d}\boldsymbol{v}\chi(\boldsymbol{x},\boldsymbol{v})J(\boldsymbol{x},\boldsymbol{v})\\&=\frac{1}{2}\int_{\mathbb{R}^3}\mathrm{d}\boldsymbol{v}\int_{\mathbb{R}^3}\mathrm{d}\boldsymbol{w}\int \mathrm{d}\Omega\sigma(\Omega)|\boldsymbol{w}-\boldsymbol{v}|(f_1f_2-f_1'f_2')(\chi_1'+\chi_2').\end{aligned} \tag{8.100}$$

Since χ is a conserved quantity, i.e., (8.95) is satisfied, (8.96) follows immediately from (8.99) and (8.100). The theorem is proved. □

Using the above theorem, we can derive the corresponding conservation laws from the Boltzmann equation by multiplying both sides of (8.54) by χ and then integrating with respect to $\boldsymbol{v}$. At this moment, by Theorem 8.1, the integral corresponding to the collision term J is zero; then we have

$$\int_{\mathbb{R}^3}\mathrm{d}\boldsymbol{v}\chi(\boldsymbol{x},\boldsymbol{v})\left(\frac{\partial}{\partial t}+\sum_{i=1}^{3}v_i\frac{\partial}{\partial x_i}+\sum_{i=1}^{3}g_i\frac{\partial}{\partial v_i}\right)f(t,\boldsymbol{x},\boldsymbol{v})=0. \tag{8.101}$$

This can be easily written as

$$\begin{aligned}&\frac{\partial}{\partial t}\int_{\mathbb{R}^3}\mathrm{d}\boldsymbol{v}\chi f+\sum_{i=1}^{3}\frac{\partial}{\partial x_i}\int_{\mathbb{R}^3}\mathrm{d}\boldsymbol{v}\chi v_i f-\sum_{i=1}^{3}\int_{\mathbb{R}^3}\mathrm{d}\boldsymbol{v}\frac{\partial\chi}{\partial x_i}v_i f\\&+\sum_{i=1}^{3}\int_{\mathbb{R}^3}\mathrm{d}\boldsymbol{v}\frac{\partial}{\partial v_i}(\chi g_i f)-\sum_{i=1}^{3}\int_{\mathbb{R}^3}\mathrm{d}\boldsymbol{v}\frac{\partial\chi}{\partial v_i}g_i f\\&-\sum_{i=1}^{3}\int_{\mathbb{R}^3}\mathrm{d}\boldsymbol{v}\chi\frac{\partial g_i}{\partial v_i}f=0.\end{aligned} \tag{8.102}$$

It is always assumed that $f(t,\boldsymbol{x},\boldsymbol{v})\to 0$ as $|\boldsymbol{v}|\to\infty$; then from Green's formula we have

$$\int_{\mathbb{R}^3}\mathrm{d}\boldsymbol{v}\frac{\partial}{\partial v_i}(\chi g_i f)=0. \tag{8.103}$$

For a quantity A related to $\boldsymbol{v}$, we define its average as

$$\langle A\rangle \stackrel{\mathrm{d}}{=} \frac{\int_{\mathbb{R}^3}\mathrm{d}\boldsymbol{v}Af}{\int_{\mathbb{R}^3}\mathrm{d}\boldsymbol{v}f}=\frac{1}{n}\int_{\mathbb{R}^3}\mathrm{d}\boldsymbol{v}Af, \tag{8.104}$$

where $f = f(t, \boldsymbol{x}, \boldsymbol{v})$ is any solution of the Boltzmann equation (8.54) and $n = n(t, \boldsymbol{x})$ is the number of molecules per unit volume at time t and position $\boldsymbol{x}$. From this we have

$$\int_{\mathbb{R}^3} \mathrm{d}\boldsymbol{v} A f = n\langle A\rangle = \langle nA\rangle. \tag{8.104$'$}$$

Noting (8.103), from (8.102) and (8.104)′ we have the following theorem.

Theorem 8.2 (conservation theorem). *For any conserved quantity χ, it always holds that*

$$\begin{aligned} &\frac{\partial}{\partial t}\langle n\chi\rangle + \sum_{i=1}^{3} \frac{\partial}{\partial x_i}\langle n v_i \chi\rangle - n\sum_{i=1}^{3}\left\langle v_i \frac{\partial \chi}{\partial x_i}\right\rangle \\ &\quad - n\sum_{i=1}^{3}\left\langle g_i \frac{\partial \chi}{\partial v_i}\right\rangle - n\sum_{i=1}^{3}\left\langle \frac{\partial g_i}{\partial v_i}\chi\right\rangle = 0, \end{aligned} \tag{8.105}$$

where f appearing in average values is any solution of the Boltzmann equation (8.54).

As usual, below we only consider the case that $\boldsymbol{g}$ is independent of $\boldsymbol{v}$; then the last term on the left-hand side of the above formula is zero, and we have

$$\frac{\partial}{\partial t}\langle n\chi\rangle + \sum_{i=1}^{3} \frac{\partial}{\partial x_i}\langle n v_i \chi\rangle - n\sum_{i=1}^{3}\left\langle v_i \frac{\partial \chi}{\partial x_i}\right\rangle - n\sum_{i=1}^{3}\left\langle g_i \frac{\partial \chi}{\partial v_i}\right\rangle = 0. \tag{8.106}$$

In the case of identical molecules, as independent conservation quantities, we have mass, momentum, and energy. Taking

$$\chi = M, \quad \chi = M v_i \quad (i = 1,2,3), \quad \text{and} \quad \chi = \frac{1}{2}M|\boldsymbol{v}|^2,$$

respectively, we may obtain corresponding conservation equations.

Taking $\chi = M$ in (8.106), we have

$$\frac{\partial}{\partial t}\langle nM\rangle + \sum_{i=1}^{3} \frac{\partial}{\partial x_i}\langle nM v_i\rangle = 0. \tag{8.107}$$

Noticing that

$$\langle nM\rangle = nM = \rho(t, \boldsymbol{x}) \tag{8.108}$$

is the mass density (see (8.4)), and

$$\langle nM v_i\rangle = nM \frac{\int_{\mathbb{R}^3} \mathrm{d}\boldsymbol{v} v_i f(t, \boldsymbol{x}, \boldsymbol{v})}{n} = \rho V_i, \tag{8.109}$$

where $\boldsymbol{V}$ is the average velocity (see (8.5)), (8.107) is reduced to

$$\frac{\partial \rho}{\partial t}+\operatorname{div}(\rho \boldsymbol{V})=0. \tag{8.110}$$

This is just the well-known conservation law of mass—the *continuity equation* (see (2.10) in Chapter 2 of Volume I).

Taking $\chi = Mv_i$ in (8.106), we have

$$\frac{\partial}{\partial t}\langle nMv_i\rangle+\sum_{j=1}^{3}\frac{\partial}{\partial x_j}\langle nMv_iv_j\rangle-nMg_i=0. \tag{8.111}$$

Using (8.108)–(8.109), and noticing

$$\begin{aligned}\langle v_iv_j\rangle &= \langle (v_i-V_i)(v_j-V_j)\rangle+\langle v_i\rangle V_j+V_i\langle v_j\rangle-V_iV_j\\ &= \langle (v_i-V_i)(v_j-V_j)\rangle+V_iV_j\end{aligned}$$

and the pressure tensor

$$\begin{aligned}p_{ij} &= \int_{\mathbb{R}^3} M(v_i-V_i)(v_j-V_j)f(t,\boldsymbol{x},\boldsymbol{v})\mathrm{d}v\\ &= \rho\langle (v_i-V_i)(v_j-V_j)\rangle,\end{aligned} \tag{8.112}$$

by (8.7), equation (8.111) can be rewritten as

$$\frac{\partial}{\partial t}(\rho V_i)+\sum_{j=1}^{3}\frac{\partial}{\partial x_j}(\rho V_iV_j)+\sum_{j=1}^{3}\frac{\partial p_{ij}}{\partial x_j}=\rho g_i \quad (i=1,2,3),$$

i.e.,

$$\frac{\partial}{\partial t}(\rho \boldsymbol{V})+\operatorname{div}(\rho \boldsymbol{V}\otimes \boldsymbol{V}+\boldsymbol{P})=\rho \boldsymbol{g}, \tag{8.113}$$

where $\boldsymbol{P}=(p_{ij})$. These are just the *conservation equations of momentum* (see (2.133) in Chapter 2 of Volume I, but here the pressure tensor $\boldsymbol{P}$ differs from that in Chapter 2 by only a minus sign).

Using the continuity equation (8.110) to simplify (8.113), we obtain

$$\frac{\partial \boldsymbol{V}}{\partial t}+\sum_{k=1}^{3}V_k\frac{\partial \boldsymbol{V}}{\partial x_k}+\frac{1}{\rho}\operatorname{div}\boldsymbol{P}=\boldsymbol{g} \tag{8.114}$$

or

$$\frac{\mathrm{d}\boldsymbol{V}}{\mathrm{d}t}+\frac{1}{\rho}\operatorname{div}\boldsymbol{P}=\boldsymbol{g}, \tag{8.115}$$

where $\frac{\mathrm{d}}{\mathrm{d}t} = \frac{\partial}{\partial t} + \sum_{k=1}^{3} V_k \frac{\partial}{\partial x_k}$. These are exactly the *Euler equations* (see (2.136) in Chapter 2 of Volume I).

Finally, taking $\chi = \frac{1}{2} M|\boldsymbol{v}|^2$ in (8.106), we have

$$\frac{1}{2}\frac{\partial}{\partial t}\langle nM|\boldsymbol{v}|^2\rangle + \frac{1}{2}\sum_{i=1}^{3}\frac{\partial}{\partial x_i}\langle nMv_i|\boldsymbol{v}|^2\rangle - n\sum_{i=1}^{3}\langle Mg_iv_i\rangle = 0. \tag{8.116}$$

From (8.10)–(8.11) we know that

$$\frac{1}{2}\langle nM|\boldsymbol{v}|^2\rangle = \frac{3}{2}nkT + \frac{1}{2}\rho|\boldsymbol{V}|^2 = \rho e + \frac{1}{2}\rho|\boldsymbol{V}|^2, \tag{8.117}$$

where e is the internal energy per unit mass. In addition,

$$n\sum_{i=1}^{3}\langle Mg_iv_i\rangle = \sum_{i=1}^{3}\rho g_i\langle v_i\rangle = \rho\boldsymbol{g}\cdot\boldsymbol{V}. \tag{8.118}$$

Now we check the second term on the left-hand side of (8.116). Obviously,

$$\frac{1}{2}\langle nMv_i|\boldsymbol{v}|^2\rangle = \frac{1}{2}\langle nM(v_i - V_i)|\boldsymbol{v}|^2\rangle + \frac{1}{2}\langle nMV_i|\boldsymbol{v}|^2\rangle. \tag{8.119}$$

From (8.117) we have

$$\frac{1}{2}\langle nMV_i|\boldsymbol{v}|^2\rangle = \left(\rho e + \frac{1}{2}\rho|\boldsymbol{V}|^2\right)V_i. \tag{8.120}$$

Moreover,

$$\begin{aligned}
&\frac{1}{2}\langle nM(v_i - V_i)|\boldsymbol{v}|^2\rangle\\
&= \frac{1}{2}nM\langle(v_i - V_i)(|\boldsymbol{v}-\boldsymbol{V}|^2 + 2(\boldsymbol{v}-\boldsymbol{V})\cdot\boldsymbol{V} + |\boldsymbol{V}|^2)\rangle\\
&= \frac{1}{2}nM\langle(v_i - V_i)|\boldsymbol{v}-\boldsymbol{V}|^2\rangle\\
&\quad + nM\langle(v_i - V_i)(\boldsymbol{v}-\boldsymbol{V})\rangle\cdot\boldsymbol{V} + \frac{1}{2}nM\langle v_i - V_i\rangle|\boldsymbol{V}|^2\\
&= \frac{1}{2}nM\langle(v_i - V_i)|\boldsymbol{v}-\boldsymbol{V}|^2\rangle + nM\sum_{j=1}^{3}\langle(v_i - V_i)(v_j - V_j)\rangle V_j\\
&= q_i + \sum_{j=1}^{3}p_{ij}V_j.
\end{aligned} \tag{8.121}$$

Here we used (8.108) and (8.112), and

$$\begin{aligned}
q_i &= \frac{1}{2}nM\langle(v_i - V_i)|\boldsymbol{v}-\boldsymbol{V}|^2\rangle\\
&= \int_{\mathbb{R}^3}\frac{M}{2}(v_i - V_i)|\boldsymbol{v}-\boldsymbol{V}|^2 f(t,\boldsymbol{x},\boldsymbol{v})\mathrm{d}\boldsymbol{v}.
\end{aligned} \tag{8.122}$$

Using the above formulas, from (8.116) we obtain the *conservation equation of energy* (see (2.139) in Chapter 2 of Volume I)

$$\begin{aligned}&\frac{\partial}{\partial t}\left(\rho e+\frac{1}{2}\rho|\boldsymbol{V}|^2\right)+\operatorname{div}\left(\left(\rho e+\frac{1}{2}\rho|\boldsymbol{V}|^2\right)\boldsymbol{V}+\boldsymbol{P}\boldsymbol{V}\right)\\&=-\operatorname{div}\boldsymbol{Q}+\rho\boldsymbol{g}\cdot\boldsymbol{V},\end{aligned}\tag{8.123}$$

where $\boldsymbol{Q}=(q_1,q_2,q_3)$ is equivalent to the heat flux density vector $-\kappa\,\mathrm{grad}\,T$.

Using the continuity equation, the above equation can be simplified to

$$\begin{aligned}&\rho\frac{\mathrm{d}}{\mathrm{d}t}\left(e+\frac{1}{2}|\boldsymbol{V}|^2\right)+\operatorname{div}(\boldsymbol{P}\boldsymbol{V})\\&=-\operatorname{div}\boldsymbol{Q}+\rho\boldsymbol{g}\cdot\boldsymbol{V},\end{aligned}\tag{8.124}$$

where $\frac{\mathrm{d}}{\mathrm{d}t}=\frac{\partial}{\partial t}+\sum_{k=1}^{3}V_k\frac{\partial}{\partial x_k}$. Furthermore, by Euler equations (8.115), (8.124) can be further reduced to

$$\rho\frac{\mathrm{d}e}{\mathrm{d}t}+\sum_{i,j=1}^{3}p_{ij}\frac{\partial V_j}{\partial x_i}=-\operatorname{div}\boldsymbol{Q}.\tag{8.125}$$

Denote

$$S_{ij}=\frac{1}{2}\left(\frac{\partial V_i}{\partial x_j}+\frac{\partial V_j}{\partial x_i}\right).\tag{8.126}$$

Noting the symmetry of $\{p_{ij}\}$, (8.125) can be written as

$$\rho\frac{\mathrm{d}e}{\mathrm{d}t}+\sum_{i,j=1}^{3}p_{ij}S_{ij}=-\operatorname{div}\boldsymbol{Q}.\tag{8.127}$$

Thus, we see that the system of fluid dynamics can be obtained formally from the conservation theorem (Theorem 8.2). However, to make this really work, we have to figure out the solution of the Boltzmann equation (8.54) and determine $\boldsymbol{P}$ and $\boldsymbol{Q}$ accordingly, so that these equations form a closed and physically meaningful system of fluid dynamics.

It is very difficult to solve the Boltzmann equation (8.54) rigorously. Next we will introduce some related approximation methods of finding solutions and give the approximations of different orders for the system of fluid dynamics.

8.5 Zero-Order Approximation

Assume that the gas under consideration is under a state not far from the equilibrium state. In particular, assume that the distribution function can be locally expressed by the Maxwell–Boltzmann distribution in a neighborhood of any point in the gas, and that the density, temperature, and average velocity are changing slowly in time and space. For this kind of gas, it is natural to make the following approximation:

$$f(t,\boldsymbol{x},\boldsymbol{v}) \approx f^{(0)}(t,\boldsymbol{x},\boldsymbol{v}),$$

where $f^{(0)}(t,\boldsymbol{x},\boldsymbol{v})$ is given by the Maxwell–Boltzmann distribution (8.87), i.e.,

$$f^{(0)}(t,\boldsymbol{x},\boldsymbol{v}) = n\left(\frac{M}{2\pi kT}\right)^{\frac{3}{2}} \exp\left(-\frac{M|\boldsymbol{v}-\boldsymbol{V}|^2}{2kT}\right), \tag{8.128}$$

where n, T, and $\boldsymbol{V}$ are slowly varying functions of t and $\boldsymbol{x}$.

Obviously, $f^{(0)}(t,\boldsymbol{x},\boldsymbol{v})$ cannot be an exact solution of the Boltzmann equation (8.54). As a matter of fact, since n, T, and $\boldsymbol{V}$ are all independent of $\boldsymbol{v}$, together with (8.22)–(8.23), we obviously have

$$\begin{aligned} J = \int_{\mathbb{R}^3} \mathrm{d}w \int \mathrm{d}\Omega\sigma(\Omega)|\boldsymbol{w}-\boldsymbol{v}|(&f^{(0)}(t,\boldsymbol{x},\boldsymbol{v}')f^{(0)}(t,\boldsymbol{x},\boldsymbol{w}') \\ &- f^{(0)}(t,\boldsymbol{x},\boldsymbol{v})f^{(0)}(t,\boldsymbol{x},\boldsymbol{w})) = 0, \end{aligned}$$

but, generally speaking,

$$\left(\frac{\partial}{\partial t} + \boldsymbol{v}\cdot\nabla_x + \boldsymbol{g}\cdot\nabla_v\right) f^{(0)}(t,\boldsymbol{x},\boldsymbol{v}) \neq 0. \tag{8.129}$$

Even so, we still assume that $f^{(0)}(t,\boldsymbol{x},\boldsymbol{v})$ is a good approximation of the distribution function and discuss its corresponding conclusions in physics.

Thanks to the assumption that $f^{(0)}(t,\boldsymbol{x},\boldsymbol{v})$ is a good approximation, being regarded as the corresponding mass, temperature, and average velocity, $nM=\rho$, T, and $\boldsymbol{V}$ should satisfy approximately corresponding conservation laws. By these conservation laws, we can put restrictions on ρ, T, and $\boldsymbol{V}$ to some extent.

To specifically obtain these conservation laws, we have to compute the corresponding pressure tensor (p_{ij}) and heat flux density vector (q_i) and denote them by $(p_{ij}^{(0)})$ and $(q_i^{(0)})$, respectively. Denoting

$$C(t,\boldsymbol{x}) = n\left(\frac{M}{2\pi kT}\right)^{3/2}, \quad A(t,\boldsymbol{x}) = \frac{M}{2kT}, \tag{8.130}$$

(8.128) can be written as

$$f^{(0)}(t,\boldsymbol{x},\boldsymbol{v}) = C(t,\boldsymbol{x})\exp(-A(t,\boldsymbol{x})|\boldsymbol{v}-\boldsymbol{V}|^2). \tag{8.131}$$

Corresponding to this distribution function, from (8.122) and (8.7) we know that its heat flux density vector and pressure tensor are

$$q_i^{(0)} = \frac{M}{2}\int_{\mathbb{R}^3}(v_i - V_i)|\boldsymbol{v}-\boldsymbol{V}|^2 C(t,\boldsymbol{x})\exp(-A(t,\boldsymbol{x})|\boldsymbol{v}-\boldsymbol{V}|^2)\mathrm{d}v = 0 \tag{8.132}$$

and

$$\begin{aligned} p_{ij}^{(0)} &= M \int_{\mathbb{R}^3} (v_i - V_i)(v_j - V_j) C(t,\boldsymbol{x}) \exp(-A(t,\boldsymbol{x})|\boldsymbol{v} - \boldsymbol{V}|^2) \mathrm{d}v \\ &= MC(t,\boldsymbol{x}) \int_{\mathbb{R}^3} v_i v_j \exp(-A(t,\boldsymbol{x})|\boldsymbol{v}|^2) \mathrm{d}v \\ &= p\delta_{ij}, \end{aligned} \tag{8.133}$$

respectively, where

$$\begin{aligned} p = p(t,\boldsymbol{x}) &= \frac{1}{3} MC(t,\boldsymbol{x}) \int_{\mathbb{R}^3} |\boldsymbol{v}|^2 \exp(-A(t,\boldsymbol{x})|\boldsymbol{v}|^2) \mathrm{d}v \\ &= \frac{1}{3} MC(t,\boldsymbol{x}) \cdot \frac{3}{2} \frac{\pi^{3/2}}{A^{5/2}} \\ &= nkT \end{aligned} \tag{8.134}$$

is the local static pressure (see (8.9)).

Plugging $q_i^{(0)}$ and $p_{ij}^{(0)}$ into equations (8.113) and (8.123) for the conservation of momentum and energy, together with equation (8.110) for the conservation of mass, we obtain now the corresponding system of conservation equations

$$\frac{\partial \rho}{\partial t} + \operatorname{div}(\rho \boldsymbol{V}) = 0, \tag{8.135}$$

$$\frac{\partial}{\partial t}(\rho \boldsymbol{V}) + \operatorname{div}(\rho \boldsymbol{V} \otimes \boldsymbol{V} + p\boldsymbol{I}) = \rho \boldsymbol{g}, \tag{8.136}$$

$$\frac{\partial}{\partial t}\left(\rho e + \frac{1}{2}\rho|\boldsymbol{V}|^2\right) + \operatorname{div}\left(\left(\rho e + \frac{1}{2}\rho|\boldsymbol{V}|^2 + p\right)\boldsymbol{V}\right) = \rho \boldsymbol{g} \cdot \boldsymbol{V}, \tag{8.137}$$

where the equation of state is

$$p = nkT = R\rho T \quad \left(\text{where } R = \frac{k}{M}\right), \tag{8.138}$$

and from (8.11) we have

$$e = \frac{3}{2}\frac{kT}{M} = C_V T \quad \left(\text{where } C_V = \frac{3}{2}R\right). \tag{8.139}$$

This implies that the gas under consideration is not only an ideal gas but also a polytropic gas (see (2.6)–(2.7) in Chapter 2 of Volume I).

Thus, as a zero-order approximation, we obtain the system of ideal fluid dynamics neglecting viscosity and heat conduction (see (2.10), (2.13), and (2.19) in Chapter 2 of Volume I), and the gas under discussion is a polytropic gas.

As stated in section 2.1 in Chapter 2 of Volume I, using the continuity equation (8.135), the momentum conservation equations (8.136) can be reduced to

$$\frac{\mathrm{d}\boldsymbol{V}}{\mathrm{d}t} + \frac{1}{\rho}\operatorname{grad} p = \boldsymbol{g}, \tag{8.140}$$

where

$$\frac{\mathrm{d}}{\mathrm{d}t} = \frac{\partial}{\partial t} + \sum_{i=1}^{3} V_i \frac{\partial}{\partial x_i}. \tag{8.141}$$

Using again the continuity equation and the above equation, the energy conservation equation (8.137) can be reduced to

$$\rho \frac{\mathrm{d}e}{\mathrm{d}t} + p \operatorname{div} \boldsymbol{V} = 0. \tag{8.142}$$

Together with (8.138) and (8.139), the above equation is reduced to

$$\frac{\mathrm{d}T}{\mathrm{d}t} + \frac{2}{3} T \operatorname{div} \boldsymbol{V} = 0. \tag{8.143}$$

8.6 First-Order Approximation

8.6.1 Chapman–Enskog Expansion

To obtain higher-order approximations of the Boltzmann equation (8.54), we introduce an approach proposed by Sydney Chapman and David Enskog, which is now called the *Chapman–Enskog expansion*.

Denote by λ the *mean free path* of molecules, i.e., the average distance traveled by a moving molecule between successive collisions. Under standard temperature and pressure (see section 8.1),

$$\lambda \approx 10^{-5} \text{ cm}.$$

Let L be the macroscopic characteristic length of the gas. Usually, $\varepsilon = \lambda / L$ is a very small quantity.

Roughly speaking, the Chapman–Enskog expansion is to expand the distribution function f in terms of powers of ε as

$$f = f^{(0)} + f^{(1)} + \cdots, \tag{8.144}$$

where $f^{(0)}$ is the zero-order approximation given in the last section, $f^{(1)}/f^{(0)}$ is of order ε, and so forth. The approximations of different orders of the Boltzmann equation can be obtained accordingly.

For a distribution function f like (8.144), we can write the Boltzmann equation (8.54) as

$$\left(\frac{\partial}{\partial t} + \boldsymbol{v} \cdot \nabla_x + \boldsymbol{g} \cdot \nabla_v\right)(f^{(0)} + f^{(1)} + \cdots)$$
$$= J^{(0)} + J^{(1)} + \cdots, \tag{8.145}$$

where

$$J^{(0)} = \int_{\mathbb{R}^3} \mathrm{d}w \int \mathrm{d}\Omega\sigma(\Omega)|\boldsymbol{w}-\boldsymbol{v}|(f_1^{(0)\prime} f_2^{(0)\prime} - f_1^{(0)} f_2^{(0)}), \tag{8.146}$$

$$J^{(1)} = \int_{\mathbb{R}^3} \mathrm{d}w \int \mathrm{d}\Omega\sigma(\Omega)|\boldsymbol{w}-\boldsymbol{v}|(f_2^{(0)\prime} f_1^{(1)\prime} - f_2^{(0)} f_1^{(1)} + f_1^{(0)\prime} f_2^{(1)\prime} - f_1^{(0)} f_2^{(1)}), \tag{8.147}$$

and so on, and $f_1^{(0)\prime} = f^{(0)}(t,\boldsymbol{x},\boldsymbol{v}')$, $f_1^{(1)} = f^{(1)}(t,\boldsymbol{x},\boldsymbol{v})$, $f_2^{(1)\prime} = f^{(1)}(t,\boldsymbol{x},\boldsymbol{w}')$, etc.

Now we analyze the order of each term in (8.145).

Assume that only when the position of $\boldsymbol{x}$ changes for a distance of L can the magnitude of a distribution function change considerably; then acting the Boltzmann operator $-\frac{\partial}{\partial t} + \boldsymbol{v}\cdot\nabla_x + \boldsymbol{g}\cdot\nabla_v$ of the left-hand side of (8.145) on a distribution function is equivalent in terms of order to multiplying this function by a factor $\overline{V}/L$, where $\overline{V}$ is the average velocity. As a matter of fact, this is obvious for the term $\boldsymbol{v}\cdot\nabla_x$, and the other two terms are assumed to be of the same order as well.

Now we check $J^{(1)}$ given on the right-hand side of (8.147). Obviously,

$$-\int_{\mathbb{R}^3} \mathrm{d}w \int \mathrm{d}\Omega\sigma(\Omega)|\boldsymbol{w}-\boldsymbol{v}| f_2^{(0)} f_1^{(1)} = -f^{(1)}(t,\boldsymbol{x},\boldsymbol{v})/\tau \approx -\varepsilon f^{(0)}/\tau, \tag{8.148}$$

where

$$\tau = \left(\int_{\mathbb{R}^3} \mathrm{d}w \int \mathrm{d}\Omega\sigma(\Omega)|\boldsymbol{w}-\boldsymbol{v}| f^{(0)}(t,\boldsymbol{x},\boldsymbol{w})\right)^{-1}. \tag{8.149}$$

From (8.38), $\int_{\mathbb{R}^3} \mathrm{d}w \int \mathrm{d}\Omega\sigma(\Omega)|\boldsymbol{w}-\boldsymbol{v}| f^{(0)}(t,\boldsymbol{x},\boldsymbol{w})$ is the number of collisions per unit time with a molecule located at $\boldsymbol{x}$ with velocity $\boldsymbol{v}$ as its target and all possible $\boldsymbol{w}$ as its incident flow. Therefore, its reciprocal τ corresponds to the collision time (the average time between successive collisions). For other terms on the right-hand side of (8.147), we have similar results. But for those terms on the right-hand side of (8.146), there is one ε factor short.

Thus, multiplying both sides of (8.145) by τ, noting $\lambda \approx \overline{V}\tau$, and letting the zero-order terms of ε on both sides be equal to each other, we obtain

$$J^{(0)} = 0. \tag{8.150}$$

This is just the equation satisfied by $f^{(0)}$ in the last section.

Now, comparing the first-order terms of ε on both sides of (8.145) after they are multiplied by τ, we get

$$\left(\frac{\partial}{\partial t} + \boldsymbol{v}\cdot\nabla_x + \boldsymbol{g}\cdot\nabla_v\right) f^{(0)} = J^{(1)}. \tag{8.151}$$

As $f^{(0)}$ was already given by (8.128), (8.151) is just the integral equation that $f^{(1)}$ should satisfy.

To simplify the above integral equation, instead of the unknown function $f^{(1)}$, we introduce a new function ϕ defined by

$$f^{(1)} = f^{(0)}\phi. \tag{8.152}$$

Plugging (8.152) into the expression (8.147) of $J^{(1)}$, and noticing that, from the definition (8.128) of $f^{(0)}$ and (8.22)–(8.23),

$$f_1^{(0)\prime} f_2^{(0)\prime} = f_1^{(0)} f_2^{(0)}, \tag{8.153}$$

we have

$$J^{(1)} = \int_{\mathbb{R}^3} \mathrm{d}\boldsymbol{w} \int \mathrm{d}\Omega\sigma(\Omega)|\boldsymbol{w}-\boldsymbol{v}| f_1^{(0)} f_2^{(0)}(\phi_1' + \phi_2' - \phi_1 - \phi_2), \tag{8.154}$$

where $\phi_1 = \phi_1(t,\boldsymbol{x},\boldsymbol{v})$, $\phi_2' = \phi(t,\boldsymbol{x},\boldsymbol{w}')$, etc.

Dividing equation (8.151) by $f_1^{(0)} = f^{(0)}(t,\boldsymbol{x},\boldsymbol{v})$, and using (8.154), we obtain

$$\left(\frac{\partial}{\partial t} + \boldsymbol{v}\cdot\nabla_x + \boldsymbol{g}\cdot\nabla_v\right)\ln f^{(0)} = -n\widetilde{J}(\phi), \tag{8.155}$$

where

$$\widetilde{J}(\phi) = \frac{1}{n}\int_{\mathbb{R}^3} \mathrm{d}\boldsymbol{w} \int \mathrm{d}\Omega\sigma(\Omega)|\boldsymbol{w}-\boldsymbol{v}| f_2^{(0)}(\phi_1 + \phi_2 - \phi_1' - \phi_2'). \tag{8.156}$$

Noting the expression (8.128) of $f^{(0)}$, we see that the integral operator $\widetilde{J}$ does not depend on either n or ρ.

Now let us first look at the left-hand side of (8.155). Since $f^{(0)}$ depends on t and $\boldsymbol{x}$ through ρ, T, and $\boldsymbol{V}$, to compute the left-hand side of (8.155), we have to first compute the partial derivatives of $f^{(0)}$ with respect to variables ρ, T, and $\boldsymbol{V}$. From (8.128) we have

$$\frac{\partial f^{(0)}}{\partial \rho} = \frac{f^{(0)}}{\rho}, \tag{8.157}$$

$$\frac{\partial f^{(0)}}{\partial T} = \frac{1}{T}\left(\frac{M|\boldsymbol{U}|^2}{2kT} - \frac{3}{2}\right) f^{(0)}, \tag{8.158}$$

$$\frac{\partial f^{(0)}}{\partial V_i} = \frac{M}{kT} U_i f^{(0)} \quad (i = 1,2,3), \tag{8.159}$$

$$\frac{\partial f^{(0)}}{\partial v_i} = -\frac{M}{kT} U_i f^{(0)} \quad (i = 1,2,3), \tag{8.160}$$

where $\boldsymbol{U} = \boldsymbol{v} - \boldsymbol{V} = (U_1, U_2, U_3)$. Thus, we obtain

$$\begin{aligned}
&\text{the left-hand side of (8.155)}\\
&= \frac{1}{f^{(0)}}\left\{\frac{\partial f^{(0)}}{\partial \rho}\left(\frac{\partial \rho}{\partial t} + \boldsymbol{v}\cdot\nabla_x \rho\right) + \frac{\partial f^{(0)}}{\partial T}\left(\frac{\partial T}{\partial t} + \boldsymbol{v}\cdot\nabla_x T\right)\right.\\
&\quad\left. + \sum_{i=1}^{3}\frac{\partial f^{(0)}}{\partial V_i}\left(\frac{\partial V_i}{\partial t} + \boldsymbol{v}\cdot\nabla_x V_i\right) + \boldsymbol{g}\cdot\nabla_v f^{(0)}\right\}\\
&= \frac{1}{\rho}\frac{\mathrm{D}\rho}{\mathrm{D}t} + \frac{1}{T}\left(\frac{M|\boldsymbol{U}|^2}{2kT} - \frac{3}{2}\right)\frac{\mathrm{D}T}{\mathrm{D}t}\\
&\quad + \frac{M}{kT}\sum_{i=1}^{3} U_i \frac{\mathrm{D}V_i}{\mathrm{D}t} - \frac{M}{kT}\boldsymbol{g}\cdot\boldsymbol{U},
\end{aligned} \tag{8.161}$$

where

$$\frac{\mathrm{D}}{\mathrm{D}t} = \frac{\partial}{\partial t} + \sum_{i=1}^{3} v_i \frac{\partial}{\partial x_i}. \tag{8.162}$$

Using the ideal fluid dynamics system of equations (8.135), (8.143), and (8.140) satisfied by the zero-order approximation, we have

$$\begin{aligned}
\frac{\mathrm{D}\rho}{\mathrm{D}t} &= -\rho \operatorname{div}\boldsymbol{V} + \boldsymbol{U}\cdot\nabla_x \rho,\\
\frac{\mathrm{D}T}{\mathrm{D}t} &= -\frac{2}{3}T \operatorname{div}\boldsymbol{V} + \boldsymbol{U}\cdot\nabla_x T,\\
\frac{\mathrm{D}V_i}{\mathrm{D}t} &= -\frac{1}{\rho}\frac{\partial p}{\partial x_i} + g_i + \boldsymbol{U}\cdot\nabla_x V_i \quad (i = 1,2,3).
\end{aligned}$$

Substituting the above formulas into (8.161) and using the equation of state (8.138), we get

$$\begin{aligned}
&\text{the left-hand side of (8.155)} = -\operatorname{div}\boldsymbol{V} + \frac{1}{\rho}\boldsymbol{U}\cdot\nabla_x \rho\\
&\quad + \frac{1}{T}\left(\frac{M|\boldsymbol{U}|^2}{2kT} - \frac{3}{2}\right)\left(-\frac{2}{3}T \operatorname{div}\boldsymbol{V} + \boldsymbol{U}\cdot\nabla_x T\right)\\
&\quad + \frac{M}{kT}\sum_{i=1}^{3} U_i\left(-\frac{1}{\rho}\frac{\partial p}{\partial x_i} + g_i + \boldsymbol{U}\cdot\nabla_x V_i\right) - \frac{M}{kT}\boldsymbol{g}\cdot\boldsymbol{U}\\
&= \frac{1}{T}\left(\frac{M}{2kT}|\boldsymbol{U}|^2 - \frac{5}{2}\right)\boldsymbol{U}\cdot\nabla_x T\\
&\quad + \frac{M}{kT}\sum_{i,j=1}^{3}\left(U_i U_j - \frac{1}{3}\delta_{ij}|\boldsymbol{U}|^2\right)S_{ij},
\end{aligned} \tag{8.163}$$

where S_{ij} is given by (8.126).

Thus, the integral equation (8.155) satisfied by ϕ can be written in the following form:

$$\begin{aligned}\widetilde{J}(\phi) = & -\frac{1}{nkT^2}\left(\frac{1}{2}M|U|^2 - \frac{5}{2}kT\right)U \cdot \nabla_x T \\ & -\frac{M}{nkT}\sum_{i,j=1}^{3}\left(U_iU_j - \frac{1}{3}\delta_{ij}|U|^2\right)S_{ij}.\end{aligned} \tag{8.164}$$

From this we can see that $\widetilde{J}(\phi)$ depends on v through U, and $\widetilde{J}(\phi)$ is the linear combination (with coefficients depending on U and T) of $\nabla_x T$ and $S = (S_{ij})$ dividing by n. Noting that from (8.156) $\widetilde{J}(\phi)$ depends linearly on ϕ, it is reasonable to suppose the following form of ϕ:

$$\phi = -\frac{1}{nkT^2}\sum_{i=1}^{3} a_i \frac{\partial T}{\partial x_i} - \frac{1}{nkT}\sum_{i,j=1}^{3} b_{ij}\frac{\partial V_i}{\partial x_j}, \tag{8.165}$$

where a_i, b_{ij} are all functions of U and T to be determined. Plugging this into (8.164) and comparing the coefficients of $\frac{\partial T}{\partial x_i}$ and $\frac{\partial V_i}{\partial x_j}$ on both sides, we get

$$\widetilde{J}(a_i) = \widetilde{a}_i, \quad \widetilde{J}(b_{ij}) = \widetilde{b}_{ij} \quad (i, j = 1, 2, 3), \tag{8.166}$$

where

$$\widetilde{a}_i = \left(\frac{1}{2}M|U|^2 - \frac{5}{2}kT\right)U_i, \quad \widetilde{b}_{ij} = M\left(U_iU_j - \frac{1}{3}\delta_{ij}|U|^2\right). \tag{8.167}$$

Equations (8.166) and (8.167) are the integral equations determining a_i and b_{ij}, respectively. As long as a_i and b_{ij} are solved by these, ϕ can be obtained by (8.165); then from (8.152) we get the first-order approximation $f^{(1)}$ of the solution of the Boltzmann equation (8.54).

8.6.2 Solvability of the Integral Equation

Suppose that $f = f(t, x, v)$ is the desired distribution function; then from (8.3), (8.5), and (8.6), we can easily see that

$$\begin{aligned} n &= \int_{\mathbb{R}^3} f \, \mathrm{d}v, \\ nV &= \int_{\mathbb{R}^3} v f \, \mathrm{d}v, \end{aligned}$$

and

$$\frac{3}{2}nkT + \frac{1}{2}nM|V|^2 = \int_{\mathbb{R}^3} \frac{1}{2}M|v|^2 f \, \mathrm{d}v.$$

From the definition of $f^{(0)}$ given in section 8.5 of this chapter, the above formulas are still true when f is replaced by $f^{(0)}$. Therefore,

$$\int_{\mathbb{R}^3} (f - f^{(0)})\chi \, \mathrm{d}v = 0 \tag{8.168}$$

for any given conserved quantity χ (see section 8.4). Supposing that $f = f^{(0)} + f^{(0)}\phi$, we have

$$\int_{\mathbb{R}^3} f^{(0)}\phi\chi \, \mathrm{d}v = 0 \tag{8.169}$$

for any given conserved quantity χ.

Denote by H the Hilbert space composed of functions of $\boldsymbol{v}$, where the inner product of elements $\phi, \psi \in H$ is defined by

$$(\phi, \psi)_H = \int_{\mathbb{R}^3} f^{(0)}\phi\psi \, \mathrm{d}v. \tag{8.170}$$

Thus, (8.169) can be rewritten as

$$(\phi, \chi)_H = 0 \tag{8.171}$$

for any given conserved quantity χ; in other words, ϕ is orthogonal to the space spanned by those five conserved quantities (i.e., mass, momentum, and energy). Noting the expression (8.165) of ϕ, from the above formula we have

$$(a_i, \chi)_H = 0, \quad (b_{ij}, \chi)_H = 0 \quad (i, j = 1,2,3) \tag{8.172}$$

for any given conserved quantity χ.

Regarding $\widetilde{J}$ as an integral operator in the above Hilbert space H, it is not hard to verify from the definition (8.156) of $\widetilde{J}$ that

$$(\widetilde{J}(\phi), \psi)_H = (\phi, \widetilde{J}(\psi))_H, \quad \forall \phi, \psi \in \text{ domain of } \widetilde{J} \tag{8.173}$$

and $\widetilde{J}$ is nonnegative, namely,

$$(\widetilde{J}(\phi), \phi)_H \geqslant 0, \quad \forall \phi \in \text{ domain of } \widetilde{J}. \tag{8.174}$$

We leave the proof of the above two formulas as exercises.

Moreover, from the definition of $\widetilde{J}$ and definition (8.95) of a conserved quantity, it is not hard to show that the kernel of operator $\widetilde{J}$ is the space spanned by the above five conserved quantities, i.e.,

$$\widetilde{J}(\chi) = 0 \Leftrightarrow \chi \text{ is a conserved quantity.} \tag{8.175}$$

In addition, using the expression (8.167) of $\widetilde{a}_i$ and $\widetilde{b}_{ij}$, it can be verified directly that

$$(\widetilde{a}_i, \chi)_H = 0, \quad (\widetilde{b}_{ij}, \chi)_H = 0 \quad (i, j = 1,2,3) \tag{8.176}$$

for any given conserved quantity χ (see Exercise 4). It can be proved that $\widetilde{J}$ is a self-adjoint operator, and the above formula is exactly the solvability condition of the self-adjoint integral equation (8.166) (see [9]).

We can see from definition (8.167) of $\widetilde{a}_i$ that $(\widetilde{a}_1,\widetilde{a}_2,\widetilde{a}_3)^{\mathrm{T}}$ is a first-order tensor (vector) in U space (see Appendix A of Volume I). Using the invariance of the differential scattering cross section with respect to the orthogonal transformation, it is not hard to verify that a solution $(a_1,a_2,a_3)^{\mathrm{T}}$ of integral equation $\widetilde{J}(a_i)=\widetilde{a}_i$ $(i=1,2,3)$ should be a first-order tensor in U space as well, and its transformation obeys completely the same law as $(\widetilde{a}_1,\widetilde{a}_2,\widetilde{a}_3)^{\mathrm{T}}$ for any given orthogonal transformation. So we have

$$a_i=\alpha(T,|\boldsymbol{U}|)\widetilde{a}_i \quad (i=1,2,3). \tag{8.177}$$

Similarly, we have

$$b_{ij}=\beta(T,|\boldsymbol{U}|)\widetilde{b}_{ij} \quad (i,j=1,2,3). \tag{8.178}$$

From the nonnegativity (8.174) of $\widetilde{J}$, both α and β are positive functions. Furthermore, since $\widetilde{J}$ is independent of ρ, both α and β are functions of just $(T,|\boldsymbol{U}|)$ and independent of ρ.

8.6.3 Coefficient of Viscosity and Coefficient of Heat Conductivity

Up to now, although we have proved the solvability of the first-order approximation of the Boltzmann equation, we have not given specifically the first-order approximate solution; however, we can give, at least formally, the pressure tensor and the heat flux density vector, and then the coefficient of viscosity and the coefficient of heat conduction accordingly.

We first discuss the heat flux density vector. Noting $\boldsymbol{U}=\boldsymbol{v}-\boldsymbol{V}$ and (8.132), from (8.122) we have

$$\begin{aligned} q_i &= \int_{\mathbb{R}^3}\frac{M}{2}U_i|\boldsymbol{U}|^2(f^{(0)}+f^{(0)}\phi)\mathrm{d}U \\ &= \int_{\mathbb{R}^3}\frac{M}{2}U_i|\boldsymbol{U}|^2 f^{(0)}\phi\mathrm{d}\boldsymbol{v} \\ &= \left(\frac{M}{2}U_i|\boldsymbol{U}|^2,\phi\right)_H. \end{aligned} \tag{8.179}$$

It easily follows from (8.178) and (8.167) that b_{ij} are even functions of $\boldsymbol{U}$; then we have

$$\left(\frac{M}{2}U_i|\boldsymbol{U}|^2,b_{ij}\right)_H=0 \quad (i,j=1,2,3). \tag{8.180}$$

Similarly, we have

$$\left(\frac{M}{2}U_i|\boldsymbol{U}|^2,a_j\right)_H=0 \quad (i\neq j) \tag{8.181}$$

and

$$\left(\frac{M}{2}U_i|\boldsymbol{U}|^2,a_i\right)_H=(\widetilde{a}_i,a_i)_H=\frac{1}{3}\sum_{j=1}^{3}(\widetilde{a}_j,a_j)_H. \tag{8.182}$$

In obtaining (8.182), we used the fact that U_i are conserved quantities and a_i is orthogonal to any given conserved quantity in space H (see (8.172)). Using expression (8.165) of ϕ and noting the above formulas, from (8.179) we get

$$q_i = -\kappa \frac{\partial T}{\partial x_i} \quad (i = 1,2,3), \tag{8.183}$$

where the coefficient of heat conduction is

$$\kappa = \frac{1}{3nkT^2} \sum_{j=1}^{3} (\widetilde{a}_j, a_j)_H. \tag{8.184}$$

Substituting $\widetilde{a}_j$ and a_j given by (8.167) and (8.177) into the above formula, and noting definition (8.170) of the inner product $(\cdot,\cdot)_H$ and expression (8.128) of $f^{(0)}$, we have

$$\begin{aligned} \kappa = {} & \frac{1}{3}\left(\frac{M}{2\pi}\right)^{\frac{3}{2}} k^{-\frac{5}{2}} T^{-\frac{7}{2}} \\ & \cdot \int_{\mathbb{R}^3} \alpha(T,|\boldsymbol{U}|)\left(\frac{1}{2}M|\boldsymbol{U}|^2 - \frac{5}{2}kT\right)^2 |\boldsymbol{U}|^2 \mathrm{e}^{-\frac{M|\boldsymbol{U}|^2}{2kT}}\, \mathrm{d}\boldsymbol{U}. \end{aligned} \tag{8.185}$$

By employing spherical coordinates, we can reduce the triple integral on the right-hand side of the above formula to a single integral, and then by a simple change of variables, we can get the coefficient of heat conduction as

$$\kappa = \frac{1}{6\sqrt{2\pi}\,M} k^2 T \int_0^{\infty} \alpha\left(T, \sqrt{\frac{kT}{M}}\, r\right)(r^2-5)^2 r^4 \mathrm{e}^{-\frac{1}{2}r^2}\, \mathrm{d}r. \tag{8.186}$$

Although the exact expression of function $\alpha(T,|\boldsymbol{U}|)$ is not given, we can see from the above formula that the heat conduction coefficient κ depends only on the absolute temperature T.

Second, we discuss the pressure tensor. Using (8.133)–(8.134), we can represent the pressure tensor $\boldsymbol{P} = (p_{ij})$ (see (8.7)) as

$$p_{ij} = p\delta_{ij} + p'_{ij}, \tag{8.187}$$

where

$$p = nkT \tag{8.188}$$

and

$$p'_{ij} = \int_{\mathbb{R}^3} MU_iU_j f^{(0)} \phi \mathrm{d}\boldsymbol{U} = (MU_iU_j, \phi)_H. \tag{8.189}$$

Similarly to (8.180), it is not difficult to verify that

$$(MU_iU_j, a_i)_H = 0 \quad (i,j = 1,2,3). \tag{8.190}$$

Thus, substituting expression (8.165) of ϕ into (8.189), we get

$$p'_{ij} = -\frac{M}{nkT}\sum_{l,m=1}^{3}(U_iU_j, b_{lm})_H \frac{\partial V_l}{\partial x_m}. \tag{8.191}$$

Noticing that $|\boldsymbol{U}|^2$ is a conserved quantity and b_{lm} is orthogonal to any given conserved quantity in space H (see (8.171)), from definition (8.166) of $\widetilde{b}_{ij}$, and noting the symmetry of (b_{lm}), we can rewrite the above formula as

$$\begin{aligned} p'_{ij} &= -\frac{1}{nkT}\sum_{l,m=1}^{3}(\widetilde{b}_{ij}, b_{lm})_H \frac{\partial V_l}{\partial x_m} \\ &= -\frac{1}{nkT}\sum_{l,m=1}^{3}(\widetilde{b}_{ij}, b_{lm})_H S_{lm}. \end{aligned} \tag{8.192}$$

Using (8.167) and (8.178), and noticing that b_{ij} is orthogonal to the conserved quantity $|\boldsymbol{U}|^2$, it is not hard to verify that

$$(\widetilde{b}_{ij}, b_{lm})_H = 0 \quad \text{if } i \neq j, (l,m) \text{ or } (m,l) \neq (i,j), \tag{8.193}$$

$$(\widetilde{b}_{ii}, b_{lm})_H = 0 \quad \text{if } l \neq m, \tag{8.194}$$

and the value of $(\widetilde{b}_{ij}, b_{ij})_H$ is independent of i, j as $i \neq j$. So, it is easy to see from (8.192) that

$$p'_{ij} = -2\mu S_{ij} \quad \text{if } i \neq j, \tag{8.195}$$

where

$$\mu = \frac{1}{nkT}(\widetilde{b}_{ij}, b_{ij})_H = \frac{M^2}{nkT}(U_iU_j, \beta U_iU_j)_H, \quad i \neq j. \tag{8.196}$$

Here, we used once again the orthogonality of b_{ij} with the conserved quantity $|\boldsymbol{U}|^2$ and the fact that $(U_iU_j, \beta|\boldsymbol{U}|^2)_H = 0$ when $i \neq j$.

Noting (8.194), similarly, from (8.192) we have

$$\begin{aligned} p'_{ii} &= -\frac{1}{nkT}\sum_{l=1}^{3}(\widetilde{b}_{ii}, b_{ll})_H S_{ll} \\ &= -\frac{1}{nkT}\sum_{l=1}^{3}(MU_i^2, b_{ll})_H S_{ll}. \end{aligned} \tag{8.197}$$

In addition, from expression (8.178) of b_{ll} and (8.167) we get

$$\begin{aligned} &\sum_{l=1}^{3}(MU_i^2, b_{ll})_H S_{ll} = (MU_i^2, M\beta U_i^2)_H S_{ii} \\ &+\sum_{l\neq i}(MU_i^2, M\beta U_l^2)_H S_{ll} - \frac{1}{3}(MU_i^2, M\beta|\boldsymbol{U}|^2)_H \operatorname{div} \boldsymbol{V}. \end{aligned} \tag{8.198}$$

It is easy to see from (8.196) that

$$(MU_i^2, M\beta U_l^2)_H = nkT\mu \quad \text{if } i \neq l \tag{8.199}$$

and

$$-(MU_i^2, M\beta|U|^2)_H = -nkT\delta - 2nkT\mu, \tag{8.200}$$

where

$$\delta = \frac{M^2}{nkT}(U_i^2, \beta U_i^2)_H, \tag{8.201}$$

and, obviously, the right-hand side of the above formula is independent of i. Thus, from (8.197) we easily get

$$\begin{aligned} p'_{ii} &= -\delta S_{ii} - \mu \sum_{l \neq i} S_{ll} + \frac{1}{3}(\delta + 2\mu)\operatorname{div}\boldsymbol{V} \\ &= -(\delta - \mu)S_{ii} + \frac{1}{3}(\delta - \mu)\operatorname{div}\boldsymbol{V}. \end{aligned} \tag{8.202}$$

Now we further give the expressions of δ and μ. Taking $i = 3$ in (8.201), and noting definition (8.170) of the inner product $(\cdot,\cdot)_H$ and definition (8.128) of $f^{(0)}$, we have

$$\delta = \frac{M}{(2\pi)^{\frac{3}{2}}}\left(\frac{M}{kT}\right)^{\frac{5}{2}} \int_{\mathbb{R}^3} \beta(T,|U|)U_3^4 \mathrm{e}^{-\frac{M|U|^2}{2kT}}\,\mathrm{d}U. \tag{8.203}$$

Computing the above integral by using the spherical coordinates $U_1 = r\cos\phi\sin\theta$, $U_2 = r\sin\phi\sin\theta$, and $U_3 = r\cos\theta$, we obtain

$$\begin{aligned} \delta &= \frac{M}{(2\pi)^{\frac{3}{2}}}\left(\frac{M}{kT}\right)^{\frac{5}{2}} \int_0^\infty \beta(T,r)r^6 \mathrm{e}^{-\frac{Mr^2}{2kT}}\,\mathrm{d}r \int_0^\pi \cos^4\theta\sin\theta \mathrm{d}\theta \int_0^{2\pi} \mathrm{d}\phi \\ &= \frac{2}{5\sqrt{2\pi}}kT\int_0^\infty \beta\left(T, \sqrt{\frac{kT}{M}}r\right) r^6 \mathrm{e}^{-\frac{r^2}{2}}\,\mathrm{d}r. \end{aligned} \tag{8.204}$$

From expressions (8.196) (taking $i = 1$ and 2, respectively) and (8.201) (taking $i = 1$, $j = 2$), it is easy to show that

$$\begin{aligned} 2(\delta - \mu) &= \frac{M^2}{nkT}\int_{\mathbb{R}^3} \beta(T,|U|)(U_1^4 + U_2^4 - 2U_1^2U_2^2)f^{(0)}\mathrm{d}U \\ &= \frac{M^2}{nkT}\int_{\mathbb{R}^3} \beta(T,|U|)(U_1 - U_2)^2(U_1 + U_2)^2 f^{(0)}\mathrm{d}U. \end{aligned} \tag{8.205}$$

Taking the orthogonal substitution

$$U_1^* = \frac{\sqrt{2}}{2}(U_1 - U_2), \quad U_2^* = \frac{\sqrt{2}}{2}(U_1 + U_2), \quad U_3^* = U_3,$$

and noting expression (8.196) of μ, we obtain

$$\begin{aligned}2(\delta-\mu) &= \frac{4M^2}{nkT}\int_{\mathbb{R}^3}\beta(T,|U^*|)U_1^{*2}U_2^{*2}f^{(0)}\mathrm{d}U^* \\ &= 4\mu.\end{aligned} \tag{8.206}$$

Then it follows immediately that

$$\mu = \frac{1}{3}\delta. \tag{8.207}$$

Together with (8.204), we get

$$\mu = \frac{2}{15\sqrt{2\pi}}kT\int_0^\infty \beta\left(T,\sqrt{\frac{kT}{M}}r\right)r^6 \mathrm{e}^{-\frac{r^2}{2}}\,\mathrm{d}r. \tag{8.208}$$

Although the exact expression of $\beta(T,|U|)$ is not given, we can see from the above formula that the viscosity coefficient μ depends only on the absolute temperature T.

Using (8.206), from (8.202) we obtain

$$p'_{ii} = -2\mu\left(S_{ii}-\frac{1}{3}\operatorname{div}\boldsymbol{V}\right) \quad (i=1,2,3). \tag{8.209}$$

Combining this with (8.195), we get

$$p'_{ij} = -2\mu\left(S_{ij}-\frac{1}{3}\delta_{ij}\operatorname{div}\boldsymbol{V}\right) \quad (i,j=1,2,3). \tag{8.210}$$

Thus, from (8.187), we obtain the constitutive equations

$$p_{ij} = p\delta_{ij}-2\mu\left(S_{ij}-\frac{1}{3}\delta_{ij}\operatorname{div}\boldsymbol{V}\right) \quad (i,j=1,2,3). \tag{8.211}$$

These are exactly the same as the constitutive equations given in (2.130) in Chapter 2 of Volume I for viscous and heat-conductive fluids with the expansive viscosity coefficient $\mu'=0$. What we need to notice is that (p_{ij}) there represents the stress tensor, while here the same notation (p_{ij}) represents the pressure tensor, and they are different by only a minus sign. Moreover, here the conclusion corresponding to the case with zero expansive viscosity coefficient is related to the assumption that gas molecules do not have internal structure.

In the above discussions, the exact expressions of functions $\alpha(T,|U|)$ and $\beta(T,|U|)$ are not given; therefore, (8.186) and (8.208) do not specifically give the formulas to compute the coefficient of viscosity and the coefficient of heat conduction. An in-depth discussion can yield approximate computing formulas with results quite close to those of experiments. We refer the interested reader to references listed at the end of this chapter (e.g., [1], [9]).

8.7 Vlasov Equation and Related Coupled Systems

8.7.1 Vlasov Equation

For a gas molecular group described by the Boltzmann equation, there is no force acting other than the collision (short-range force) between particles (molecules). Then the state of motion of any given particle is completely independent when no collision occurs. If there exist long-range interactions between particles in the particle group under study, while collisions between them occur rarely and can be neglected, the Boltzmann equation is no longer applicable. For instance, in a plasma (see Chapter 3 of Volume I), since the Coulomb force between charged particles is a kind of long-range force, any given charged particle interacts with other particles at any moment, so the independent state of motion cannot exist. Moreover, there is a similar story for the universe, since stars interact with each other because of (long-range) gravitation. None of these fit with conditions for satisfying the Boltzmann equation. However, in many situations, the effect of interactions between particles can be regarded approximately as introducing an additional average force field: each particle moves under the action of this field, and in the meantime this average field is determined by the state of motion of all particles. Thus, replacing the long-range interactions between particles with an additional average force field enables us to continue using the basic framework of the Boltzmann equation to handle related problems.

Now we discuss a particle group in which the interactions between particles are long range, and we assume that no collision occurs. Suppose that $\boldsymbol{g}(t,\boldsymbol{x})$ is the force (per unit mass) acting on each particle from the average force field determined by the state of motion of all particles. From the derivation in section 8.2 we know that the distribution function $f(t,\boldsymbol{x},\boldsymbol{v})$ of particles should satisfy the following equation:

$$\frac{\partial f}{\partial t}+\boldsymbol{v}\cdot\nabla_x f+\boldsymbol{g}\cdot\nabla_v f=0. \tag{8.212}$$

This equation is called the *Vlasov equation*. As a special form of the Boltzmann equation, it is one of the fundamental equations in the study of plasmas.

8.7.2 Vlasov–Poisson System

Now we consider the motion of a charged particle group (e.g., the electron group in a plasma). Assume that the process of motion is not too fast so that the electromagnetic effect can be neglected, i.e., it can be approximated as an electrostatic process. Suppose that the mass of one particle is M and its charge q, and that the (average) electric field generated by these charged particles is $\boldsymbol{E}(t,\boldsymbol{x})$. Thus, since the magnetic field has been ignored, the electric field force per unit mass received by a particle is $\boldsymbol{g}=\frac{q}{M}\boldsymbol{E}(t,\boldsymbol{x})$. So the distribution function $f(t,\boldsymbol{x},\boldsymbol{v})$ of these charged particles should satisfy the Vlasov equation

$$\frac{\partial f}{\partial t}+\boldsymbol{v}\cdot\nabla_x f+\frac{q}{M}\boldsymbol{E}\cdot\nabla_v f=0. \tag{8.213}$$

From the Maxwell equations, the Gauss theorem still holds (see (1.54) in Chapter 1 of Volume I), although $\boldsymbol{E}$ is not an electrostatic field now, i.e.,

$$\operatorname{div}\boldsymbol{E}=\frac{\rho}{\varepsilon_0}, \tag{8.214}$$

where ε_0 is the dielectric constant in a vacuum and ρ is the charge density. By the definition of the distribution function $f(t,\boldsymbol{x},\boldsymbol{v})$, the charge density, i.e., the charge per unit volume, is

$$\rho(t,\boldsymbol{x})=q\int_{\mathbb{R}^3} f(t,\boldsymbol{x},\boldsymbol{v})\mathrm{d}\boldsymbol{v}. \tag{8.215}$$

Under the electrostatic approximation, we can denote by $\phi(t,\boldsymbol{x})$ the potential of the electric field $\boldsymbol{E}$:

$$\boldsymbol{E} = -\nabla_x \phi; \tag{8.216}$$

then, noting (8.215), equations (8.213)–(8.214) can be written as

$$\frac{\partial f}{\partial t} + \boldsymbol{v} \cdot \nabla_x f - \frac{q}{M} \nabla_x \phi \cdot \nabla_v f = 0 \tag{8.217}$$

and

$$-\Delta \phi = \frac{q}{\varepsilon_0} \int_{\mathbb{R}^3} f(t,\boldsymbol{x},\boldsymbol{v}) \mathrm{d}\boldsymbol{v}. \tag{8.218}$$

This system of equations is called the *Vlasov–Poisson system*.

Now we investigate specifically the Vlasov–Poisson system (8.217)–(8.218) when it is applied in a high-temperature plasma without an extra magnetic field. If the frequency of the process is not too low, cations can be regarded as motionless and generate a background of positive charges uniformly distributed in space. Suppose that $f(t,\boldsymbol{x},\boldsymbol{v})$ represents the distribution function of electrons in the plasma, and $q = -e$ for an electron. In consideration of the above uniformly distributed positive charge background, the charges to determine the electric field are provided not only by the electrons but also by the positive charges induced by the cations. The density of positive charges is

$$e \int_{\mathbb{R}^3} f_0(t,\boldsymbol{x},\boldsymbol{v}) \mathrm{d}\boldsymbol{v}, \tag{8.219}$$

where $f_0(t,\boldsymbol{x},\boldsymbol{v})$ is the Maxwell–Boltzmann distribution given by (8.87) (where $\boldsymbol{v}_0 = 0$). Thus, the total density of charges should be

$$\rho(t,\boldsymbol{x}) = -e \int_{\mathbb{R}^3} (f(t,\boldsymbol{x},\boldsymbol{v}) - f_0(t,\boldsymbol{x},\boldsymbol{v})) \mathrm{d}\boldsymbol{v}. \tag{8.220}$$

So now the Vlasov–Poisson system (8.217)–(8.218) is modified to

$$\frac{\partial f}{\partial t} + \boldsymbol{v} \cdot \nabla_x f + \frac{e}{M} \nabla_x \phi \cdot \nabla_v f = 0 \tag{8.221}$$

and

$$\Delta \phi = \frac{e}{\varepsilon_0} \int_{\mathbb{R}^3} (f(t,\boldsymbol{x},\boldsymbol{v}) - f_0(t,\boldsymbol{x},\boldsymbol{v})) \mathrm{d}\boldsymbol{v}. \tag{8.222}$$

For the particle group under discussion, if there exist only (long-range) gravitational interactions between particles, its distribution function f of particles and potential ϕ of the gravitational field also satisfy a Vlasov–Poisson system similar to (8.217)–(8.218) (see Exercise 5).

8.7.3 Vlasov–Maxwell System

Generally speaking, the magnetic field effect cannot be ignored (see Chapter 3 of Volume I) in a group of charged particles in motion (e.g., a plasma). Under this situation, this group of charged particles is in an electromagnetic field generated by the macroscopic motion of all particles. In this case, the Vlasov–Poisson system stated above should be modified.

We still suppose that the mass of a single particle is M and its charge q. Suppose furthermore that the electric field intensity and the magnetic induction intensity are $\boldsymbol{E}(t,\boldsymbol{x})$ and $\boldsymbol{B}(t,\boldsymbol{x})$, respectively. From the Lorentz force formula (see (1.64) in Chapter 1 of Volume I) in the electromagnetic field, the force received by a charged particle is

$$q(\boldsymbol{E}+\boldsymbol{v}\times\boldsymbol{B}). \tag{8.223}$$

Therefore, the Vlasov equation (8.212) satisfied by the distribution function $f(t,\boldsymbol{x},\boldsymbol{v})$ of particles reduces to

$$\frac{\partial f}{\partial t}+\boldsymbol{v}\cdot\nabla_x f+\frac{q}{M}(\boldsymbol{E}+\boldsymbol{v}\times\boldsymbol{B})\cdot\nabla_v f=0. \tag{8.224}$$

Meanwhile, for simplicity, we assume that the electric field intensity $\boldsymbol{E}$ and the magnetic induction intensity $\boldsymbol{B}$ satisfy the following Maxwell equations in a vacuum (see (1.54)–(1.57) in Chapter 1 of Volume I):

$$\varepsilon_0\mu_0\frac{\partial \boldsymbol{E}}{\partial t}=\operatorname{rot}\boldsymbol{B}-\mu_0\boldsymbol{j}, \tag{8.225}$$

$$\frac{\partial \boldsymbol{B}}{\partial t}=-\operatorname{rot}\boldsymbol{E}, \tag{8.226}$$

$$\operatorname{div}\boldsymbol{E}=\frac{\rho}{\varepsilon_0}, \tag{8.227}$$

$$\operatorname{div}\boldsymbol{B}=0, \tag{8.228}$$

where ρ is the charge density, $\boldsymbol{j}$ is the current density, and ε_0 and μ_0 are the dielectric constant and the magnetic conductivity in vacuum, respectively. The charge density ρ can be given by the following formula using the distribution function f:

$$\rho(t,\boldsymbol{x})=q\int_{\mathbb{R}^3}f(t,\boldsymbol{x},\boldsymbol{v})\mathrm{d}\boldsymbol{v}. \tag{8.229}$$

Now we use the distribution function f again to determine the current density $\boldsymbol{j}$. The current is produced by the macroscopic transport of charged particles, and the macroscopic velocity of charged particles is exactly the average velocity of particles (see (8.5))

$$\boldsymbol{V}(t,\boldsymbol{x})=\frac{1}{n}\int_{\mathbb{R}^3}\boldsymbol{v}f(t,\boldsymbol{x},\boldsymbol{v})\mathrm{d}\boldsymbol{v}.$$

By definition (see section 1.2.2.1 in Chapter 1 of Volume I), the direction of the current density $\boldsymbol{j}$ is the same as that of $\boldsymbol{V}$, while the magnitude is equal to the quantity of charges passing across a unit area vertical to the $\boldsymbol{V}$ direction in unit time. Since the number of particles passing across a unit area vertical to the $\boldsymbol{V}$ direction in unit time is $n|\boldsymbol{V}|$, it is easy to show that the current density is

$$\boldsymbol{j}=q\int_{\mathbb{R}^3}\boldsymbol{v}f(t,\boldsymbol{x},\boldsymbol{v})\mathrm{d}\boldsymbol{v}. \tag{8.230}$$

The system constituted by equations (8.224)–(8.230) is a combination of the Vlasov equation describing the conservation of particle number and the Maxwell equations describing the change of electromagnetic field called the *Vlasov–Maxwell system*.

Before discussing the formulation of well-posed problems of the Vlasov–Maxwell system, we first prove two properties related to solutions of this system according to physical knowledge.

Assume that particles occupy the whole space $\mathbb{R}^3$; then the total number of particles is

$$N = \int_{\mathbb{R}^3}\int_{\mathbb{R}^3} f(t,\boldsymbol{x},\boldsymbol{v})\mathrm{d}x\mathrm{d}v. \tag{8.231}$$

From the conservation of particle number, this quantity should not change with time t. To be specific, we have the following theorem.

Theorem 8.3. *Suppose that* $(f,\boldsymbol{E},\boldsymbol{B})$ *is a* C^1 *solution of the Vlasov–Maxwell system* (8.224)–(8.230) *on* $t \geqslant 0$ *and is compactly supported with respect to variables* $\boldsymbol{x}$ *and* $\boldsymbol{v}$*; then*

$$N = constant, \quad \forall t \geqslant 0, \tag{8.232}$$

where N *is given by* (8.231).

Proof. Using (8.224), from (8.231) we get

$$\frac{\mathrm{d}N}{\mathrm{d}t} = -\int_{\mathbb{R}^3}\int_{\mathbb{R}^3}\Big(\boldsymbol{v}\cdot\nabla_x f + \frac{q}{M}(\boldsymbol{E}+\boldsymbol{v}\times\boldsymbol{B})\cdot\nabla_v f\Big)\mathrm{d}x\mathrm{d}v. \tag{8.233}$$

Since f is compactly supported with respect to $\boldsymbol{x}$ and $\boldsymbol{v}$, using Green's formula and noticing that $\boldsymbol{E}$ and $\boldsymbol{B}$ are functions only of t and $\boldsymbol{x}$, we obtain

$$\int_{\mathbb{R}^3}\int_{\mathbb{R}^3} \boldsymbol{v}\cdot\nabla_x f\mathrm{d}x\mathrm{d}v = 0, \tag{8.234}$$

$$\int_{\mathbb{R}^3}\int_{\mathbb{R}^3} \boldsymbol{E}\cdot\nabla_v f\mathrm{d}x\mathrm{d}v = 0, \tag{8.235}$$

and

$$\begin{aligned}&\int_{\mathbb{R}^3}\int_{\mathbb{R}^3}(\boldsymbol{v}\times\boldsymbol{B})\cdot\nabla_v f\mathrm{d}x\mathrm{d}v\\ &= -\int_{\mathbb{R}^3}\int_{\mathbb{R}^3}\mathrm{div}_v(\boldsymbol{v}\times\boldsymbol{B})f\mathrm{d}x\mathrm{d}v\\ &= -\int_{\mathbb{R}^3}\int_{\mathbb{R}^3}(\boldsymbol{B}\cdot\mathrm{rot}_v\,\boldsymbol{v})f\mathrm{d}x\mathrm{d}v = 0,\end{aligned} \tag{8.236}$$

where div_v and rot_v stand for the divergence and rotation with respect to variable $\boldsymbol{v}$, respectively. In obtaining (8.236), we used the vector analysis formula

$$\mathrm{div}(\boldsymbol{a}\times\boldsymbol{b}) = \boldsymbol{b}\cdot\mathrm{rot}\,\boldsymbol{a} - \boldsymbol{a}\cdot\mathrm{rot}\,\boldsymbol{b} \tag{8.237}$$

and $\mathrm{rot}_v \boldsymbol{v} = 0$. Substituting (8.235)–(8.236) into the right-hand side of (8.233), we get

$$\frac{\mathrm{d}N}{\mathrm{d}t} = 0, \quad \forall t \geqslant 0.$$

This proves the theorem. □

Now we check the total energy of the charged particle group. From (8.10), the average total kinetic energy per unit volume is

$$\int_{\mathbb{R}^3} \frac{M}{2} |\boldsymbol{v}|^2 f(t,\boldsymbol{x},\boldsymbol{v})\mathrm{d}v,$$

while, from section 1.4.1 in Chapter 1 of Volume I, the electromagnetic energy per unit volume, namely, the electromagnetic energy density, is

$$\frac{1}{2}\left(\varepsilon_0 |\boldsymbol{E}|^2 + \frac{1}{\mu_0}|\boldsymbol{B}|^2\right).$$

Then the total energy is

$$\mathcal{E} = \int_{\mathbb{R}^3}\int_{\mathbb{R}^3} \frac{M}{2}|\boldsymbol{v}|^2 f(t,\boldsymbol{x},\boldsymbol{v})\mathrm{d}x\mathrm{d}v + \int_{\mathbb{R}^3} \frac{1}{2}\left(\varepsilon_0 |\boldsymbol{E}|^2 + \frac{1}{\mu_0}|\boldsymbol{B}|^2\right)\mathrm{d}x. \tag{8.238}$$

By the conservation of energy, this quantity should not change with time either. Specifically, we have the following.

Theorem 8.4. *Suppose that $(f,\boldsymbol{E},\boldsymbol{B})$ is a C^1 solution of the Vlasov–Maxwell system* (8.224)–(8.230) *on $t \geqslant 0$ and is compactly supported with respect to variables $\boldsymbol{x}$ and $\boldsymbol{v}$; then*

$$\mathcal{E} = \textit{constant}, \quad \forall t \geqslant 0, \tag{8.239}$$

where $\mathcal{E}$ is given by (8.238).

Proof. Multiplying both sides of equation (8.224) by $|\boldsymbol{v}|^2$, and integrating over the whole space with respect to $\boldsymbol{x}$ and $\boldsymbol{v}$, we have

$$\begin{aligned} &\int_{\mathbb{R}^3}\int_{\mathbb{R}^3}\left(|\boldsymbol{v}|^2\frac{\partial f}{\partial t} + |\boldsymbol{v}|^2(\boldsymbol{v}\cdot\nabla_x f)\right)\mathrm{d}x\mathrm{d}v \\ &= -\int_{\mathbb{R}^3}\int_{\mathbb{R}^3}\frac{q}{M}|\boldsymbol{v}|^2(\boldsymbol{E}+\boldsymbol{v}\times\boldsymbol{B})\cdot\nabla_v f\mathrm{d}x\mathrm{d}v. \end{aligned} \tag{8.240}$$

Noting that f is compactly supported with respect to $\boldsymbol{x}$ and $\boldsymbol{v}$ and using Green's formula, we get

$$\int_{\mathbb{R}^3}\int_{\mathbb{R}^3}|\boldsymbol{v}|^2(\boldsymbol{v}\cdot\nabla_x f)\mathrm{d}x\mathrm{d}v = 0 \tag{8.241}$$

and

$$
\begin{aligned}
&\int_{\mathbb{R}^3}\int_{\mathbb{R}^3}\frac{q}{M}|\boldsymbol{v}|^2(\boldsymbol{E}+\boldsymbol{v}\times\boldsymbol{B})\cdot\nabla_v f\,\mathrm{d}x\mathrm{d}v\\
&=-\int_{\mathbb{R}^3}\int_{\mathbb{R}^3}2\frac{q}{M}\boldsymbol{v}\cdot(\boldsymbol{E}+\boldsymbol{v}\times\boldsymbol{B})f\,\mathrm{d}x\mathrm{d}v\\
&\quad-\int_{\mathbb{R}^3}\int_{\mathbb{R}^3}\frac{q}{M}|\boldsymbol{v}|^2\,\mathrm{div}_v(\boldsymbol{v}\times\boldsymbol{B})f\,\mathrm{d}x\mathrm{d}v\\
&=-2\frac{q}{M}\int_{\mathbb{R}^3}\int_{\mathbb{R}^3}\boldsymbol{E}\cdot\boldsymbol{v}f\,\mathrm{d}x\mathrm{d}v\\
&=-\frac{2}{M}\int_{\mathbb{R}^3}\boldsymbol{E}\cdot\boldsymbol{j}\mathrm{d}x, \qquad (8.242)
\end{aligned}
$$

in which we used

$$\boldsymbol{v}\cdot(\boldsymbol{v}\times\boldsymbol{B})=0,\quad \mathrm{div}_v(\boldsymbol{v}\times\boldsymbol{B})=0,$$

and equation (8.230). Plugging (8.241) and (8.242) into (8.240) and using equation (8.226) yields

$$
\begin{aligned}
&\frac{\mathrm{d}}{\mathrm{d}t}\int_{\mathbb{R}^3}\int_{\mathbb{R}^3}|\boldsymbol{v}|^2 f\,\mathrm{d}x\mathrm{d}v\\
&=\int_{\mathbb{R}^3}\int_{\mathbb{R}^3}|\boldsymbol{v}|^2\frac{\partial f}{\partial t}\mathrm{d}x\mathrm{d}v\\
&=\frac{2}{M}\int_{\mathbb{R}^3}\boldsymbol{E}\cdot\boldsymbol{j}\mathrm{d}x\\
&=\frac{2}{M}\int_{\mathbb{R}^3}\boldsymbol{E}\cdot\left(\frac{1}{\mu_0}\mathrm{rot}\,\boldsymbol{B}-\varepsilon_0\frac{\partial\boldsymbol{E}}{\partial t}\right)\mathrm{d}x\\
&=-\frac{1}{M}\frac{\mathrm{d}}{\mathrm{d}t}\int_{\mathbb{R}^3}\varepsilon_0|\boldsymbol{E}|^2\mathrm{d}x+\frac{2}{M}\int_{\mathbb{R}^3}\frac{1}{\mu_0}\boldsymbol{E}\cdot\mathrm{rot}\,\boldsymbol{B}\mathrm{d}x. \qquad (8.243)
\end{aligned}
$$

Since $\boldsymbol{E}$ and $\boldsymbol{B}$ are compactly supported with respect to $\boldsymbol{x}$, using the vector analysis formula (8.237) and Green's formula, it is not hard to verify that

$$\int_{\mathbb{R}^3}\boldsymbol{E}\cdot\mathrm{rot}\,\boldsymbol{B}\mathrm{d}x=\int_{\mathbb{R}^3}\mathrm{rot}\,\boldsymbol{E}\cdot\boldsymbol{B}\mathrm{d}x.$$

Thus, noting (8.226), we can rewrite (8.243) as

$$
\begin{aligned}
&\frac{\mathrm{d}}{\mathrm{d}t}\left(\int_{\mathbb{R}^3}\int_{\mathbb{R}^3}M|\boldsymbol{v}|^2 f\,\mathrm{d}x\mathrm{d}v+\int_{\mathbb{R}^3}\varepsilon_0|\boldsymbol{E}|^2\mathrm{d}x\right)\\
&=2\int_{\mathbb{R}^3}\frac{1}{\mu_0}\mathrm{rot}\,\boldsymbol{E}\cdot\boldsymbol{B}\mathrm{d}x\\
&=-2\int_{\mathbb{R}^3}\frac{1}{\mu_0}\frac{\partial\boldsymbol{B}}{\partial t}\cdot\boldsymbol{B}\mathrm{d}x\\
&=-\frac{\mathrm{d}}{\mathrm{d}t}\int_{\mathbb{R}^3}\frac{1}{\mu_0}|\boldsymbol{B}|^2\mathrm{d}x.
\end{aligned}
$$

It follows immediately from the above formula that

$$\frac{\mathrm{d}\mathcal{E}}{\mathrm{d}t} = 0, \quad \forall t \geqslant 0. \tag{8.244}$$

This proves the theorem. □

8.7.4 Mathematical Structures and Well-Posed Problems of Vlasov Coupled Systems

In the Vlasov–Poisson system, equation (8.217) is a first-order hyperbolic partial differential equation of unknown function f with coefficients depending on the gradient of the other unknown function ϕ with respect to $\boldsymbol{x}$, while equation (8.218) is a Poisson equation of unknown function ϕ with the right-hand side given by an integral of unknown function f. Therefore, this system is a nonlinear hyperbolic–elliptic coupled integrodifferential system.

Now we focus on the Vlasov–Maxwell system. First we point out that equations (8.227) and (8.228) in this system can be reduced to requirements on the initial data. Precisely, if at $t = 0$

$$f = f^0(\boldsymbol{x}, \boldsymbol{v}), \quad \boldsymbol{E} = \boldsymbol{E}^0(\boldsymbol{x}), \quad \boldsymbol{B} = \boldsymbol{B}^0(\boldsymbol{x}), \tag{8.245}$$

and these initial data satisfy compatibility conditions

$$\text{div } \boldsymbol{E}^0 = \frac{q}{\varepsilon_0} \int_{\mathbb{R}^3} f^0(\boldsymbol{x}, \boldsymbol{v}) \mathrm{d}v \tag{8.246}$$

and

$$\text{div } \boldsymbol{B}^0 = 0, \tag{8.247}$$

then when $t > 0$, it always holds that

$$\text{div } \boldsymbol{E} = \frac{q}{\varepsilon_0} \int_{\mathbb{R}^3} f(t, \boldsymbol{x}, \boldsymbol{v}) \mathrm{d}v \tag{8.248}$$

and

$$\text{div } \boldsymbol{B} = 0. \tag{8.249}$$

In fact, applying the divergence operator div on both sides of equation (8.226) and noticing that div rot $\boldsymbol{E} = 0$, we have

$$\frac{\partial}{\partial t} \text{div } \boldsymbol{B} = 0. \tag{8.250}$$

Then (8.247) leads to (8.249) immediately. Applying again the divergence operator div on both sides of equation (8.226) and noting (8.230), similarly we have

$$\begin{aligned}
\varepsilon_0 \mu_0 \frac{\partial}{\partial t} \text{div } \boldsymbol{E} &= -\mu_0 \, \text{div } \boldsymbol{j} \\
&= -\mu_0 q \int_{\mathbb{R}^3} \text{div}_x (\boldsymbol{v} f) \mathrm{d}v \\
&= -\mu_0 q \int_{\mathbb{R}^3} \boldsymbol{v} \cdot \nabla_x f \mathrm{d}v.
\end{aligned} \tag{8.251}$$

Using the Vlasov equation (8.224), we obtain

$$\varepsilon_0\mu_0\frac{\partial}{\partial t}\operatorname{div}\boldsymbol{E}=\mu_0 q\int_{\mathbb{R}^3}\frac{\partial f}{\partial t}\mathrm{d}v+\mu_0\frac{q^2}{M}\int_{\mathbb{R}^3}(\boldsymbol{E}+\boldsymbol{v}\times\boldsymbol{B})\cdot\nabla_v f\mathrm{d}v. \tag{8.252}$$

Since we always assume that f goes to zero at a sufficiently fast speed as $|\boldsymbol{v}|\to\infty$, similarly to (8.235) and (8.236), using Green's formula, we can easily verify that

$$\int_{\mathbb{R}^3}(\boldsymbol{E}+\boldsymbol{v}\times\boldsymbol{B})\cdot\nabla_v f\mathrm{d}v=0. \tag{8.253}$$

Then (8.252) can be rewritten as

$$\frac{\partial}{\partial t}\left(\operatorname{div}\boldsymbol{E}-\frac{q}{\varepsilon_0}\int_{\mathbb{R}^3}f(t,\boldsymbol{x},\boldsymbol{v})\mathrm{d}v\right)=0. \tag{8.254}$$

Thus, (8.250) follows immediately from (8.246).

Therefore, for the Vlasov–Maxwell system, we only need to consider the system constituted by equations (8.224)–(8.226) and (8.230). As shown before, the Vlasov equation (8.224) is a first-order hyperbolic partial differential equation of unknown function f with coefficients depending on unknown functions $\boldsymbol{E}$ and $\boldsymbol{B}$, while the Maxwell equations (8.226)–(8.226) constitute a first-order symmetric hyperbolic system of unknown functions $\boldsymbol{E}$ and $\boldsymbol{B}$ (see section 1.5 in Chapter 1 of Volume I), where $\boldsymbol{j}$ is given by an integral of unknown function f as in (8.230). Thus, the Vlasov–Maxwell system is a *system of first-order nonlinear hyperbolic integro-partial differential equations*.

Finally, we discuss the formulation of well-posed problems of the above coupled systems. Since in principle well-posed conditions on the unknown function ϕ in the Vlasov–Poisson system can be given according to the case of the Poisson equation, and well-posed conditions on unknown functions $\boldsymbol{E}$ and $\boldsymbol{B}$ can be given according to the case of the Maxwell equations (see Chapter 1 of Volume I), here we only need to focus on how to impose well-posed conditions on the Vlasov equation (8.212).

Well-posed problems of the Vlasov equation (8.212) fall into two types: the initial value problem and the initial-boundary value problem.

(1) *Initial value problem (Cauchy problem)*: Let $T>0$. Find a solution $f=f(t,\boldsymbol{x},\boldsymbol{v})$ of equation (8.212) in $(0,T)\times\mathbb{R}^3\times\mathbb{R}^3$ such that it satisfies at $t=0$ the initial condition

$$f(0,\boldsymbol{x},\boldsymbol{v})=f^0(\boldsymbol{x},\boldsymbol{v}),\quad\forall(\boldsymbol{x},\boldsymbol{v})\in\mathbb{R}^3\times\mathbb{R}^3, \tag{8.255}$$

where $f^0(\boldsymbol{x},\boldsymbol{v})$ is a given function.

Any solution $\boldsymbol{x}=\boldsymbol{x}(t)$, $\boldsymbol{v}=\boldsymbol{v}(t)$ of the ordinary differential equations

$$\frac{\mathrm{d}\boldsymbol{x}}{\mathrm{d}t}=\boldsymbol{v},\quad\frac{\mathrm{d}\boldsymbol{v}}{\mathrm{d}t}=\boldsymbol{g} \tag{8.256}$$

is called the characteristic curve of the Vlasov equation (8.212). From (8.212), the distribution function $f(t,\boldsymbol{x},\boldsymbol{v})$ remains constant along any characteristic curve. Therefore, the distribution function $f(t,\boldsymbol{x},\boldsymbol{v})$ has the same upper and lower bounds as the initial distribution $f^0(\boldsymbol{x},\boldsymbol{v})$. For the Vlasov equation (8.224), the characteristic equations (8.256) turn into

$$\frac{\mathrm{d}\boldsymbol{x}}{\mathrm{d}t}=\boldsymbol{v},\quad\frac{\mathrm{d}\boldsymbol{v}}{\mathrm{d}t}=\frac{q}{M}(\boldsymbol{E}+\boldsymbol{v}\times\boldsymbol{B}). \tag{8.257}$$

These are exactly the kinetic equations of charged particles moving in an electromagnetic field (see (1.64) in Chapter 1 of Volume I). By the way, in many practical problems, the velocity of particles may be quite large, which requires us to modify the Vlasov equation in the relativistic regime (see Exercises 13 and 14 in Chapter 9).

(2) *Initial-boundary value problem*: Let $T > 0$ and $\Omega \subset \mathbb{R}^3$. Find a solution $f = f(t, \boldsymbol{x}, \boldsymbol{v})$ of equation (8.212) in $(0,T) \times \Omega \times \mathbb{R}^3$ such that it satisfies at $t = 0$ the initial condition

$$f(0,\boldsymbol{x},\boldsymbol{v}) = f^0(\boldsymbol{x},\boldsymbol{v}), \quad \forall(\boldsymbol{x},\boldsymbol{v}) \in \Omega \times \mathbb{R}^3 \tag{8.258}$$

and satisfies a certain boundary condition on part of the boundary $\Gamma = \partial\Omega \times \mathbb{R}^3$. Here we point out that the variable $\boldsymbol{v}$ varies in the whole space $\mathbb{R}^3$.

To illustrate on which part of boundary Γ we should give boundary conditions and what kind of boundary conditions we should prescribe, we start by considering a simple example from the mathematical point of view.

Example 8.1. Consider the following one-dimensional Vlasov equation:

$$\frac{\partial f}{\partial t} + v\frac{\partial f}{\partial x} + g\frac{\partial f}{\partial v} = 0, \quad t > 0, x > 0, v \in \mathbb{R}. \tag{8.259}$$

We prescribe at $t = 0$ the initial condition

$$f(0,x,v) = f^0(x,v), \quad x > 0, v \in \mathbb{R}, \tag{8.260}$$

where $f^0(x,v)$ is a known function.

A characteristic curve $x = x(t)$, $v = v(t)$ of equation (8.259) satisfies

$$\frac{\mathrm{d}x}{\mathrm{d}t} = v, \quad \frac{\mathrm{d}v}{\mathrm{d}t} = g. \tag{8.261}$$

By (8.259), along any characteristic curve, we have

$$\frac{\mathrm{d}f}{\mathrm{d}t} = 0. \tag{8.262}$$

For simplicity, we assume that $g \equiv 0$. Thus, the characteristic curve passing through point $(0,\alpha,\beta)$ in (t,x,v) space is a straight line

$$x = \beta t + \alpha, \quad v = \beta, \tag{8.263}$$

and along this characteristic line, we have

$$f = f^0(\alpha,\beta) = f^0(x - vt, v). \tag{8.264}$$

Now we check the direction of this characteristic line as t increases. It is not hard to see from (8.263) that if $\beta > 0$, the characteristic line points in the direction along which x increases, and then points to the interior of the domain of the solution on the boundary $x = 0$, while if $\beta < 0$, the characteristic line points in the direction along which x decreases, and then points to the exterior of the domain of the solution on the boundary $x = 0$. Therefore, from (8.264), using the initial conditions given for $x > 0$ and $v \in \mathbb{R}$, we can determine the solution

$$f = f^0(x - vt, v) \tag{8.265}$$

in and only in domains

$$t > 0, \quad x > 0, \quad v \leqslant 0 \tag{8.266}$$

and

$$t > 0, \quad x > vt, \quad v > 0. \tag{8.267}$$

To determine the value of solution f in domain

$$t > 0, \quad 0 \leqslant x < vt, \quad v > 0, \tag{8.268}$$

appropriate boundary conditions should be prescribed on the boundary $x = 0$, $v > 0$. For instance, we assume a boundary condition on this part of the boundary as

$$f(t,0,v) = g^0(t,v) \quad \text{for } v > 0. \tag{8.269}$$

Noting that the characteristic curve through $t = \tau$, $x = 0$, $v = \beta$ is

$$x = \beta(t-\tau), \quad v = \beta, \tag{8.270}$$

and along this characteristic curve

$$f = g^0(\tau,\beta), \tag{8.271}$$

we can determine the value of f in domain (8.268) as

$$f = g^0\left(t - \frac{x}{v}, v\right). \tag{8.272}$$

Thus, we obtain the solution of the well-posed problem (8.259)–(8.260) and (8.269). Of course, to find a continuous and even smooth solution, $f^0(x,v)$ and $g^0(t,v)$ should satisfy some compatibility conditions. ■

From the above example, the boundary condition of form (8.269) should only be given on the part of the boundary satisfying

$$\boldsymbol{v} \cdot \boldsymbol{n} < 0, \tag{8.273}$$

where $\boldsymbol{n}$ stands for the outward unit normal vector on the boundary $\partial\Omega$ of domain Ω.

Moreover, it is not hard to show by further discussion of the above example that on the other part of the boundary under discussion, i.e., the boundary satisfying $v < 0$, we prescribe a boundary condition of the following form:

$$f(t,0,v) = f(t,0,-v) \quad \text{as } v < 0, \tag{8.274}$$

while, giving no boundary condition on the part of the boundary corresponding to $v > 0$, we can still determine uniquely the solution f in the whole domain. A boundary condition like (8.274) corresponds to a specular reflection of particles from the boundary (see (8.279) below). Certainly, this kind of reflection occurs only on the part of the boundary satisfying

$$\boldsymbol{v} \cdot \boldsymbol{n} > 0. \tag{8.275}$$

Now we discuss the general case. Denote

$$\Gamma_- = \{(\boldsymbol{x},\boldsymbol{v}) \in \Gamma, \boldsymbol{v} \cdot \boldsymbol{n} < 0\}, \tag{8.276}$$

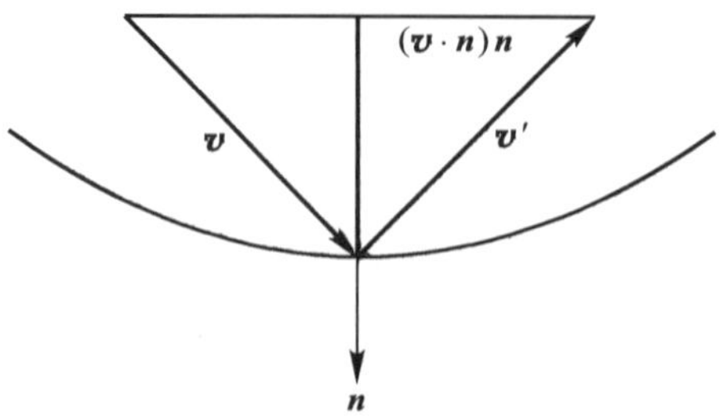

Figure 8.7

where $\boldsymbol{n}$ is the outward unit normal vector on the boundary $\partial\Omega$ of domain Ω. Since the velocity $\boldsymbol{v}$ of particles on Γ_- points to the interior of the domain of the solution, there exist particles entering the interior of the domain of the solution from outside. At this time, the distribution function of incoming particles should be given, so the boundary condition should take the following form:

$$f(t,\boldsymbol{x},\boldsymbol{v}) = f_b(t,\boldsymbol{x},\boldsymbol{v}), \quad t > 0, (\boldsymbol{x},\boldsymbol{v}) \in \Gamma_-, \tag{8.277}$$

where $f_b(t,\boldsymbol{x},\boldsymbol{v})$ is a given function.

Now we consider the case that the boundary is a fixed wall. Generally speaking, the boundary layer of gases (or other particle groups) is very complicated in the case of a fixed-wall boundary. Gas molecules could be reflected and absorbed on the boundary layer and could even form chemical bonds to turn into ions. Their states depend not only on the temperature of the boundary but also on how bright and clean the boundary is. Either the theoretical analyses or the experimental data are insufficient about this. An ideal situation is the specular reflection of gas from the fixed-wall boundary. Assume that the velocity of a molecule with an incident velocity $\boldsymbol{v}$ turns into

$$\boldsymbol{v}' = \boldsymbol{v} - 2(\boldsymbol{v}\cdot\boldsymbol{n})\boldsymbol{n} \tag{8.278}$$

after a specular reflection from the boundary (see Figure 8.7). Since the incident velocity $\boldsymbol{v}$ and the reflected velocity $\boldsymbol{v}'$ always appear in pairs, the corresponding boundary conditions should be

$$f(t,\boldsymbol{x},\boldsymbol{v}) = f(t,\boldsymbol{x},\boldsymbol{v} - 2(\boldsymbol{v}\cdot\boldsymbol{n})\boldsymbol{n}), \quad t > 0, (\boldsymbol{x},\boldsymbol{v}) \in \Gamma_+, \tag{8.279}$$

where

$$\Gamma_+ = \{(\boldsymbol{x},\boldsymbol{v}) \in \Gamma, \boldsymbol{v}\cdot\boldsymbol{n} > 0\}. \tag{8.280}$$

The well-posed problem of the Vlasov–Poisson system and the Vlasov–Maxwell system has been a hot topic in the research of applied partial differential equations over the past few decades. For the Vlasov–Maxwell system (including both relativistic and nonrelativistic cases), the existence of global classical solutions to the Cauchy problem with small initial data was proved (see [5]), but it is still an open and deep problem for the case of general initial data. The existence of global weak solutions was also obtained (see [8]), but the uniqueness of weak solutions is still open. For the reader who is interested in the existence of global weak solutions to the Cauchy problem of the Boltzmann equation with general collision term, we refer to [4].

Exercises

1. Suppose that the gas molecules are hard spheres of radius r. Find the differential scattering cross section of the molecular hard-sphere collision and prove that the total cross section is $\sigma_{\text{tot}} = 4\pi r^2$.
2. Suppose that $f(t,\boldsymbol{x},\boldsymbol{v})$ is a C^1 solution of the Boltzmann equation
$$\frac{\partial f}{\partial t} + \boldsymbol{v} \cdot \nabla_x f = J$$
and f is compactly supported with respect to variables $\boldsymbol{x}$ and $\boldsymbol{v}$ for any given $t \geqslant 0$. Prove that the integral
$$\int_{\mathbb{R}^3} \int_{\mathbb{R}^3} f(t,\boldsymbol{x},\boldsymbol{v}) |\boldsymbol{x} - t\boldsymbol{v}|^2 \mathrm{d}x \mathrm{d}v$$
is a constant independent of $t \geqslant 0$.
3. Suppose that $\widetilde{J}$ is an integral operator defined by (8.156) in the Hilbert space H. Prove that this operator is a nonnegative self-adjoint operator.
4. Using expression (8.167) of $\widetilde{a}_i$ and $\widetilde{b}_{ij}$, prove directly that (8.176) is true for any given conserved quantity χ.
5. Suppose that particles in a particle group interact with each other by gravitational force (long-range interaction), and collisions are ignored. Prove that the distribution function $f(t,\boldsymbol{x},\boldsymbol{v})$ of particles and the potential $\phi(t,\boldsymbol{x})$ of the gravitational field satisfy the following Vlasov–Poisson system:
$$\frac{\partial f}{\partial t} + \boldsymbol{v} \cdot \nabla_x f + \nabla_x \phi \cdot \nabla_v f = 0,$$
$$-\Delta_x \phi = \int_{\mathbb{R}^3} f(t,\boldsymbol{x},\boldsymbol{v}) \mathrm{d}v.$$
6. Suppose that $f(t,\boldsymbol{x},\boldsymbol{v})$ is a continuous and differentiable distribution function of particles determined by the Vlasov–Maxwell system and f is compactly supported with respect to variables $\boldsymbol{x}$ and $\boldsymbol{v}$ for any given $t \geqslant 0$. Additionally, suppose that ϕ is a smooth function and $\phi(0) = 0$. Prove that the integral
$$\int_{\mathbb{R}^3} \int_{\mathbb{R}^3} \phi(f(t,\boldsymbol{x},\boldsymbol{v})) \mathrm{d}x \mathrm{d}v$$
is a constant that does not change with time $t \geqslant 0$.
7. Solve the following Cauchy problem of the one-dimensional Vlasov equation:
$$\frac{\partial f}{\partial t} + v\frac{\partial f}{\partial x} + g\frac{\partial f}{\partial v} = 0, \quad t > 0, (x,v) \in \mathbb{R}^2,$$
$$f(0,x,v) = f^0(x,v),$$
where g is a constant and $f^0(x,v)$ is a suitably smooth given function.

8. Solve the following initial-boundary value problem of the one-dimensional Vlasov equation:

$$\begin{aligned}
&\frac{\partial f}{\partial t} + v\frac{\partial f}{\partial x} = 0, \quad t > 0, x > 0, v \in \mathbb{R},\\
&f(0,x,v) = f^0(x,v), \quad x > 0, v \in \mathbb{R},\\
&f(t,0,v) = f(t,0,-v), \quad t > 0, v < 0,
\end{aligned}$$

where $f^0(x,v)$ is a suitably smooth given function satisfying the compatibility condition

$$f^0(0,v) = f^0(0,-v), \quad v < 0.$$

Bibliography

[1] Bond J. W. Jr., Watson K. M., Welch J. A. Jr. *Atomic Theory of Gas Dynamics*. Reading, MA: Addison-Wesley, 1965.

[2] Chapman S., Cowling T. G. *The Mathematical Theory of Non-Uniform Gases*. Cambridge: Cambridge University Press, 1958.

[3] Cercignani C. *The Boltzmann Equation and Its Applications*. New York: Springer–Verlag, 1988.

[4] DiPerna R., Lions P. L. *On the Cauchy problem for Boltzmann equations: Global existence and weak stability results*. Annals of Math., 130 (1989), 321–366.

[5] Glassey R. T., Strauss W. A. *Absence of shocks in an initial dilute, collisionless plasma*. Comm. Math. Phys., 113 (1987), 191–208.

[6] Huo Y. P., Zheng J. R. *Non-equilibrium Statistical Physics* (in Chinese). Beijing: Science Press, 1987.

[7] Lions P. L. *Modèles Mathématiques des Phénomènes de Transport*. Paris: Ecole Polytechnique, 1994.

[8] Lions P. L., DiPerna R. *Global solutions of Vlasov–Maxwell system*. Comm. Pure Appl. Math., 42 (1989), 729–757.

[9] McLennan J. A. *Introduction to Non-Equilibrium Statistical Mechanics*. Englewood Cliffs, NJ: Prentice Hall, 1989.

[10] Strauss W. A. *Nonlinear Wave Equations*. Providence, RI: American Mathematical Society, 1989.

Chapter 9

Special Relativity and Relativistic Fluid Dynamics

9.1 Introduction

9.1.1 The Principle of Relativity and the Galilean Transformation

Manifestations of physical phenomena in different reference frames may be different. For example, when discussing an electromagnetic field generated by a point charge, we must be sure whether this point charge is at rest or in motion. If this point charge is at rest, it only generates an electrostatic field, while if it is in motion, it also generates a magnetic field in addition to an electric field. In order to determine whether this point charge is at rest or in motion, we must first determine in which reference frame it is. However, a point charge at rest in one reference frame may be moving in another reference frame. Therefore, the above physical phenomenon manifests differently in different reference frames. Since the selection of a reference frame can be quite arbitrary, and the same physical phenomenon can take different forms in different reference frames, should physical laws that reflect objective laws of physical phenomena take different forms in different reference frames? If this is true, the forms of physical laws describing the same physical phenomenon will be of a wide variety and will lose their objective criteria, which is clearly unreasonable. Therefore, physical laws should take the same form in all corresponding reference frames—this is the *principle of relativity*.

First let us look at classical Newtonian mechanics. We know that the manifestations of Newton's laws are valid in all inertial frames of reference, and a so-called inertial frame of reference means a frame in which Newton's first law (the law of inertia) is valid. Any reference frame that is in a state of rectilinear motion at a constant velocity with respect to a certain inertial frame is still an inertial frame. In this way, Newton's mechanics satisfies the following *principle of relativity*: physical laws take the same form in all inertial frames. It can be described mathematically as follows: the form of physical laws remains the same when an inertial frame is converted to another under a space-time transformation. As to the form of the space-time transformation between inertial frames, it depends on our concepts about time and space. In Newtonian mechanics, time and space are considered in an absolute sense: they do not depend on the reference frame chosen. In other words, both the time interval between two events and the spatial distance between two points on a rigid body are independent of the reference frame chosen. At this point, the space-time transformation between inertial frames of reference is called the *Galilean transformation*, and the corresponding principle of relativity is called the *Galilean principle of relativity*.

Suppose that K and $\overline{K}$ are two inertial frames, where the time-space coordinates are (t, x_1, x_2, x_3) and $(\overline{t}, \overline{x}_1, \overline{x}_2, \overline{x}_3)$, respectively, and that $\overline{K}$ is in rectilinear motion at a constant velocity $\boldsymbol{V}$ with respect to K. For simplicity, we may assume that the axis directions are the same in these two reference frames. Thus, as long as we choose appropriate origins of time and space, the Galilean transformation can be written as

$$\begin{cases} \overline{t} = t, \\ \overline{\boldsymbol{x}} = \boldsymbol{x} - \boldsymbol{V}t, \end{cases} \tag{9.1}$$

where $\boldsymbol{x} = (x_1, x_2, x_3)$, and so forth.

Here we show that the form of Newton's equation (Newton's second law) remains the same under the Galilean transformation (9.1). Suppose that the form of Newton's equation in reference frame K is

$$m\frac{\mathrm{d}^2\boldsymbol{x}}{\mathrm{d}t^2} = \boldsymbol{f}. \tag{9.2}$$

Under the Galilean transformation (9.1), we have, obviously,

$$\frac{\mathrm{d}^2\overline{\boldsymbol{x}}}{\mathrm{d}\overline{t}^2} = \frac{\mathrm{d}^2\boldsymbol{x}}{\mathrm{d}t^2}. \tag{9.3}$$

In addition, in Newtonian mechanics, it is also assumed that the particle mass is independent of the reference frame, that is,

$$\overline{m} = m, \tag{9.4}$$

where $\overline{m}$ stands for the particle mass in reference frame $\overline{K}$. As for the force $\boldsymbol{f}$ acting on the particle, it is in general a function of time (e.g., the periodic force), relative distance (e.g., universal gravitation), or relative speed (e.g., friction). Under a Galilean transformation, the time and the relative distance clearly remain unchanged, and it is easy to verify that the relative speed remains unchanged as well. Therefore, the acting force is the same in different inertial frames, that is,

$$\overline{\boldsymbol{f}} = \boldsymbol{f}, \tag{9.5}$$

where $\overline{\boldsymbol{f}}$ stands for the force in reference frame $\overline{K}$. Combining (9.3)–(9.5), we have

$$\overline{m}\frac{\mathrm{d}^2\overline{\boldsymbol{x}}}{\mathrm{d}\overline{t}^2} = \overline{\boldsymbol{f}}. \tag{9.6}$$

This indicates that Newton's equation takes the same form in inertial frame $\overline{K}$ as in inertial frame K. This is also the case for Newton's third law. Therefore, Newtonian mechanics satisfies the Galilean principle of relativity, and equations in classical mechanics preserve their forms under a Galilean transformation.

9.1.2 Maxwell's Theory of Electromagnetic Fields and the Galilean Principle of Relativity

Now we investigate Maxwell's equations (see Chapter 1 of Volume I) describing the laws of electromagnetic fields. It is known that in the absence of charge and current, the electric field intensity $\boldsymbol{E}$ and the magnetic induction intensity $\boldsymbol{B}$ in a vacuum satisfy the wave equations

$$\frac{1}{c^2}\frac{\partial^2 \boldsymbol{E}}{\partial t^2} - \Delta \boldsymbol{E} = 0 \tag{9.7}$$

and

$$\frac{1}{c^2}\frac{\partial^2 \boldsymbol{B}}{\partial t^2} - \Delta \boldsymbol{B} = 0 \tag{9.8}$$

(see (1.111) and (1.112) in Chapter 1 of Volume I). These two equations tell us that the electromagnetic field propagates in the form of waves at velocity $c = 3 \times 10^8$ m/s in a vacuum. Suppose that the electromagnetic equations in inertial frame K take the above form; then we can verify that the original form of equations (9.7) and (9.8) will not be preserved any longer under the Galilean transformation (9.1), which converts K to another inertial frame $\overline{K}$. We can conclude the same by a direct physical analysis. As a matter of fact, because of the symmetry of equations (9.7) and (9.8) with respect to space variables x_1, x_2, x_3, electromagnetic waves propagate in all directions at the same speed c in inertial frame K, and the propagation speed c is isotropic. But, according to the Galilean transformation (9.1), when the reference frame $\overline{K}$ moves at the velocity $\boldsymbol{V}$ with respect to K, from the superposition of velocities, electromagnetic waves should propagate at different speeds along different directions in reference frame $\overline{K}$ (along the direction of $\boldsymbol{V}$, the propagation speed is $c - |\boldsymbol{V}|$, and along the opposite direction of $\boldsymbol{V}$, the propagation speed is $c + |\boldsymbol{V}|$). This feature must be reflected in equations in reference frame $\overline{K}$; thus Maxwell's equations cannot still take the forms (9.7) and (9.8) in $\overline{K}$. This implies that Maxwell's equations have different forms in different inertial frames and so do not satisfy the Galilean principle of relativity.

Thus, we are faced with the following three possible options:

1. *Maxwell's equations are not correct.* Correct electromagnetic field equations should be invariant under the Galilean transformation.

2. *The Galilean principle of relativity applies only to classical mechanics, but not to electromagnetics.* In electromagnetics, there exists a special and preferred reference frame, and electromagnetic field equations in this reference frame take the form of Maxwell's equations.

3. *The principle of relativity remains a universally applicable physical law, but fundamental changes should be made to the notions of time and space.* We cannot consider time and space in an absolute sense; on the contrary, both time and length are likely to change with a transformation from one inertial frame to another. In other words, transformations between inertial frames are no longer Galilean transformations.

Since the time when Maxwell established the electromagnetic equations in 1861 to the end of the 19th century, this theory has won amazing success thanks to the efforts of H. R. Hertz, H. A. Lorentz, and others, and has been generally accepted by people, so it is almost impossible to deny it. Therefore, we are left with options 2 and 3. Before A. Einstein, an overwhelming majority of physicists accepted option 2, trying to amend the existing theory. Then what is this special and preferred reference frame in which Maxwell's equations are valid?

We know that mechanical waves propagate in media. Let us take a certain one-dimensional small longitudinal vibration of an elastic rod as an example; it satisfies the equation

$$\frac{\partial^2 u}{\partial t^2} - a^2 \frac{\partial^2 u}{\partial x^2} = 0 \tag{9.9}$$

(see, e.g., [7]), where $a^2 = E/\rho$, E is the elastic modulus of the material, and ρ is the linear density of that rod. At that time physicists, regarding electromagnetic waves as mechanical waves, thought that there existed a medium called the ether in a vacuum through which electromagnetic waves propagate at speed c, and Maxwell's equations are satisfied only in this inertial frame of the ether. At this point, since the value of c is very large, the ether should be a transparent medium with very large elastic modulus and very small density, in which light propagates at speed c as an electromagnetic wave.

However, if this assumption is accepted, for bodies (e.g., the earth) moving relative to the ether, the propagation speed of light (electromagnetic wave) should be different from c. In addition, by the difference between light speeds measured on the earth and in the ether, the speed of the earth relative to the ether can also be deduced. But the actual measurement result is quite different from these expectations. In 1887, an experiment by A. A. Michelson and E. W. Morley denied the possibility of measuring the speed of the earth relative to the ether based on the difference of light speed.

Hence, 3 becomes the only possible option. That is, the principle of relativity is valid not only for classical mechanics but also for electromagnetic phenomena (including light). Thus, Maxwell's equations should take the same form in all inertial frames, and the light speed c in a vacuum takes the same value in all inertial frames. As mentioned before, the latter conclusion is completely opposite to the traditional Newtonian notions of time and space, which initiated Einstein's theory of special relativity.

9.2 Fundamental Principles in Special Relativity; the Lorentz Transformation

9.2.1 Fundamental Postulates of Special Relativity

In 1905, at the young age of 26, when he worked as an examiner at the Swiss Patent Office, Albert Einstein analyzed the achievement of physics at that time and published an epochal paper entitled "On the Electrodynamics of Moving Bodies" in the most prestigious journal, *Annalen der Physik*, of Germany. This paper, together with another of his papers published in the same year, laid the foundation for and marked the birth of special relativity.

Einstein's theory depends on the following two fundamental postulates:

(1) *The principle of relativity*: Physical laws take the same form in all inertial frames.

This principle and the Galilean principle of relativity are not really different in form. But in the Galilean principle of relativity, the corresponding change of time and space coordinates under which an inertial frame is converted to another is given by a Galilean transformation. This is because absolute time and space are assumed to exist; that is, the time interval and length (distance between two points) do not change with different choices of inertial frames. Now since this principle of relativity is adopted, Maxwell's equations hold in all inertial frames and, naturally, Einstein replaced the absoluteness of time and space with the invariance of the speed of light, which violates classical Newtonian mechanics.

(2) *The principle of invariance of light speed*: Light propagates in a vacuum at a definite velocity c along any direction in all inertial frames.

9.2.2 Lorentz Transformation

The Lorentz transformation, which is the core and mathematical expression of special relativity, was proposed by Dutch physicist H. A. Lorentz in 1904 after generalizing his research results on electromagnetic fields over many years. The famous French scientist H. Poincaré esteemed Lorentz's work greatly and called this transformation the Lorentz transformation.

Under the above two postulates of Einstein's, we now figure out how the corresponding space-time coordinates (t,x_1,x_2,x_3) and $(\overline{t},\overline{x}_1,\overline{x}_2,\overline{x}_3)$ are related when an inertial frame K is converted to another inertial frame $\overline{K}$. Suppose

$$\begin{cases} \overline{t} = f_0(t,x_1,x_2,x_3), \\ \overline{x}_i = f_i(t,x_1,x_2,x_3) \quad (i=1,2,3). \end{cases} \tag{9.10}$$

We have the following conclusions:

(1) *Transformation* (9.10) *is linear.* Only in this way can we ensure any body in uniform rectilinear motion with respect to K is also in uniform rectilinear motion with respect to $\overline{K}$. This reflects the homogeneity of time and space. Thus, except for the translation, (9.10) should be an invertible linear transformation of the following form:

$$\overline{x}_\alpha = \sum_{\beta=0}^{3} a_{\alpha\beta} x_\beta \quad (\alpha = 0,\ldots,3), \tag{9.11}$$

where $a_{\alpha\beta}$ are constants, and for simplicity of expression, t and $\overline{t}$ are denoted by x_0 and $\overline{x}_0$, respectively.

(2) $\sum_{i=1}^{3}(\Delta x_i)^2-(c\Delta t)^2$ *is an invariant under the above transformation.* In fact, according to the principle of invariance of light speed, the wavefront, at the instant $t+\Delta t$, of the light emitting at the instant t from a point located at (x_1,x_2,x_3) in the inertial frame K is a sphere with radius $c\Delta t$ centered at (x_1,x_2,x_3):

$$\sum_{i=1}^{3}(\Delta x_i)^2-(c\Delta t)^2=0, \tag{9.12}$$

where $\Delta x_i = X_i - x_i$ $(i=1,2,3)$, and $(X_1,X_2,X_3)=(x_1+\Delta x_1, x_2+\Delta x_2, x_3+\Delta x_3)$ is the moving point. When converted to the frame $\overline{K}$, $(t,x_1,x_2,x_3)\to(\overline{t},\overline{x}_1,\overline{x}_2,\overline{x}_3)$, $(t+\Delta t, x_1+\Delta x_1, x_2+\Delta x_2, x_3+\Delta x_3)\to(\overline{t}+\Delta\overline{t}, \overline{x}_1+\Delta\overline{x}_1, \overline{x}_2+\Delta\overline{x}_2, \overline{x}_3+\Delta\overline{x}_3)$, and according to the principle of invariance of light speed, the wavefront now is the sphere

$$\sum_{i=1}^{3}(\Delta\overline{x}_i)^2-(c\Delta\overline{t})^2=0. \tag{9.13}$$

Considering that transformation (9.11) is linear, and wavefront should correspond to wavefront, from (9.12) and (9.13) we get

$$\sum_{i=1}^{3}(\Delta\overline{x}_i)^2-(c\Delta\overline{t})^2=k\left(\sum_{i=1}^{3}(\Delta x_i)^2-(c\Delta t)^2\right), \tag{9.14}$$

where k is a constant. Since the roles of the two inertial frames K and $\overline{K}$ are completely symmetric, we have similarly

$$\sum_{i=1}^{3}(\Delta x_i)^2-(c\Delta t)^2=k\left(\sum_{i=1}^{3}(\Delta\overline{x}_i)^2-(c\Delta\overline{t})^2\right).$$

Then $k^2=1$, i.e., $k=\pm1$. Noticing that (9.14) is a relation between two quadratic forms under the invertible transformation (9.11), from the inertial theorem of quadratic forms, we have $k=1$. This proves conclusion (2).

Since (9.11) is a homogeneous linear transformation, conclusion (2) reduces equivalently to the conclusion that

$$\sum_{i=1}^{3}x_i^2-(ct)^2 \tag{9.15}$$

is an invariant of transformation (9.11).

A linear transformation (9.11) satisfying (2), i.e., with an invariant (9.15), is called a *Lorentz transformation*.

To obtain the general form of the Lorentz transformation, we first check a special case: x_2 and x_3 remain unchanged in the transformation, i.e., we suppose that the transformation is

$$\begin{cases}\overline{x}_0=a_{00}x_0+a_{01}x_1,\\ \overline{x}_1=a_{10}x_0+a_{11}x_1,\\ \overline{x}_2=x_2,\\ \overline{x}_3=x_3.\end{cases} \tag{9.16}$$

Substituting the above transformation into the expression (9.15) of the invariant, we get

$$(a_{10}x_0+a_{11}x_1)^2-c^2(a_{00}x_0+a_{01}x_1)^2=x_1^2-c^2x_0^2.$$

Comparing the coefficients of x_0^2, x_0x_1, and x_1^2 on both sides of the above formula, respectively, we have

$$\begin{cases}a_{10}^2-c^2a_{00}^2=-c^2,\\ a_{10}a_{11}-c^2a_{00}a_{01}=0,\\ a_{11}^2-c^2a_{01}^2=1.\end{cases} \tag{9.17}$$

It is not hard to see from (9.16) that the reference frame $\overline{K}$ determined by this transformation is in uniform rectilinear motion with speed

$$V=-\frac{a_{10}}{a_{11}} \tag{9.18}$$

along the x_1 direction with respect to the reference frame K. Using the second formula in (9.17), from its first and third formulas we have

$$a_{00}^2 = a_{11}^2,$$
$$a_{01}^2 = \frac{1}{c^4} a_{10}^2.$$

Using (9.18) and the above two formulas, we can solve from the first and the third formulas in (9.17) that

$$a_{00}^2 = a_{11}^2 = \gamma^2,$$
$$a_{01}^2 = \frac{1}{c^4} a_{10}^2 = \frac{\gamma^2 V^2}{c^4},$$

where $\gamma = 1\Big/\sqrt{1-\left(\frac{V}{c}\right)^2}$. From this, V defined by (9.18) must satisfy $|V| < c$. Since we may take reflection transformations of x_0 and x_1 (they are all Lorentz transformations) if necessary, we can always assume $a_{00} > 0$ and $a_{11} > 0$. Thus, we have

$$a_{00} = \gamma, \quad a_{11} = \gamma.$$

Using (9.18) and the second formula in (9.17), we further have

$$a_{10} = -\gamma V, \quad a_{01} = -\gamma \frac{V}{c^2}.$$

Substituting the above formula into (9.16), we obtain a special kind of Lorentz transformation,

$$\begin{cases} \overline{x}_0 = \gamma\left(x_0 - \dfrac{V}{c^2} x_1\right), \\ \overline{x}_1 = \gamma(-V x_0 + x_1), \\ \overline{x}_2 = x_2, \\ \overline{x}_3 = x_3, \end{cases} \tag{9.19}$$

where V is a real constant satisfying $|V| < c$ and $\gamma = 1\Big/\sqrt{1-\left(\frac{V}{c}\right)^2}$. Conversely, for any given real constant V satisfying $|V| < c$ (the corresponding value of γ can be inferred), (9.19) yields a Lorentz transformation.

Once we have a Lorentz transformation of special form (9.19), we can give general Lorentz transformations by the following theorem.

Theorem 9.1. *Any Lorentz transformation is a composition of an orthogonal transformation, a transformation of the form* (9.19)*, and a reflection transformation (a transformation that changes the sign of a certain variable) with respect to variables* x_1, x_2, x_3 *(*x_0 *is kept unchanged).*

Proof. Suppose that a Lorentz transformation is given by (9.11).

We first check the case that not all of a_{01}, a_{02}, and a_{03} are zero. Apply an orthogonal transformation to variables x_1, x_2, and x_3: $(x_1, x_2, x_3) \to (y_1, y_2, y_3)$, such that

$$a_{01}x_1 + a_{02}x_2 + a_{03}x_3 = \alpha y_1, \tag{9.20}$$

where α is a suitably chosen nonzero constant. This can always be done. Let $y_0 = x_0$. Since $x_1^2 + x_2^2 + x_3^2$ is an invariant under orthogonal transformations, the transformation

$$(x_0, x_1, x_2, x_3) \to (y_0, y_1, y_2, y_3)$$

is a Lorentz transformation.

By the above Lorentz transformation, (9.11) reduces to

$$\begin{cases} \overline{x}_0 = a_{00} y_0 + \alpha y_1, \\ \overline{x}_1 = a_{10} y_0 + b_{11} y_1 + b_{12} y_2 + b_{13} y_3, \\ \overline{x}_2 = a_{20} y_0 + b_{21} y_1 + b_{22} y_2 + b_{23} y_3, \\ \overline{x}_3 = a_{30} y_0 + b_{31} y_1 + b_{32} y_2 + b_{33} y_3. \end{cases} \tag{9.21}$$

This is still a Lorentz transformation. Now we have

$$c^2 \alpha^2 < a_{00}^2. \tag{9.22}$$

In fact, from the definition of the Lorentz transformation, we have

$$\overline{x}_1^2 + \overline{x}_2^2 + \overline{x}_3^2 - c^2 \overline{x}_0^2 = y_1^2 + y_2^2 + y_3^2 - c^2 y_0^2. \tag{9.23}$$

In particular, take $\overline{x}_0 = 0$. From the first formula of (9.21) we have $y_0 = -\frac{\alpha}{a_{00}} y_1$. Plugging this into (9.23), we get

$$\overline{x}_1^2 + \overline{x}_2^2 + \overline{x}_3^2 = \left(1 - \frac{c^2 \alpha^2}{a_{00}^2}\right) y_1^2 + y_2^2 + y_3^2.$$

The right-hand side of the above formula must be positive as $y_1^2 + y_2^2 + y_3^2 > 0$. Otherwise, we have $\overline{x}_1 = \overline{x}_2 = \overline{x}_3 = 0$. Noting furthermore that $\overline{x}_0 = 0$, we get $y_1 = y_2 = y_3 = 0$, a contradiction. Therefore, we have

$$1 - \frac{c^2 \alpha^2}{a_{00}^2} > 0,$$

which is just (9.22).

Let $V = -c^2 \alpha / a_{00}$. We know from (9.22) that $|V| < c$. To this given V (thus the corresponding γ), apply a Lorentz transformation of form (9.19):

$$\begin{cases} \overline{y}_0 = \gamma \left(y_0 - \frac{V}{c^2} y_1 \right), \\ \overline{y}_1 = \gamma(-V y_0 + y_1), \\ \overline{y}_2 = y_2, \\ \overline{y}_3 = y_3. \end{cases} \tag{9.24}$$

Noticing that the first formula of (9.24) can be written as

$$\overline{y}_0 = \frac{\gamma}{a_{00}} (a_{00} y_0 + \alpha y_1),$$

and plugging (9.24) into (9.21), we obtain

$$\begin{cases} \overline{x}_0 = c_{00}\overline{y}_0, \\ \overline{x}_1 = c_{10}\overline{y}_0 + c_{11}\overline{y}_1 + c_{12}\overline{y}_2 + c_{13}\overline{y}_3, \\ \overline{x}_2 = c_{20}\overline{y}_0 + c_{21}\overline{y}_1 + c_{22}\overline{y}_2 + c_{23}\overline{y}_3, \\ \overline{x}_3 = c_{30}\overline{y}_0 + c_{31}\overline{y}_1 + c_{32}\overline{y}_2 + c_{33}\overline{y}_3, \end{cases} \tag{9.25}$$

where $c_{00} = a_{00}/\gamma$. It is still a Lorentz transformation.

Therefore, if a_{01}, a_{02}, and a_{03} in (9.11) are not all zero, (9.11) can always be reduced to a transformation of form (9.25) by an orthogonal transformation with respect to x_1, x_2, and x_3 and a Lorentz transformation of form (9.19), while if $a_{01} = a_{02} = a_{03} = 0$ in (9.11), (9.11) already has the form of (9.25). Thus, we need only discuss the Lorentz transformation of form (9.25).

Taking in (9.25) particularly $\overline{y}_0 = 1$ and $\overline{y}_1 = \overline{y}_2 = \overline{y}_3 = 0$, we have

$$\overline{x}_0 = c_{00}, \quad \overline{x}_1 = c_{10}, \quad \overline{x}_2 = c_{20}, \quad \overline{x}_3 = c_{30}.$$

Plugging them into (9.15), we get

$$c^2 c_{00}^2 - c_{10}^2 - c_{20}^2 - c_{30}^2 = c^2. \tag{9.26}$$

Then it follows immediately that

$$c_{00}^2 \geqslant 1. \tag{9.27}$$

On the other hand, taking particularly $\overline{x}_0 = 1$ and $\overline{x}_1 = \overline{x}_2 = \overline{x}_3 = 0$ in (9.25) leads to $\overline{y}_0 = 1/c_{00}$, and taking corresponding values $\widetilde{y}_1$, $\widetilde{y}_2$, $\widetilde{y}_3$ of $\overline{y}_1$, $\overline{y}_2$, $\overline{y}_3$ and plugging them into (9.15), we obtain

$$\frac{c^2}{c_{00}^2} - \widetilde{y}_1^2 - \widetilde{y}_2^2 - \widetilde{y}_3^2 = c^2.$$

So we also have

$$c_{00}^2 \leqslant 1. \tag{9.28}$$

Inequalities (9.27) and (9.28) yield $c_{00}^2 = 1$. Then (9.26) gives

$$c_{10} = c_{20} = c_{30} = 0.$$

Thus, transformation (9.25) should actually take the following form:

$$\begin{cases} \overline{x}_0 = \pm\overline{y}_0, \\ \overline{x}_1 = c_{11}\overline{y}_1 + c_{12}\overline{y}_2 + c_{13}\overline{y}_3, \\ \overline{x}_2 = c_{21}\overline{y}_1 + c_{22}\overline{y}_2 + c_{23}\overline{y}_3, \\ \overline{x}_3 = c_{31}\overline{y}_1 + c_{32}\overline{y}_2 + c_{33}\overline{y}_3. \end{cases} \tag{9.29}$$

From this special form of the Lorentz transformation we know that $\overline{y}_1^2+\overline{y}_2^2+\overline{y}_3^2$ is an invariant; i.e., the transformation $(\overline{y}_1,\overline{y}_2,\overline{y}_3)\to(\overline{x}_1,\overline{x}_2,\overline{x}_3)$ given by the last three formulas in (9.29) is an orthogonal transformation, and the first formula with the negative sign in (9.29) is a reflection with respect to the variable $\overline{y}_0$. The theorem is proved. □

Using the above theorem, it is not difficult to verify that the wave equation

$$\frac{1}{c^2}\frac{\partial^2 u}{\partial t^2}-\Delta u=0 \tag{9.30}$$

takes the same form under Lorentz transformations. Conversely, we can prove that any invertible linear transformation (9.11) preserving the form of the wave equation (9.30) must be a composition of a Lorentz transformation and a similarity transformation $\overline{x}_i=ax_i$ $(i=0,1,2,3)$ $(a\neq 0$ is a constant) (see [14]).

As stated above, Lorentz transformation (9.19) is a transformation of space-time coordinates between the inertial frame $\overline{K}$ (with space-time coordinates $(\overline{x}_0,\overline{x}_1,\overline{x}_2,\overline{x}_3)$) and the inertial frame K (with space-time coordinates (x_0,x_1,x_2,x_3)) when $\overline{K}$ is in uniform rectilinear motion along the x_1 direction at speed V with respect to K. Since we will need it in section 9.5 below, now we further discuss what the form is of the transformation of space-time coordinates between the inertial frames $\overline{K}$ and K when $\overline{K}$ is in uniform rectilinear motion at velocity $\boldsymbol{V}$ with respect to K. The following theorem answers this question.

Theorem 9.2. *Suppose that the space-time coordinates of the inertial frames K and $\overline{K}$ coincide with each other at a certain instant (e.g., $x_0=0$ $(t=0)$), and $\overline{K}$ is in rectilinear motion at a constant velocity $\boldsymbol{V}=(V_1,V_2,V_3)$ with respect to K. Denote by (x_0,x_1,x_2,x_3) and $(\overline{x}_0,\overline{x}_1,\overline{x}_2,\overline{x}_3)$ the space-time coordinates of K and $\overline{K}$, respectively (see Figure 9.1); then the relation between them is given by the following Lorentz transformation:*

$$\begin{aligned}\overline{x}_0&=\gamma\left(x_0-\frac{V_1}{c^2}x_1-\frac{V_2}{c^2}x_2-\frac{V_3}{c^2}x_3\right),\\ \overline{x}_i&=-\gamma V_i x_0+\sum_{j=1}^{3}\left(\frac{\gamma-1}{V^2}V_iV_j+\delta_{ij}\right)x_j \quad (i=1,2,3),\end{aligned} \tag{9.31}$$

where $\gamma=1/\sqrt{1-\left(\frac{V}{c}\right)^2}$, $V=|\boldsymbol{V}|$, and δ_{ij} is the Kronecker symbol.

Proof. First apply a rotation transformation to the space coordinates while keeping the time coordinate x_0 unchanged, such that the positive y_1 direction is the same as that of $\boldsymbol{V}$. This transformation can be written as

$$\begin{cases} y_0=x_0,\\ y_i=\displaystyle\sum_{j=1}^{3}a_{ij}x_j \quad (i=1,2,3), \end{cases} \tag{9.32}$$

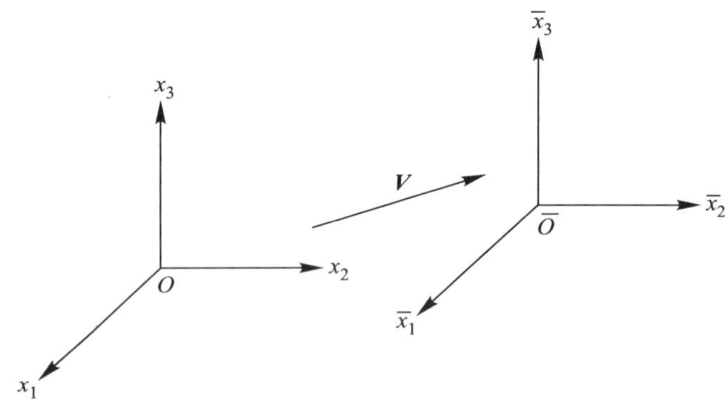

Figure 9.1

where (a_{ij}) is an orthogonal matrix,

$$\det(a_{ij}) = 1, \tag{9.33}$$

and

$$a_{11} = \frac{V_1}{V}, \quad a_{12} = \frac{V_2}{V}, \quad a_{13} = \frac{V_3}{V}. \tag{9.34}$$

Take the transformation of form (9.32) in the inertial frames K and $\overline{K}$, and denote the obtained space-time coordinates by (y_0, y_1, y_2, y_3) and $(\overline{y}_0, \overline{y}_1, \overline{y}_2, \overline{y}_3)$, respectively. Under the new coordinates, $\overline{K}$ is in uniform rectilinear motion with respect to K along the y_1 direction at speed V. From (9.19) and (9.32) we have

$$\begin{cases} \overline{y}_0 = \gamma\left(y_0 - \dfrac{V}{c^2}y_1\right) = \gamma\left(x_0 - \dfrac{V}{c^2}\displaystyle\sum_{j=1}^{3} a_{1j}x_j\right), \\ \overline{y}_1 = \gamma(-Vy_0 + y_1) = \gamma\left(-Vx_0 + \displaystyle\sum_{j=1}^{3} a_{1j}x_j\right), \\ \overline{y}_2 = y_2 = \displaystyle\sum_{j=1}^{3} a_{2j}x_j, \\ \overline{y}_3 = y_3 = \displaystyle\sum_{j=1}^{3} a_{3j}x_j. \end{cases} \tag{9.35}$$

Now transform $(\overline{y}_0, \overline{y}_1, \overline{y}_2, \overline{y}_3)$ back to the original $(\overline{x}_0, \overline{x}_1, \overline{x}_2, \overline{x}_3)$ space-time coordinates. From the corresponding (9.32), and noticing that the inverse matrix of an orthogonal matrix is its transpose, we obtain

$$
\begin{aligned}
\overline{x}_i &= \sum_{j=1}^{3} a_{ji}\overline{y}_j \\
&= a_{1i}\gamma\left(-Vx_0 + \sum_{j=1}^{3} a_{1j}x_j\right) \\
&\quad + a_{2i}\sum_{j=1}^{3} a_{2j}x_j + a_{3i}\sum_{j=1}^{3} a_{3j}x_j \\
&= -\gamma a_{1i}Vx_0 + (\gamma-1)\sum_{j=1}^{3} a_{1i}a_{1j}x_j + \sum_{j,k=1}^{3} a_{ki}a_{kj}x_j \\
&= -\gamma V_i x_0 + \sum_{j=1}^{3}\left(\frac{\gamma-1}{V^2}V_iV_j + \delta_{ij}\right)x_j \quad (i = 1,2,3).
\end{aligned} \tag{9.36}
$$

Here, we used (9.34) and the orthogonality of matrix (a_{ij}). This proves the second formula of (9.31).

Noting $\overline{x}_0 = \overline{y}_0$ and (9.34), the first formula of (9.31) follows immediately from the first formula of (9.35). □

It is easy to show that the Lorentz transformation (9.31) reduces to the Galilean transformation (9.1) as $c \to \infty$.

9.3 Space-Time Theory in Special Relativity

9.3.1 Relativity of Simultaneity

In classical Newtonian space-time, time is absolute. Sir Isaac Newton wrote in his famous book *Mathematical Principles of Natural Philosophy* that "Absolute, true and mathematical time, of itself, and from its own nature flows equably without regard to anything external." According to this notion of time, it has clear meaning at least in theory saying that two events occur in different locations at the same time. In relativity, is it possible to say in an absolute sense (without the reference frame) that two distinct events occurring at different points in space are simultaneous?

We first point out that the order of priority in time (including simultaneity) for two events to occur at the same point in space in a certain reference frame remains invariant in any reference frame. This can easily be shown by the Lorentz transformation formula (9.19) (see also formula (9.37) below).

For two events that are at different points in space, we check starting from an example. Suppose that a train moves at a constant speed V along the rail. Do two events occurring simultaneously with respect to the railbed (e.g., flashes of light given off simultaneously from A and B) still occur simultaneously with respect to the train?

As a matter of fact, when we say that the pulses of light emitted from A and B are simultaneous with respect to the railbed, it means that the pulses of light emitted from A and B meet at the center point M of the interval $[A,B]$ on the railbed. Suppose that A

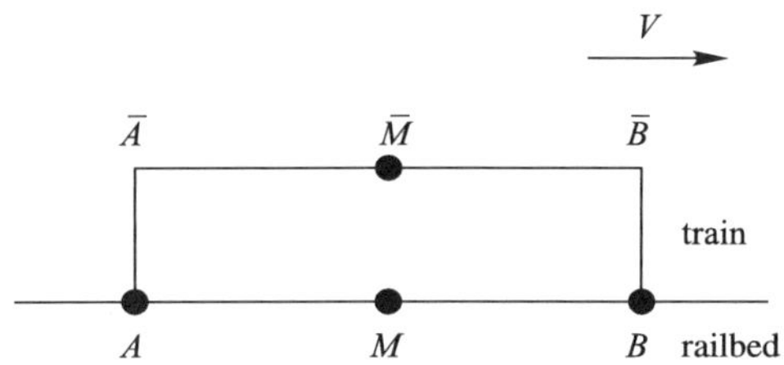

Figure 9.2

and B just coincide with points $\overline{A}$ and $\overline{B}$ on the train, respectively, at the instant when the pulses of light are emitted, and that $\overline{M}$ is the center point of the interval $[\overline{A},\overline{B}]$ on the train. In the reference frame of the train, if these two events still occur simultaneously, according to the constancy of the speed of light, the pulses of light emitted from A and B should meet at $\overline{M}$. But in the reference frame of the railbed, $\overline{M}$ moves to the right at speed V (see Figure 9.2), facing the light ray coming from B. Therefore, the light ray emitted from B should arrive at $\overline{M}$ first. Since the order of priority in time for two events to occur at the same point in space in a certain reference frame remains invariant in any reference frame, if we take the train as a reference frame, the occurrence of the light pulse at B precedes that at A, so these two events are not simultaneous.

Now we analyze the above phenomenon by using the Lorentz transformation. Assume that we have two reference frames K and $\overline{K}$, with $\overline{K}$ moving with respect to K along the x_1 direction at a constant speed V. Suppose the time and space coordinates of A and B are (t^A,x^A) and (t^B,x^B) in reference frame K, respectively, and $(\bar{t}^A,\bar{x}^A)$ and $(\bar{t}^A,\bar{x}^B)$ in $\overline{K}$, respectively. Since the motion is along the x_1 direction, for simplicity of presentation, we omit the other two space coordinates x_2 and x_3, and denote x_1 by x. From the Lorentz transformation (9.19) we have

$$\bar{t}^A = \gamma\left(t^A - \frac{V}{c^2}x^A\right),$$
$$\bar{t}^B = \gamma\left(t^B - \frac{V}{c^2}x^B\right).$$

So

$$\bar{t}^B - \bar{t}^A = \gamma\left(t^B - t^A - \frac{V}{c^2}(x^B - x^A)\right). \tag{9.37}$$

If the two events A and B are simultaneous in K, i.e., $t^A = t^B$, we have

$$\bar{t}^B - \bar{t}^A = \gamma\frac{V}{c^2}(x^A - x^B). \tag{9.38}$$

Let $V > 0$. From the above formula we can see that if $x^B > x^A$, then $\bar{t}^B < \bar{t}^A$; thus for frame $\overline{K}$, event B occurs earlier than A, while if $x^B < x^A$, then $\bar{t}^A < \bar{t}^B$; thus for frame $\overline{K}$, event A occurs earlier than B. This is consistent with the analysis in the train example.

From the above analysis we can see that there is no common time for different reference frames; the simultaneity of two events is meaningless in an absolute sense. This indicates that two events happening in two different locations and occurring simultaneously in one inertial frame may occur nonsimultaneously in another inertial frame. This is just the relativity of simultaneity, which is the foundation of Einstein's theory of time and space.

We further ask if the following phenomenon can happen: event A occurs earlier than B in one inertial frame K, while event B occurs earlier than A in another inertial frame $\overline{K}$. If this is true, then does the phenomenon of causal order reversing happen?

Now we show that if two events are causally connected, i.e., one event causes the other, then the causal order between them is preserved. We discuss this problem by using the Lorentz transformation. In formula (9.37) suppose that $t^A < t^B$, i.e., A occurs earlier than B in frame K; then $\bar{t}^A > \bar{t}^B$ holds, i.e., event B occurs earlier than A in frame $\overline{K}$ if and only if

$$t^B - t^A < \frac{V}{c^2}(x^B - x^A), \tag{9.39}$$

i.e.,

$$\frac{x^B - x^A}{t^B - t^A} V > c^2. \tag{9.40}$$

Assume that the speed of transmission of all matters, actions, or effects does not exceed the speed of light c (see section 9.3.5); then for two causally connected events A and B, we have

$$\left| \frac{x^B - x^A}{t^B - t^A} \right| \leqslant c. \tag{9.41}$$

Moreover, we have

$$|V| \leqslant c.$$

So (9.40) cannot be true.

The above result can be revealed more clearly by using the notion of a light cone. For a given event (t^A, x^A), we can see from condition (9.41) that an event (t, x) that may be causally connected with this event is given by

$$(x - x^A)^2 - c^2(t - t^A)^2 \leqslant 0. \tag{9.42}$$

The domain determined by this is the domain (see the shaded domain in Figure 9.3) lying between the two characteristic lines

$$x - x^A = \pm c(t - t^A) \tag{9.43}$$

passing through point (t^A, x^A) of the following wave equation on the (t, x) plane:

$$\frac{1}{c^2}\frac{\partial^2 u}{\partial t^2} - \frac{\partial^2 u}{\partial x^2} = 0.$$

Events represented by points inside domain I must occur earlier than the event represented by point (t^A, x^A), so they can cause event (t^A, x^A), while events represented by points inside domain II must occur later than the event represented by point (t^A, x^A), so they can be caused by event (t^A, x^A). Those events represented by points outside the domain are not causally connected with event (t^A, x^A): their time orders with respect to event (t^A, x^A) are meaningful only in a relative sense, and time order may be reversed when transforming from one inertial frame to another.

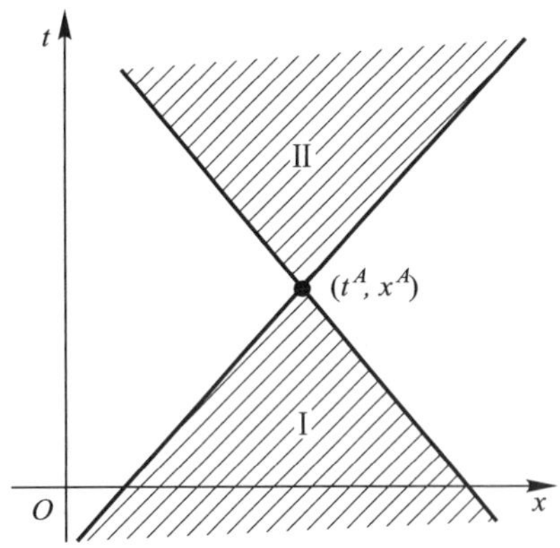

Figure 9.3

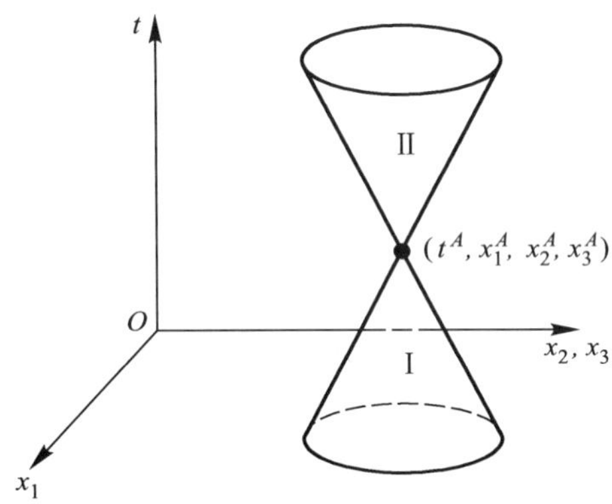

Figure 9.4

In three-dimensional space, given an event A represented by space-time coordinates $(t^A, x_1^A, x_2^A, x_3^A)$, the characteristic cone passing through this point of wave equation (9.30) is (see [7])

$$(x_1 - x_1^A)^2 + (x_2 - x_2^A)^2 + (x_3 - x_3^A)^2 - c^2(t - t^A)^2 = 0. \tag{9.44}$$

Points in domains I and II (see Figure 9.4) correspond to events occurring earlier than or later than A, respectively, that is, events that may cause or be caused by event A. The characteristic cone (9.44) is called the *light cone*.

9.3.2 Time Dilation of Motion

The absoluteness of time in classical physics, i.e., the time interval between two events is independent of the motion state of the reference frame, is one of the bases on which the Galilean transformation is derived. The relativity of simultaneity discussed above shows that there does not exist absolute time without the reference frame according to the space-time theory in relativity. Now we further discuss the impact of motion on time (interval).

Consider a physical process, e.g., the elastic vibration of a crystal. Suppose that the period of vibration of a crystal at rest is T_0; we check the impact of the motion of the crystal on the period of vibration. Assume that the crystal moves with respect to the inertial frame K at a constant speed V along the x_1 (denoted by x) direction, and remains at rest with respect to the inertial frame $\overline{K}$, i.e., $\overline{K}$ is the comoving reference frame with respect to the crystal. Assume furthermore that two successive events at which the amplitude of vibration of the crystal reaches the maximum are represented by the space-time coordinates (t^A, x^A) and (t^B, x^B) in the frame K, respectively, and by $(\bar{t}^A, \bar{x}^A)$ and $(\bar{t}^B, \bar{x}^B)$ in frame $\overline{K}$, respectively. Since the crystal is at rest with respect to the frame $\overline{K}$, then $\bar{x}^A = \bar{x}^B$. From the first and second formulas in the Lorentz transformation (9.19), we find that

$$t = \gamma\left(\bar{t} + \frac{V}{c^2}\bar{x}\right). \tag{9.45}$$

This can also be obtained by the assumption that K moves with respect to $\overline{K}$ along the x_1 direction at speed V. From (9.45) we have

$$t^A = \gamma\left(\bar{t}^A + \frac{V}{c^2}\bar{x}^A\right),$$

$$t^B = \gamma\left(\bar{t}^B + \frac{V}{c^2}\bar{x}^B\right).$$

Subtracting the first formula above from the second, and noting $\bar{x}^A = \bar{x}^B$, we get

$$t^B - t^A = \gamma(\bar{t}^B - \bar{t}^A). \tag{9.46}$$

Since the crystal is at rest in the inertial frame $\overline{K}$, we have $\bar{t}^B - \bar{t}^A = T_0$. Then the period of vibration of the crystal when it is moving at speed V, i.e., the period of vibration of the crystal measured in the inertial frame K, is

$$T = t^B - t^A = \gamma T_0. \tag{9.47}$$

Because $\gamma > 1$, the above formula implies that the period of vibration of the crystal in motion is dilated compared to the crystal at rest.

We have similar results for the time interval between any two events. The time dilation of moving bodies, known as "moving clock running slow," is the basic nature of space-time for any physical process taking place on moving bodies, which has been verified by a great deal of experiments in elementary particle physics.

9.3.3 Length Contraction along the Direction of Motion

Another important foundation of the classical Newtonian theory of space-time is that the distance between two points on a rigid body, namely, the length, does not change with the motion state of the body and does not depend on the choice of reference frame either, which is also a fundamental basis for deriving the Galilean transformation. In the theory of relativity, is this conclusion still true?

Assume that a rigid rod with a proper length (i.e., the length of the rod at rest) l_0 moves with respect to the inertial frame K at a constant speed V along the direction parallel to its length, and $\overline{K}$ is the comoving reference frame, i.e., the rod is at rest with respect to $\overline{K}$. Now we check if the measurements of the length of the rod are the same in the inertial frames K and $\overline{K}$.

Without loss of generality, suppose that the direction of the rod, i.e., the direction along which the rod is moving with respect to the frame K, is the direction of the x_1 (denoted by x) axis, and that the space-time coordinates of the two endpoints of the rod measured in the frame K at the same time t are (t, x^A) and (t, x^B), respectively. Thus, the length of the rod in the frame K is

$$l = x^B - x^A. \tag{9.48}$$

Now we check in the frame $\overline{K}$. Suppose that the two events corresponding to the two endpoints of the rod in the above measurement can be represented by the space-time coordinates $(\bar{t}^A, \bar{x}^A)$ and $(\bar{t}^B, \bar{x}^B)$ in $\overline{K}$, respectively. From the relativity of simultaneity, these two events may not occur simultaneously in $\overline{K}$, namely, $\bar{t}^A = \bar{t}^B$ may not be true; However, the rod is at rest in the frame $\overline{K}$, so we should have

$$l_0 = \bar{x}^B - \bar{x}^A. \tag{9.49}$$

From the second formula in the Lorentz transformation (9.19) we get

$$\bar{x}^A = \gamma(-Vt + x^A)$$

and

$$\bar{x}^B = \gamma(-Vt + x^B);$$

then

$$\bar{x}^B - \bar{x}^A = \gamma(x^B - x^A).$$

Noting (9.48) and (9.49), we have

$$l = \gamma^{-1} l_0. \tag{9.50}$$

This implies that when the rod is moving along the direction of its length, the length of the rod is shortened to $\frac{1}{\gamma}$ (< 1) times the length at rest. Similarly, when the rod is not aligned along the direction of motion, the length of the rod along the direction of motion is still shortened to $\frac{1}{\gamma}$ (< 1) times the corresponding length at rest. In particular, if the rod is still aligned along the direction of the x_1 axis, and its direction of movement is along the x_2 axis or the x_3 axis, then from the Lorentz transformation (9.19), we can easily see that the length of the rod is not contracted (but its thickness is contracted). Thus, the length contraction along the direction of motion, known as "moving ruler getting shorter," is also a basic nature of space-time.

9.3.4 Minkowski Four-Space-Time

In 1907, H. Minkowski, a famous mathematician who taught Einstein at the Swiss Federal Polytechnic in Zurich, started to deliver some academic lectures on the theory of relativity. He thought that "Einstein's profound theory was mathematically imprecise" and developed special relativity in the form of tensors. Minkowski's theoretical expressions created a powerful tool for the development of Einstein's special relativity to general relativity later.

Here we give a brief introduction to the basic concepts of Minkowski four-space-time, and more detailed discussions will be given in the next section.

The Lorentz transformation is a linear transformation in a four-dimensional real vector space. In this real vector space, defining the inner product of its elements $x = (x_0, x_1, x_2, x_3)$ and $y = (y_0, y_1, y_2, y_3)$ by the law

$$\langle x, y\rangle = c^2 x_0 y_0 - x_1 y_1 - x_2 y_2 - x_3 y_3, \tag{9.51}$$

we obtain the *Minkowski four-space-time*, denoted by M.

We need to point out that the inner product $\langle\cdot,\cdot\rangle$ of Minkowski four-space-time M, defined by (9.51), does not satisfy the positivity condition in the usual sense; in other words, the quadratic form $\langle x, x\rangle$ is not positive definite.

An element (point) in Minkowski four-space-time M is called an *event*. Suppose that $x = (x_0, x_1, x_2, x_3)$, $y = (y_0, y_1, y_2, y_3) \in M$ are two events; the Minkowski distance between them is defined as

$$r = \sqrt{\langle y-x, y-x\rangle} = \sqrt{c^2(y_0-x_0)^2 - \sum_{i=1}^{3}(y_i - x_i)^2}. \tag{9.52}$$

Of course, this distance does not always assume real values. From the definition of the Lorentz transformation we know that the Minkowski distance between two events in Minkowski four-space-time is an invariant under the Lorentz transformation. Moreover, from the previous causality analysis of two events we can see that two events in Minkowski four-space-time are causally connected if and only if their Minkowski distance is a real value.

For simplicity, an event is denoted by a capital Latin letter in what follows.

Theorem 9.3. *Suppose that A, B, and C are three events causally connected with each other in the Minkowski space-time M (as shown in Figure 9.5); then we have*

$$r_{AC} \geqslant r_{AB} + r_{BC}, \tag{9.53}$$

where r_{AB} stands for the Minkowski distance between events A and B, etc.

The proof of Theorem 9.3 is standard and is left as an exercise for the reader.

This theorem shows that in the Minkowski space-time, the sum of any two sides of a triangle is shorter than the third one. Formula (9.53) is also called the *reversed triangle inequality*.

Now we discuss the physical meaning of the Minkowski distance. Suppose that A and B are two events in M, and B is caused by A. For simplicity, let the space-time coordinates of A and B be $(0,0,0,0)$ and (t, x_1, x_2, x_3), respectively. Thus, the Minkowski distance between A and B is

$$r_{AB} = c\tau_{AB}, \tag{9.54}$$

where

$$\tau_{AB} = t\sqrt{1 - \sum_{i=1}^{3}\left(\frac{x_i}{ct}\right)^2}. \tag{9.55}$$

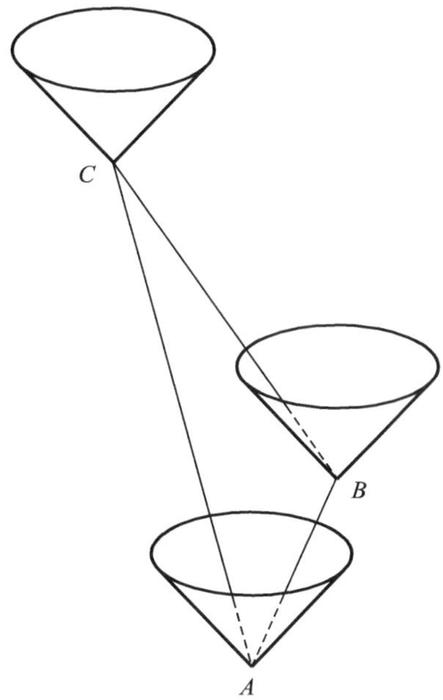

Figure 9.5

In the chosen reference frame, t is the time difference between events A and B, and $\sqrt{\sum_{i=1}^{3} x_i^2}$ is the spatial distance between A and B (hereafter, we use the same letter to describe the event and the corresponding point in space as long as no confusion occurs). Therefore, if a particle starts in uniform rectilinear motion from point A at the instant 0 along the direction of AB at the speed

$$V = \frac{\sqrt{\sum_{i=1}^{3} x_i^2}}{t}, \tag{9.56}$$

then this particle arrives at point B at the instant t. Then (9.55) reads as

$$\tau_{AB} = t\sqrt{1-\left(\frac{V}{c}\right)^2}. \tag{9.57}$$

By (9.46) and the definition of γ therein, we know that τ_{AB} given by the above formula is just the time spent on traveling from A to B in the reference frame of the moving particle.

Theorem 9.3 and the physical meaning of the Minkowski distance can be used to explain the so-called twin paradox: one of the twin brothers makes a journey to some other planet at close to the speed of light and returns home. It is found that the twin who remained on the earth has aged more.

In fact, let A represent the event that the traveling twin starts from the earth, while B represents the event that he arrives at another planet by a rocket at speed V (> 0), and the event that he returns to earth by this rocket at speed V is denoted by C. From the perspective of people on the earth (the stay-at-home twin), the time spent on this interplanetary round trip is τ_{AC}; but from the standpoint of the traveling twin (in his own reference frame), the time spent on this trip is calculated as $\tau_{AB}+\tau_{BC}$. By Theorem 9.3, we know that

$$\tau_{AC} \geqslant \tau_{AB}+\tau_{BC},$$

and it is easy to verify that the above inequality is strict on this occasion. We actually have

$$\tau_{AC} = \gamma(\tau_{AB}+\tau_{BC}), \tag{9.58}$$

where $\gamma = 1/\sqrt{1-\left(\frac{V}{c}\right)^2}$ (see Exercise 7). This implies the correctness of the twin paradox.

The twin paradox was proposed by A. Einstein in a paper in 1905, about which there had always been vigorous debate. People who disagreed with this thought that since the traveling twin moves relative to the earth-based twin, then according to the relativity of motion, the earth-based twin also moves relative to the traveling twin, so the above conclusion cannot be inferred. But in fact, since the traveling twin leaves the earth for a space trip and returns home, he must undergo acceleration, so the twins' situations are not symmetric.

Until 1966, twins travel experiments were done in the laboratory using muons: let a muon move around a circular orbit of diameter 14 m back to the starting point. This experiment shows that the traveling muon is indeed younger than the at-rest muon. If one day humans are capable of manufacturing a carrier with a speed close to the speed of light, then the fantasy of "a few days here, a thousand years there" will be brought to life.

9.3.5 Prohibition of Motion Faster than Light in a Vacuum

The Lorentz transformation can be regarded as the relation of the space-time coordinates between two inertial frames. Since the space-time coordinates all take real values, the coefficients of the Lorentz transformation must be real, which requires that $|V| < c$. This implies that the relative velocity between two inertial frames will not be greater than the speed of light c. Due to the arbitrariness of the choice of reference frame, the speed of motion of any object and the propagation speed of any signal, and so on, will not be greater than the speed of light c. As stated in section 9.3.1, this preserves the order of causality between two events.

The conclusion that any speed cannot be greater than the speed of light is fundamental to special relativity. Therefore, if a faster-than-light event is found, that would deny special relativity. Since special relativity was founded, people have kept trying to find faster-than-light events, and there have been some positive reports. However, no faster-than-light event has been confirmed so far by the scientific community.

9.4 Relativistic Dynamics

Just as we stated in section 9.1, the classical Newtonian equations are invariant under the Galilean transformation. Then what is the case under the Lorentz transformation? For instance, is the Newtonian equation

$$m\frac{\mathrm{d}^2\boldsymbol{x}}{\mathrm{d}t^2} = \boldsymbol{f}$$

invariant under the Lorentz transformation? To obtain the relevant conclusion, we only need to do some intuitive analysis without examining directly the behavior of the above equation under the Lorentz transformation. In fact, from the Newtonian equation, the acceleration speed of an object is proportional to the external force; then, under the action of a constant force, the speed of motion of this object will increase linearly and will finally be greater than the speed of light c, which is not permitted in relativity. This tells us that the Newtonian equation cannot have Lorentz invariance.

To establish a relativistic dynamics equation, at least the following two kinds of conditions should be required: one is that the form of this equation is invariant under the Lorentz transformation, and the other is that this equation is consistent with the classical one when the speed of motion of the object is far less than the speed of light c. The latter requirement is proposed because the correctness of the classical mechanics equations has been sufficiently verified by various experiments and our daily experiences. That is, we need to reconstruct relativistic mechanics equations on the basis of classical mechanics equations. To this end, the physical quantities involved should be scalars, vectors, and tensors with respect to the Lorentz transformation in the Minkowski four-space-time M.

9.4.1 Tensors in Minkowski Four-Space-Time

Different from the Cartesian tensor (see Appendix A in Volume I), tensors in the Minkowski four-space-time M can be classified as contravariant tensors, covariant tensors, and mixed tensors; refer to Appendix C in this volume. Although the usage of the above tensors will make the discussion of relativistic fluid dynamics convenient, we introduce only contravariant tensors here to avoid involving too much about tensors.

For convenience, just as in general references about the theory of relativity, denote

$$x^0 = ct, \quad x^i = x_i \quad (i = 1,2,3) \tag{9.59}$$

(comparing with notations $x_0 = t$ and x_i $(i = 1,2,3)$ in section 9.2, we have $x^0 = cx_0$). Thus, the inner product for $x = (x^0, x^1, x^2, x^3)$ and $y = (y^0, y^1, y^2, y^3) \in M$ is defined as (cf. (9.51))

$$\langle x, y\rangle = x^0 y^0 - x^1 y^1 - x^2 y^2 - x^3 y^3 \tag{9.60}$$

or

$$\langle x, y\rangle = g_{\alpha\beta} x^\alpha x^\beta, \tag{9.61}$$

where

$$(g_{\alpha\beta}) = \begin{pmatrix} 1 & & & \\ & -1 & & \\ & & -1 & \\ & & & -1 \end{pmatrix}, \tag{9.62}$$

and all the unwritten elements in this matrix are zeros (hereafter, we always use this convention), and the Einstein summation notation is adopted: repeated superscript and subscript imply summation. Hereafter, we use the common convention that the Greek indices $\alpha, \beta, \dots$ take values from 0 to 3, and the Latin indices $i, j, \dots$ take values from 1 to 3.

If $x, y \in M$ and $\langle x, y\rangle = 0$, then x and y are said to be *orthogonal* to each other. If $x \in M$ and $\langle x, x\rangle = \pm 1$, then we say that x is a *unit vector* in M.

Let L be a linear transformation from M to M. If the inner product of any two elements in M is invariant under this transformation, i.e.,

$$\langle Lx, Ly \rangle = \langle x, y \rangle, \quad \forall x, y \in M, \tag{9.63}$$

then the linear transformation L is called an *orthogonal transformation* in the Minkowski space-time M.

We can see immediately from the definition of the Lorentz transformation that an orthogonal transformation in the Minkowski space-time M must be a Lorentz transformation. Conversely, it is not hard to prove that a Lorentz transformation must be an orthogonal transformation in M (see Exercise 3). Therefore, the inner product in M is invariant under the Lorentz transformation.

Suppose that the linear transformation L is given by the matrix $A = (a^\alpha_\beta)$:

$$(Lx)^\alpha = a^\alpha_\beta x^\beta \quad (\alpha = 0, \ldots, 3), \tag{9.64}$$

where $(Lx)^\alpha$ represents the αth component of Lx. By the definition (9.63) of the orthogonal transformation and using (9.61), it is easy to verify that L is an orthogonal transformation if and only if its corresponding matrix A satisfies

$$A^{\mathrm{T}} \begin{pmatrix} 1 & & & \\ & -1 & & \\ & & -1 & \\ & & & -1 \end{pmatrix} A = \begin{pmatrix} 1 & & & \\ & -1 & & \\ & & -1 & \\ & & & -1 \end{pmatrix} \tag{9.65}$$

(see Exercise 4).

Suppose that $\{e_0, e_1, e_2, e_3\}$ is an orthonormal basis in M, i.e.,

$$\langle e_\alpha, e_\beta \rangle = g_{\alpha\beta} \quad (\alpha, \beta = 0, \ldots, 3), \tag{9.66}$$

where $g_{\alpha\beta}$ is given by (9.62). For any given orthogonal transformation L in M, denote

$$\overline{e}_\alpha = L e_\alpha \quad (\alpha = 0, \ldots, 3). \tag{9.67}$$

Obviously, $\{\overline{e}_0, \overline{e}_1, \overline{e}_2, \overline{e}_3\}$ is also an orthonormal basis in M. In fact, just as in the Euclidean space, it is easy to prove that for any two given orthonormal bases $\{e_0, e_1, e_2, e_3\}$ and $\{\overline{e}_0, \overline{e}_1, \overline{e}_2, \overline{e}_3\}$, there exists an orthogonal transformation L such that (9.67) is satisfied.

Suppose that the relation between the two orthonormal bases $\{e_0, e_1, e_2, e_3\}$ and $\{\overline{e}_0, \overline{e}_1, \overline{e}_2, \overline{e}_3\}$ in M is given by

$$e_\alpha = a^\beta_\alpha \overline{e}_\beta \quad (\alpha = 0, \ldots, 3) \tag{9.68}$$

(comparing with (A.3) in Appendix A of Volume I). Representing by the above two orthonormal bases, an element x in M can be denoted, respectively, as

$$x = x^\alpha e_\alpha \tag{9.69}$$

and

$$x = \overline{x}^\beta \overline{e}_\beta. \tag{9.70}$$

Plugging (9.68) into the right-hand side of (9.69), and comparing with (9.70), we immediately have

$$\overline{x}^{\beta} = a^{\beta}_{\alpha} x^{\alpha} \quad (\beta = 0, \ldots, 3) \tag{9.71}$$

(comparing with (A.5) in Appendix A of Volume I).

Definition 9.1. *If a quantity can be represented by four real numbers* (x^0, x^1, x^2, x^3) *under the orthonormal basis* $\{e_0, e_1, e_2, e_3\}$ *in* M*, and these components can be transformed according to* (9.71) *under the transformation* (9.68) *of orthonormal bases, then this quantity is called a (contravariant) vector in the Minkowski space-time* M*, or a (contravariant) four-vector for short. If a quantity* P *can be represented by* 16 *components* $p^{\alpha\beta}$ $(\alpha, \beta = 0, 1, 2, 3)$ *under an orthonormal basis in* M*, and these components can be transformed, under the transformation* (9.68) *of orthonormal bases, according to the laws*

$$\overline{p}^{\alpha\beta} = a^{\alpha}_{\gamma} a^{\beta}_{\delta} p^{\gamma\delta} \quad (\alpha, \beta = 0, \ldots, 3) \tag{9.72}$$

into 16 *components* $\overline{p}^{\alpha\beta}$ $(\alpha, \beta = 0, \ldots, 3)$ *under the orthonormal basis* $\{\overline{e}_0, \overline{e}_1, \overline{e}_2, \overline{e}_3\}$*, then* P *is called a second-order (contravariant) tensor in the Minkowski space-time* M*, or a second-order (contravariant) four-tensor for short.*

As for a scalar in the Minkowski space-time M (a four-scalar for short), it is a quantity that does not change under the transformation of orthonormal bases in M.

9.4.2 Four-Velocity and Energy–Momentum Four-Vector

A point in the Minkowski four-space-time M is called an event. The history of a particle is represented in M by a continuous sequence of events, called the *world line* of this particle. Suppose that the time and space coordinates of an event on the world line of a particle in inertial frames K and $\overline{K}$ (corresponding to the orthonormal bases $\{e_0, e_1, e_2, e_3\}$ and $\{\overline{e}_0, \overline{e}_1, \overline{e}_2, \overline{e}_3\}$, respectively) are (x^0, x^1, x^2, x^3) and $(\overline{x}^0, \overline{x}^1, \overline{x}^2, \overline{x}^3)$, respectively; then from (9.71) we have

$$\mathrm{d}\overline{x}^{\alpha} = a^{\alpha}_{\beta} \mathrm{d}x^{\beta} \quad (\alpha = 0, \ldots, 3). \tag{9.73}$$

Therefore, $\mathrm{d}x = (\mathrm{d}x^0, \mathrm{d}x^1, \mathrm{d}x^2, \mathrm{d}x^3)$ is a (contravariant) vector in M. But since the differential $\mathrm{d}t$ of time t is not a four-scalar, $\frac{\mathrm{d}x}{\mathrm{d}t}$ is not a (contravariant) four-vector, and thus it cannot be regarded as a four-velocity.

By (9.73) and noting the invariance of the inner product in M with respect to the Lorentz transformation, we have

$$(\mathrm{d}x^0)^2 - \sum_{i=1}^{3} (\mathrm{d}x^i)^2 = (\mathrm{d}\overline{x}^0)^2 - \sum_{i=1}^{3} (\mathrm{d}\overline{x}^i)^2. \tag{9.74}$$

Suppose that the velocities of the particle with respect to inertial frames K and $\overline{K}$ are v and $\overline{v}$, respectively. Since $\sum_{i=1}^{3} (\mathrm{d}x^i)^2$ and $\sum_{i=1}^{3} (\mathrm{d}\overline{x}^i)^2$ are the square of motion distances of the particle in the time intervals $\mathrm{d}t$ and $\mathrm{d}\bar{t}$ observed in K and $\overline{K}$, respectively, we have

$$\sum_{i=1}^{3}\left(\frac{\mathrm{d}x^i}{\mathrm{d}t}\right)^2 = v^2, \quad \sum_{i=1}^{3}\left(\frac{\mathrm{d}\bar{x}^i}{\mathrm{d}\bar{t}}\right)^2 = \bar{v}^2.$$

Thus, (9.74) can be rewritten as

$$(\mathrm{d}x^0)^2 - (v\mathrm{d}t)^2 = (\mathrm{d}\bar{x}^0)^2 - (\bar{v}\mathrm{d}\bar{t})^2.$$

Noting that $x^0 = ct$ and $\bar{x}^0 = c\bar{t}$, the above formula leads to

$$\sqrt{1-\left(\frac{v}{c}\right)^2}\,\mathrm{d}t = \sqrt{1-\left(\frac{\bar{v}}{c}\right)^2}\,\mathrm{d}\bar{t}. \tag{9.75}$$

Denote

$$\mathrm{d}\tau = \sqrt{1-\left(\frac{v}{c}\right)^2}\,\mathrm{d}t. \tag{9.76}$$

From (9.75), $\mathrm{d}\tau$ is a four-scalar called the *proper time interval.*

Dividing the (contravariant) four-vector $\mathrm{d}x$ by the scalar $\mathrm{d}\tau$ in M, we still get a contravariant four-vector $\frac{\mathrm{d}x}{\mathrm{d}\tau}$ called the *four-velocity*, denoted by

$$(u^0,u^1,u^2,u^3) = \left(\frac{\mathrm{d}x^0}{\mathrm{d}\tau}, \frac{\mathrm{d}x^1}{\mathrm{d}\tau}, \frac{\mathrm{d}x^2}{\mathrm{d}\tau}, \frac{\mathrm{d}x^3}{\mathrm{d}\tau}\right). \tag{9.77}$$

Noting the definition (9.76) of $\mathrm{d}\tau$, we have

$$(u^0,u^1,u^2,u^3) = \gamma(c,v^1,v^2,v^3), \tag{9.78}$$

where $\gamma = 1/\sqrt{1-(\frac{v}{c})^2}$, $v = |\boldsymbol{v}|$, and

$$\boldsymbol{v} = (v^1,v^2,v^3) = \left(\frac{\mathrm{d}x^1}{\mathrm{d}t}, \frac{\mathrm{d}x^2}{\mathrm{d}t}, \frac{\mathrm{d}x^3}{\mathrm{d}t}\right) \tag{9.79}$$

is the velocity in the classical sense. Please note that here $\boldsymbol{v}$ is the motion velocity of the particle, not that of the reference frame $\overline{K}$ with respect to the reference frame K. Therefore, the γ in (9.78) has a different meaning from the γ in the Lorentz transformation mentioned previously.

Similarly, we define

$$\left(\frac{\mathrm{d}u^0}{\mathrm{d}\tau}, \frac{\mathrm{d}u^1}{\mathrm{d}\tau}, \frac{\mathrm{d}u^2}{\mathrm{d}\tau}, \frac{\mathrm{d}u^3}{\mathrm{d}\tau}\right) \tag{9.80}$$

as the *four-acceleration.*

We denote by m_0 the mass of the particle at rest, called the *rest mass.* The rest mass m_0 of the particle can be regarded reasonably as a four-scalar whose multiplication by a four-velocity is also a (contravariant) four-vector, called the *four-momentum*, denoted by (p^0,p^1,p^2,p^3), i.e.,

$$(p^0,p^1,p^2,p^3) = m_0(u^0,u^1,u^2,u^3). \tag{9.81}$$

Denoting by $\boldsymbol{p} = (p^1, p^2, p^3)$ the spatial components of the four-momentum, from (9.78) we have

$$\boldsymbol{p} = m\boldsymbol{v}, \tag{9.82}$$

where

$$m = \gamma m_0 \tag{9.83}$$

is called the *inertial mass* of the particle. We need to notice that the inertial mass m is not an invariant under the Lorentz transformation, namely, it is not a four-scalar.

Now we examine the time component p^0 of the four-momentum. From (9.81), (9.78), and (9.83) we have

$$p^0 = \frac{1}{c}E, \tag{9.84}$$

where

$$E = mc^2 \tag{9.85}$$

is called the *total energy*.

Thus, the four-momentum can be written as

$$(p^0, p^1, p^2, p^3) = \left(\frac{1}{c}E, \boldsymbol{p}\right). \tag{9.86}$$

It provides both the momentum $\boldsymbol{p}$ and the total energy E of the particle. Therefore, the four-momentum is also called the *energy–momentum four-vector*.

For any given (contravariant) four-vector, the difference between its time component and the sum of its spatial components is a four-scalar. In particular, for the four-velocity, from (9.78) we have

$$(u^0)^2 - (u^1)^2 - (u^2)^2 - (u^3)^2 = c^2. \tag{9.87}$$

Together with the definition of the four-momentum, we get

$$(p^0)^2 - (p^1)^2 - (p^2)^2 - (p^3)^2 = \frac{1}{c^2}E_0^2, \tag{9.88}$$

where

$$E_0 = m_0 c^2 \tag{9.89}$$

is called the *rest energy* of the particle, which is a scalar in M. Noting the expression (9.84) of p^0, it follows from (9.88) that

$$E^2 - c^2|\boldsymbol{p}|^2 = E_0^2. \tag{9.90}$$

This implies that the difference between the square of the particle's total energy and the square of the particle's momentum (multiplied by c^2 for dimensional consistency) is a scalar in M that is equal to the square of the rest energy. $E^2 - c^2|\boldsymbol{p}|^2$ is called the *energy–momentum invariant*, and (9.90) is called the *energy–momentum formula*.

Einstein's mass–energy equivalence relation (9.85) is one of the most important formulas in special relativity (this result was published in 1905 in his second paper about special relativity). It states that mass generates energy. Even for an object at rest, the magnitude of its rest energy E_0 is breathtaking since it has rest mass m_0 and c is a very large quantity. For example, the rest energy produced by 1 kilogram of rest mass is approximately equivalent to three years' electricity production of a 1 million kilowatt power plant. Einstein's mass–energy relation has been verified by plenty of experimental facts and played an important role in the development of the atomic bomb and nuclear power industries.

9.4.3 Conservation Laws of Momentum and Energy

Taking derivatives of the particle's four-momentum (p^0, p^1, p^2, p^3) with respect to the proper time τ, we obtain a (contravariant) four-vector called the *four-force* and denoted by (g^0, g^1, g^2, g^3):

$$(g^0, g^1, g^2, g^3) = \left(\frac{\mathrm{d}p^0}{\mathrm{d}\tau}, \frac{\mathrm{d}p^1}{\mathrm{d}\tau}, \frac{\mathrm{d}p^2}{\mathrm{d}\tau}, \frac{\mathrm{d}p^3}{\mathrm{d}\tau}\right). \tag{9.91}$$

The four-force is also called the *Minkowski force*.

The spatial components of the four-force are

$$g^i = \frac{\mathrm{d}p^i}{\mathrm{d}\tau} = \gamma \frac{\mathrm{d}p^i}{\mathrm{d}t} \quad (i = 1, 2, 3), \tag{9.92}$$

where $\frac{\mathrm{d}p^i}{\mathrm{d}t}$ $(i = 1, 2, 3)$ are the components of the force in the classical sense, denoted by f^i $(i = 1, 2, 3)$; namely, the classical force vector is

$$\boldsymbol{f} = (f^1, f^2, f^3) = \frac{\mathrm{d}\boldsymbol{p}}{\mathrm{d}t}. \tag{9.93}$$

We need to notice that the classical force $\boldsymbol{f}$ is not the spatial component of the four-force.

From (9.86), the time component of the four-force is

$$g^0 = \frac{\mathrm{d}p^0}{\mathrm{d}\tau} = \frac{\gamma}{c}\frac{\mathrm{d}E}{\mathrm{d}t}. \tag{9.94}$$

Differentiating both sides of (9.90) with respect to t, and noting $\frac{\mathrm{d}E_0}{\mathrm{d}t} = 0$, we get

$$E\frac{\mathrm{d}E}{\mathrm{d}t} = c^2 \boldsymbol{p} \cdot \frac{\mathrm{d}\boldsymbol{p}}{\mathrm{d}t}. \tag{9.95}$$

Using (9.82) and (9.85), and noting (9.93), we obtain

$$\frac{\mathrm{d}E}{\mathrm{d}t} = \boldsymbol{v} \cdot \boldsymbol{f}. \tag{9.96}$$

This shows that the rate of change $\frac{\mathrm{d}E}{\mathrm{d}t}$ of the particle's total energy is equal to the rate of work $\boldsymbol{v} \cdot \boldsymbol{f}$ done by the classical force on the particle. This is why $E = mc^2$ is called the total energy. From (9.96), the expression (9.94) of the time component of the four-force can also be written as

$$g^0 = \frac{\gamma}{c}\boldsymbol{v} \cdot \boldsymbol{f}. \tag{9.97}$$

Combining (9.92)–(9.93) and (9.97), we obtain the expression of the four-force

$$(g^0, g^1, g^2, g^3) = \gamma \left(\frac{1}{c} \boldsymbol{v} \cdot \boldsymbol{f}, \boldsymbol{f} \right). \tag{9.98}$$

Equation (9.93), namely,

$$\frac{\mathrm{d}\boldsymbol{p}}{\mathrm{d}t} = \boldsymbol{f} \tag{9.99}$$

is formally the same as Newton's second law in classical mechanics, but here we only define $\frac{\mathrm{d}\boldsymbol{p}}{\mathrm{d}t}$ as the force. If we regard (9.99) as the motion equation of the particle just as we do in classical mechanics, the force $\boldsymbol{f}$ must be given independently, and the transformation relation of $\boldsymbol{f}$ under the Lorentz transformation must be the same as that of $\frac{\mathrm{d}\boldsymbol{p}}{\mathrm{d}t}$, that is, $\gamma(\frac{1}{c}\boldsymbol{v} \cdot \boldsymbol{f}, \boldsymbol{f})$ must be a (contravariant) four-vector. There is not a ray of hope for this kind of relativistic correction to extend those common forces in Newtonian mechanics, such as universal gravitation and so on, to their four-dimensional forms such that the corresponding transformation requirements are satisfied for them. As a matter of fact, according to Newton's law of universal gravitation, the gravitational force between two objects depends on the distance between them; therefore, if one object moves, the force on the other object changes immediately. This implies that the gravitational effect should propagate at infinite speed, which violates the conclusion that any speed in special relativity is less than the speed of light c in a vacuum. Einstein tried during 1908 to 1914 to establish a gravitational theory consistent with special relativity but failed. He proposed a revolutionary idea in 1915 that the gravitational force, unlike any other kind of force, is caused by the fact that space-time is not flat; based on this, he established general relativity.

Fortunately, the Lorentz force

$$\boldsymbol{f} = \rho(\boldsymbol{E} + \boldsymbol{v} \times \boldsymbol{B}) \tag{9.100}$$

(see (1.64) in Chapter 1 of Volume I), received by a charged particle moving in an electromagnetic field, satisfies the above-required transformation law (see section 9.7).

When the four-force $(g^0, g^1, g^2, g^3) = 0$, it follows from (9.91) that the four-momentum (p^0, p^1, p^2, p^3) is conserved. According to (9.86), the four-momentum includes the momentum $\boldsymbol{p}$ and the total energy E. Therefore, when the four-momentum is conserved, both the momentum and the energy are conserved. Thus, two conservation laws for momentum and energy come down to one conservation law—the conservation of four-momentum.

Now we consider a particle system composed of a number of particles.

In classical Newtonian mechanics, the momentum conservation of the system is obtained from Newton's second law with the aid of Newton's third law. But in special relativity, since the action and reaction between two separate particles propagate at a finite speed less than the speed of light in a vacuum, Newton's third law is no longer valid, so that it cannot be used to derive the conservation of momentum. Even so, the conservation laws of momentum and energy, as the most general conservation laws in physics, have been verified by all experiments and observations so far, so it is reasonable to assume that these two general conservation laws are still valid in special relativity.

The *conservation law of momentum* for a many-particle system can be described as follows: **The total momentum of a particle system subject to no external action is conserved in any given inertial frame.**

In the above law, the total momentum contains not only the momentum of the particle form but also the momentum of the field (e.g., electromagnetic field) form. In the special case that only the momentum of the particle form is involved, the conservation law of momentum can be given by

$$\sum_i \boldsymbol{p}_i = \text{constant}, \tag{9.101}$$

where $\boldsymbol{p}_i$ is the momentum of the ith particle.

We need to notice that the total number of particles does not remain unchanged during the movements of this particle system. For example, in a system composed of two particles, particle 1 and particle 2, these two particles may be combined into a new particle after a collision called a *completely inelastic collision*. They may also generate two (or more) other particles, particle 3 and particle 4, after a collision; then we have

$$\gamma_1 m_{10} \boldsymbol{v}_1 + \gamma_2 m_{20} \boldsymbol{v}_2 = \gamma_3 m_{30} \boldsymbol{v}_3 + \gamma_4 m_{40} \boldsymbol{v}_4, \tag{9.102}$$

where m_{i0} and $\boldsymbol{v}_i$ represent the rest mass and the velocity of the ith particle, respectively, $\gamma_i = 1/\sqrt{1 - \frac{|\boldsymbol{v}_i|^2}{c^2}}$ $(i = 1,2,3,4)$. Generally speaking,

$$m_{10} + m_{20} \neq m_{30} + m_{40}. \tag{9.103}$$

This case is called an *inelastic collision*. The collision after which they are still the original two particles is called a *completely elastic collision*; at this moment, $m_{30} = m_{10}$, $m_{40} = m_{20}$.

Experiments show that for a particle system, the following conservation law of energy is satisfied: **The total energy of a particle system subject to no external action is conserved in any given inertial frame.**

In this description, the total energy contains not only the total energy of the particle form but also the total energy of the field form, i.e.,

$$\sum_i m_i c^2 + E_f = \text{constant}, \tag{9.104}$$

where m_i represents the inertial mass of the ith particle, and E_f the field energy. In the special case that only the particle form is involved, the energy conservation law (9.104) is reduced to

$$\sum_i m_i c^2 = \text{constant}. \tag{9.105}$$

We make some remarks about (9.105). Even for a system involving only particle form, since there may exist decay and combining phenomena during evolution, the total number of particles and the rest mass of particles do not always remain unchanged. Equation (9.105) also contains the mutual transformation between the kinetic energy and the internal energy of particles inside the system. This internal energy means the energy measured by $m_0 c^2$ inside the particle, not the internal energy produced by the irregular thermal motion of a large number of particles in thermodynamics.

9.5 Relativistic Fluid Dynamics

When the fluid is in motion at a speed close to the speed of light, the relativistic effect has to be taken into consideration. Meanwhile, we will see that even if the macroscopic speed of the fluid is not so large that the relativistic effect has to be taken into account, if the microscopic speed of the fluid particles is very large, the relativistic effect cannot be ignored.

As early as when he established special relativity, Einstein pointed out that (see [5]) the Euler equations for a frictionless adiabatic fluid can be established by the field equations, which proposed tasks for the study of the motion law of relativistic fluids. But the study of relativistic fluid dynamics actually started much later; it was until 1970 that the first international workshop on relativistic fluid dynamics was held. Since then, significant progress has been made in this research to satisfy the needs of the development of plasma physics and nuclear physics. In fact, the system of relativistic fluid dynamics can be regarded as a good mathematical model of high-energy astrophysical plasma and plays an important role in the reaction analysis of heavy ions in nuclear physics.

The object of study in this section is restricted to the ideal fluid. As for the fluid with dissipation effects of viscosity and heat conduction, although some constitutive relations of the energy–momentum tensor have been established—for instance, L. D. Landau and E. M. Lifshitz as well as C. Ekart introduced a certain model—there has been no satisfactory result so far.

As stated in Chapter 2 of Volume I, an important step in establishing the system of classical fluid dynamics is to give the mass flux vector, the momentum flux tensor, and the energy flux vector. Thanks to these, it is convenient to establish a system reflecting the conservation of mass, momentum, and energy, which, in the absence of external force, reads as

$$\frac{\partial \rho}{\partial t}+\sum_{k=1}^{3}\frac{\partial}{\partial x_k}(\rho u_k)=0, \tag{9.106}$$

$$\frac{\partial}{\partial t}(\rho u_i)+\sum_{k=1}^{3}\frac{\partial}{\partial x_k}(\rho u_i u_k+p\delta_{ik})=0 \quad (i=1,2,3), \tag{9.107}$$

$$\frac{\partial}{\partial t}\left(\rho e+\frac{1}{2}\rho u^2\right)+\sum_{k=1}^{3}\frac{\partial}{\partial x_k}\left(\left(\rho e+\frac{1}{2}\rho u^2+p\right)u_k\right)=0 \tag{9.108}$$

(see (2.10), (2.14), and (2.15) in Chapter 2 of Volume I). To reconstruct the above system in relativistic form, the corresponding energy–momentum tensor has to be determined first.

9.5.1 Energy–Momentum Tensor

We make two remarks before establishing the energy–momentum tensor. First, mass is no longer conserved in relativistic mechanics; it is only part of the total energy. The mass conservation equation (9.106) and the energy conservation equation (9.108) should be considered along with each other in relativistic fluid dynamics. In addition, in Chapter 2 of Volume I, tensor $\rho \boldsymbol{u}\otimes \boldsymbol{u}$ is called the momentum flux tensor. For convenience, below we merge the impulse produced by the pressure into this tensor and regard the tensor $\rho \boldsymbol{u}\otimes \boldsymbol{u}+pI$ as the momentum flux tensor.

Our desired *energy–momentum tensor* $(T^{\alpha\beta})$ is a second-order (contravariant) four-tensor. If we represent the conservation law of energy by the 0th conservation equation, then T^{00} stands for the energy per unit volume and (T^{01}, T^{02}, T^{03}) the energy passing through unit area in unit time, meaning that the energy passing through the area element $\mathrm{d}S$ along the normal direction $\boldsymbol{n}$ in unit time is

$$c(T^{01}\cos(\boldsymbol{n},x_1)+T^{02}\cos(\boldsymbol{n},x_2)+T^{03}\cos(\boldsymbol{n},x_3))\mathrm{d}S,$$

where the factor c appears due to $x^0 = ct$. As long as we have the expression of $(T^{00}, T^{01}, T^{02}, T^{03})$, the equation in divergence form

$$\frac{\partial}{\partial x^\alpha}T^{0\alpha} = 0 \tag{9.109}$$

yields the conservation law of energy. Correspondingly, if we represent the conservation laws of momentum by the other three equations (the first, the second, and the third equations), then $\frac{1}{c}(T^{10}, T^{20}, T^{30})^{\mathrm{T}}$ stands for the momentum per unit volume, and (T^{ij}) $(i, j = 1, 2, 3)$ represents the momentum passing through unit area in unit time.

Now we determine the energy–momentum tensor $(T^{\alpha\beta})$ of the fluid. For any given fluid element, we adopt a comoving reference frame such that this element is at rest in this reference frame. This kind of reference frame is called the *proper reference frame*, or the *proper frame* for short. Denoting the proper frame by $\overline{K}$, and representing the tensor $(T^{\alpha\beta})$ in frame $\overline{K}$ by $(\overline{T}^{\alpha\beta})$, we have

$$\overline{T}^{00} = \mu, \tag{9.110}$$

where μ stands for the proper internal energy density, i.e., the internal energy density measured in the proper frame, which is at rest with respect to the fluid. Moreover, in addition to the thermodynamical energy, such as the heat energy, the chemical energy, and so on, the internal energy should also include the relativistic energy corresponding to the mass. Moreover, obviously we have

$$\overline{T}^{01} = \overline{T}^{02} = \overline{T}^{03} = 0. \tag{9.111}$$

As for the momentum, in the proper frame $\overline{K}$, obviously we have

$$(\overline{T}^{10}, \overline{T}^{20}, \overline{T}^{30})^{\mathrm{T}} = 0, \tag{9.112}$$

and the only momentum passing through any given area element $\mathrm{d}S$ in the fluid field in unit time is the impulse of the fluid pressure, i.e.,

$$(\overline{T}^{ij}) = \begin{pmatrix} p & & \\ & p & \\ & & p \end{pmatrix}. \tag{9.113}$$

Combining the above formulas, we get, in the proper frame $\overline{K}$, that

$$(\overline{T}^{\alpha\beta}) = \begin{pmatrix} \mu & & & \\ & p & & \\ & & p & \\ & & & p \end{pmatrix}. \tag{9.114}$$

Now we use the Lorentz transformation to obtain the expression of the energy–momentum tensor in general inertial reference frame K from its expression (9.114) in the proper frame $\overline{K}$. Suppose that the macroscopic velocity of the fluid element in the reference frame K is $\boldsymbol{v} = (v_1, v_2, v_3)$, since $\overline{K}$ is the comoving frame; then its velocity with respect to K is just $\boldsymbol{v}$. Correspondingly, the velocity of reference frame K with respect to $\overline{K}$ is $\boldsymbol{V} = -\boldsymbol{v}$. Suppose that the space-time coordinates in frames K and $\overline{K}$ are (x^0, x^1, x^2, x^3) and $(\overline{x}^0, \overline{x}^1, \overline{x}^2, \overline{x}^3)$, respectively; then by Theorem 9.2, and noticing that in (9.31) $x_0 = t$ instead of $x^0 = ct$, it is easy to show that

$$x^\alpha = a^\alpha_\beta \overline{x}^\beta \quad (\alpha = 0,1,2,3), \tag{9.115}$$

where

$$a^0_0 = \gamma, \quad a^0_i = \gamma \frac{v_i}{c} \quad (i = 1,2,3), \tag{9.116}$$

$$a^i_0 = \gamma \frac{v_i}{c}, \quad a^i_j = \frac{\gamma - 1}{v^2} v_i v_j + \delta_{ij} \quad (i, j = 1,2,3), \tag{9.117}$$

and $\gamma = 1/\sqrt{1 - \left(\frac{v}{c}\right)^2}$, $v = |\boldsymbol{v}|$.

According to the definition of the second-order (contravariant) tensor (see (9.72)), it follows from (9.115) that in K we have

$$T^{00} = a^0_\alpha a^0_\beta \overline{T}^{\alpha\beta} = \frac{\mu + p\dfrac{v^2}{c^2}}{1 - \dfrac{v^2}{c^2}}, \tag{9.118}$$

$$\begin{aligned} T^{0i} &= T^{i0} = a^i_\alpha a^0_\beta \overline{T}^{\alpha\beta} \\ &= \gamma^2 \mu \frac{v_i}{c} + \sum_{j=1}^{3} \gamma p \left(\frac{\gamma - 1}{v^2} v_i v_j + \delta_{ij} \right) \frac{v_j}{c} \\ &= \frac{(\mu + p) v_i}{c\left(1 - \dfrac{v^2}{c^2}\right)} \quad (i = 1,2,3), \end{aligned} \tag{9.119}$$

$$\begin{aligned} T^{ij} &= a^i_\alpha a^j_\beta \overline{T}^{\alpha\beta} \\ &= \gamma^2 \mu \frac{v_i v_j}{c^2} + \sum_{k=1}^{3} p \left(\frac{\gamma - 1}{v^2} v_i v_k + \delta_{ik} \right) \times \left(\frac{\gamma - 1}{v^2} v_j v_k + \delta_{jk} \right) \\ &= \frac{(\mu + p) v_i v_j}{c^2 \left(1 - \dfrac{v^2}{c^2}\right)} + p\delta_{ij} \quad (i, j = 1,2,3). \end{aligned} \tag{9.120}$$

Noting the four-velocity $(u^0, u^1, u^2, u^3) = \gamma(c, v_1, v_2, v_3)$ (see (9.78)), (9.118)–(9.120) can be unified as

$$T^{\alpha\beta} = \frac{1}{c^2} (\mu + p) u^\alpha u^\beta - p g^{\alpha\beta} \quad (\alpha, \beta = 0, \ldots, 3), \tag{9.121}$$

where

$$(g^{\alpha\beta}) = \begin{pmatrix} 1 & & & \\ & -1 & & \\ & & -1 & \\ & & & -1 \end{pmatrix}. \tag{9.122}$$

Equation (9.121) is just the expression of the energy–momentum tensor in general inertial reference frame K.

In (9.121), μ is the proper internal energy density, which certainly does not change with the change of reference frame. In addition, p is also measured in the proper frame, but it can be proved that the pressure p is invariant under the Lorentz transformation (see [16]).

9.5.2 System of Conservation Laws

After obtaining the energy–momentum tensor $(T^{\alpha\beta})$, we can immediately get the system describing the conservation of energy and momentum in relativistic fluid dynamics as follows:

$$\frac{\partial}{\partial x^\beta} T^{\alpha\beta} = 0 \quad (\alpha = 0, \ldots, 3). \tag{9.123}$$

The left-hand side of the above formula is the (covariant) divergence of a second-order (contravariant) tensor, so it is a first-order (contravariant) vector (see Appendix C). Therefore, as long as (9.123) is satisfied in an inertial reference frame, it is always satisfied in any given inertial reference frame.

Plugging (9.118)–(9.120) into the conservation system (9.123), and noting $x^0 = ct$ and $x^i = x_i$ $(i = 1, 2, 3)$, we obtain

$$\frac{\partial}{\partial t}\left(\frac{\mu + p}{c^2 - v^2} - \frac{p}{c^2}\right) + \sum_{k=1}^{3} \frac{\partial}{\partial x_k}\left(\frac{\mu + p}{c^2 - v^2} v_k\right) = 0 \tag{9.124}$$

and

$$\frac{\partial}{\partial t}\left(\frac{\mu + p}{c^2 - v^2} v_i\right) + \sum_{k=1}^{3} \frac{\partial}{\partial x_k}\left(\frac{\mu + p}{c^2 - v^2} v_i v_k + p\delta_{ik}\right) = 0$$
$$(i = 1, 2, 3). \tag{9.125}$$

Denote by ρ the proper mass density, namely, the rest mass per unit volume of fluid in the proper frame, and by e the thermodynamical internal energy per unit rest mass; then in the above formulas,

$$\mu = \rho c^2 + \rho e. \tag{9.126}$$

Just as in classical fluid dynamics, only two of the thermodynamical quantities ρ, p, and e are independent, and they are related by the equation of state.

We have only four equations in system (9.124)–(9.125), but five unknown functions; generally we need an additional conservation equation.

In relativistic fluid dynamics, mass is no longer a conserved quantity, but the particle number of the fluid is conserved. Denote by n the proper number density of particles, i.e., the number of particles per unit volume of fluid in the comoving proper frame. Thus, the particle flux (four-)vector in the proper frame should be represented by $(cn, 0, 0, 0)$, where the 0th component is the number of particles per unit volume in this frame multiplied by c (noting $x^0 = ct$), and the last three components represent the number of particles passing through unit area in unit time. By the definition of the (contravariant) vector (see (9.71)) and (9.116)–(9.117), it is easy to get the following expression of the particle flux vector in general reference frame K:

$$(nu^0, nu^1, nu^2, nu^3), \tag{9.127}$$

where (u^0, u^1, u^2, u^3) is the four-velocity. Thus, we obtain the following conservation equation of particle numbers:

$$\frac{\partial}{\partial x^\alpha}(nu^\alpha) = 0. \tag{9.128}$$

Noting that

$$\rho = nm_0, \tag{9.129}$$

where m_0 is the average rest mass of an individual particle, (9.128) can also be written in the following form:

$$\frac{\partial}{\partial x^\alpha}(\rho u^\alpha) = 0. \tag{9.130}$$

Equation (9.128) or (9.130) describing the conservation of particle numbers is also called the *continuity equation*.

Noting $x^0 = ct$, $x^i = x_i$ $(i = 1, 2, 3)$, and the expression (9.78) of the four-velocity, equation (9.130) can also be rewritten as

$$\frac{\partial}{\partial t}\left(\frac{\rho}{\sqrt{1-\frac{v^2}{c^2}}}\right) + \sum_{k=1}^{3}\frac{\partial}{\partial x_k}\left(\frac{\rho}{\sqrt{1-\frac{v^2}{c^2}}}v_k\right) = 0. \tag{9.131}$$

Equations (9.124)–(9.125) and (9.131), together with the corresponding equation of state in thermodynamics, constitute the *system of relativistic ideal fluid dynamics*.

Now we further discuss the change of the fluid entropy. Plugging the expression (9.121) of tensor $(T^{\alpha\beta})$ into the conservation system (9.123), we get

$$\frac{1}{c^2}u^\alpha \frac{\partial}{\partial x^\beta}((\mu + p)u^\beta) + \frac{1}{c^2}(\mu + p)u^\beta \frac{\partial u^\alpha}{\partial x^\beta} - g^{\alpha\beta}\frac{\partial p}{\partial x^\beta} = 0$$
$$(\alpha = 0, \ldots, 3). \tag{9.132}$$

Denote

$$u_0 = u^0, \quad u_i = -u^i \quad (i = 1, 2, 3). \tag{9.133}$$

In fact, the quantities on the left-hand side of the above formulas are first-order covariant components of a contravariant tensor (u^0, u^1, u^2, u^3) (see Appendix C). But for the time being, it does not cause any trouble in understanding the following if you have no idea what a covariant tensor is. Multiplying both sides of equation (9.132) by u_α and summing over 0 to 3 with respect to α, we get

$$\frac{\partial}{\partial x^\beta}((\mu + p)u^\beta) - u^\beta \frac{\partial p}{\partial x^\beta} = 0. \tag{9.134}$$

In obtaining the above formula, we used

$$u_\alpha u^\alpha = c^2 \tag{9.135}$$

(this is just (9.87)),

$$u_\alpha \frac{\partial u^\alpha}{\partial x^\beta} = 0, \tag{9.136}$$

and

$$u_\alpha g^{\alpha\beta} = u^\beta. \tag{9.137}$$

It is very easy to verify the above formulas directly.

Noticing that $\mu = \rho(c^2 + e)$ (see (9.126)) and $i = e + \frac{p}{\rho}$ is the enthalpy per unit rest mass of fluid (see (B.35) in Appendix B of Volume I), namely, the proper specific enthalpy, using the continuity equation (9.130), (9.134) can be reduced to

$$\frac{\partial}{\partial x^\beta}(i\rho u^\beta) - u^\beta \frac{\partial p}{\partial x^\beta} = 0. \tag{9.138}$$

Using again the continuity equation (9.130), the above formula can also be written as

$$\rho u^\beta \left(\frac{\partial i}{\partial x^\beta} - \frac{1}{\rho} \frac{\partial p}{\partial x^\beta} \right) = 0. \tag{9.139}$$

In the proper frame,

$$\mathrm{d}i = \frac{1}{\rho}\mathrm{d}p + \theta \mathrm{d}S \tag{9.140}$$

(see (B.38) in Appendix B of Volume I), where θ denotes the absolute temperature of fluid in the proper frame, namely, the proper temperature, and S is the entropy per unit rest mass, namely, the proper entropy. By using (9.140), equation (9.139) is then reduced to

$$\rho\theta u^\beta \frac{\partial S}{\partial x^\beta} = 0,$$

i.e.,

$$u^\beta \frac{\partial S}{\partial x^\beta} = 0. \tag{9.141}$$

This states that the proper entropy remains unchanged along the world line, and, naturally, the entropy of an individual particle remains unchanged.

By using the continuity equation (9.130), equation (9.141) can also be written in the following conservation law form:

$$\frac{\partial}{\partial x^\beta}(\rho S u^\beta) = 0. \tag{9.142}$$

The above two equations imply that the flow of the fluid is adiabatic, which is consistent with the consideration that inner friction and heat conduction are not taken into account in the energy–momentum for the ideal fluid.

Being transformed into arguments t, x_1, x_2, x_3 and the velocity $\boldsymbol{v}$ in the classical sense, equations (9.141) and (9.142) can be reduced to, respectively,

$$\frac{\partial S}{\partial t} + \sum_{k=1}^{3} v_k \frac{\partial S}{\partial x_k} = 0 \tag{9.143}$$

and

$$\frac{\partial}{\partial t}\left(\frac{\rho S}{\sqrt{1-\frac{v^2}{c^2}}}\right) + \sum_{k=1}^{3} \frac{\partial}{\partial x_k}\left(\frac{\rho S v_k}{\sqrt{1-\frac{v^2}{c^2}}}\right) = 0. \tag{9.144}$$

The form of equation (9.143) is exactly the same as that in classical fluid dynamics (see (2.25) in Chapter 2 of Volume I).

Thus, we now have six conservation equations altogether, (9.124)–(9.125), (9.131), and (9.144), for five unknown functions altogether, including v_1, v_2, v_3, and any two given independent thermodynamical quantities. Just as in classical fluid dynamics, in a smooth fluid field, the entropy equation (9.144) is equivalent to the energy conservation equation (9.124), so only five of these six equations are independent. But if a shock wave appears, then the energy, momentum, and number of the fluid particles are still conserved when passing across the shock wave, while the entropy increases.

Now we make further reductions to the above system (9.124)–(9.125) and (9.131) of relativistic fluid dynamics. From (9.125) we have

$$\begin{aligned} &v_i \frac{\partial}{\partial t}\left(\frac{\mu+p}{c^2-v^2}\right) + v_i \sum_{k=1}^{3} \frac{\partial}{\partial x_k}\left(\frac{\mu+p}{c^2-v^2} v_k\right) \\ &+ \frac{\mu+p}{c^2-v^2}\frac{\partial v_i}{\partial t} + \frac{\mu+p}{c^2-v^2}\sum_{k=1}^{3} v_k \frac{\partial v_i}{\partial x_k} + \frac{\partial p}{\partial x_i} = 0 \\ &\qquad (i = 1,2,3). \end{aligned}$$

By using equation (9.124), the above formula can also be written as

$$\begin{aligned} &\frac{\mu+p}{c^2-v^2}\frac{\partial v_i}{\partial t} + \frac{\mu+p}{c^2-v^2}\sum_{k=1}^{3} v_k \frac{\partial v_i}{\partial x_k} + \frac{v_i}{c^2}\frac{\partial p}{\partial t} + \frac{\partial p}{\partial x_i} = 0 \\ &\qquad (i = 1,2,3); \end{aligned}$$

it follows immediately that

$$\frac{\partial v_i}{\partial t}+\sum_{k=1}^{3} v_k \frac{\partial v_i}{\partial x_k}+\frac{c^2-v^2}{\mu+p}\left(\frac{1}{c^2} v_i \frac{\partial p}{\partial t}+\frac{\partial p}{\partial x_i}\right)=0 \quad (i=1,2,3). \tag{9.145}$$

These are just the relativistic *Euler equations*.

From the expression (9.126) of μ (but noting that ρ is the proper mass density, not the mass density in the reference frame under consideration) we can see that, as $c \to \infty$, system (9.145) is reduced formally to the classical system of Euler equations (see (2.15) in Chapter 2 of Volume I). From (9.145) we can see that, even if the macroscopic velocity v of the fluid is not very large, but the average microscopic velocity of fluid particles is very large, since the thermodynamical internal energy e may be very large, and thanks to (9.126), the relativistic effect cannot be ignored in general.

9.5.3 Simplified Equations of State

It is known that only two of all the thermodynamical quantities are independent. In order to close the above-stated relativistic fluid equations, the equations of state have to be given, e.g.,

$$e=e(\rho, p), \tag{9.146}$$

and so on.

Assume that the fluid is composed of some particles without internal structure, and no interaction other than elastic collisions exists among these particles. On this occasion, we can approximately give simple equations of state for two extreme cases.

1. *Cold nonrelativistic fluid.* For this kind of fluid, the average microscopic velocity of its particles is much less than the speed of light. Then it can be proved that (see [16], [18])

$$e \approx \frac{3p}{2\rho}, \tag{9.147}$$

so it follows approximately that

$$\mu=\rho c^2+\frac{3}{2} p. \tag{9.148}$$

2. *Hot extremely relativistic fluid.* For this kind of fluid, the average microscopic velocity of its particles is close to the speed of light. Then it can be proved that (see [16], [18])

$$e \approx \frac{3p}{\rho} \tag{9.149}$$

and

$$e \gg c^2, \tag{9.150}$$

so it follows approximately that

$$\mu=3p. \tag{9.151}$$

In fact, (9.149) can be obtained from the Stefan–Boltzmann law of thermal radiation (see [1]), and (9.150) shows that the energy density of thermal radiation is much larger than the relativistic total energy density of the fluid. This situation is very significant in celestial mechanics.

The study of cosmogony indicates that at the beginning when dim stars explode to form supernovas, a lot of high-temperature matter is ejected. This is one of the important sources of cosmic radiation. The fluid at this moment belongs to hot extremely relativistic fluids.

Formulas (9.147) and (9.149) can be unified into the following form:

$$e = \frac{p}{(\gamma - 1)\rho}, \tag{9.152}$$

where

$$\gamma = \begin{cases} \dfrac{5}{3} & \text{for nonrelativistic fluids,} \\ \dfrac{4}{3} & \text{for extremely relativistic fluids.} \end{cases} \tag{9.153}$$

In form, (9.152) is exactly the same as the equation of state for polytropic gases in classical fluid dynamics (see (65) in Appendix B of Volume I). But it should be pointed out that (9.152) is only a kind of approximation, and in relativistic fluid dynamics, γ is generally not a constant, nor is it an adiabatic exponent.

For an extremely relativistic fluid, $\mu = 3p$. Then the four equations (9.124)–(9.125) describing the conservation of energy and momentum are already closed and can be solved independently.

9.6 Mathematical Structure of the System of Relativistic Fluid Dynamics

9.6.1 An Alternative Form of the Entropy Equation

Just like in many other references about relativistic fluid dynamics, we introduce the mass–energy density of the fluid

$$\varepsilon = \frac{\mu}{c^2}, \tag{9.154}$$

which is the proper energy density measured by the dimension of mass. Suppose that the equation of state is

$$p = p(\varepsilon, S). \tag{9.155}$$

It can be proved that the square of the propagation speed of small perturbation in a static uniform relativistic fluid is $\frac{\partial p}{\partial \varepsilon}$ (see [16], [18]), so it is reasonable to assume that

$$\frac{\partial p}{\partial \varepsilon} \overset{\mathrm{d}}{=} a^2 > 0, \tag{9.156}$$

where a is called the local speed of sound.

Using the expression $i = e + \frac{p}{\rho}$ of the enthalpy and the definition (9.154) of ε, and noting (9.126), we have

$$i = \frac{\varepsilon c^2 + p}{\rho} - c^2. \tag{9.157}$$

Therefore,

$$\begin{aligned}\rho \frac{\partial i}{\partial x^\beta} &= \frac{\partial}{\partial x^\beta}(\varepsilon c^2 + p) - \frac{\varepsilon c^2 + p}{\rho}\frac{\partial \rho}{\partial x^\beta} \\ &= c^2 \frac{\partial \varepsilon}{\partial p}\frac{\partial p}{\partial x^\beta} + c^2 \frac{\partial \varepsilon}{\partial S}\frac{\partial S}{\partial x^\beta} + \frac{\partial p}{\partial x^\beta} - \frac{\varepsilon c^2 + p}{\rho}\frac{\partial \rho}{\partial x^\beta}.\end{aligned}$$

Thanks to (9.156), we have $\frac{\partial \varepsilon}{\partial p}(p,S) = a^{-2}$. Plugging this into (9.139) and using the entropy equation (9.141), we get

$$\left(\frac{c}{a}\right)^2 u^\beta \frac{\partial p}{\partial x^\beta} - \frac{\varepsilon c^2 + p}{\rho} u^\beta \frac{\partial \rho}{\partial x^\beta} = 0, \tag{9.158}$$

i.e.,

$$\frac{\partial p}{\partial t} - \frac{\varepsilon c^2 + p}{\rho c^2} a^2 \frac{\partial \rho}{\partial t} + \sum_{k=1}^{3} v_k \frac{\partial p}{\partial x_k} - \frac{\varepsilon c^2 + p}{\rho c^2} a^2 \sum_{k=1}^{3} v_k \frac{\partial \rho}{\partial x_k} = 0. \tag{9.159}$$

This is just another form of the entropy equation.

9.6.2 Mathematical Structure of the System of Relativistic Fluid Dynamics

The Euler equations (9.145), the entropy equation (9.159), and the continuity equation (9.131)—these five equations constitute a system that reads as

$$\frac{\partial v_i}{\partial t} + \frac{b}{c^2} v_i \frac{\partial p}{\partial t} + \sum_{k=1}^{3} v_k \frac{\partial v_i}{\partial x_k} + b \frac{\partial p}{\partial x_i} = 0 \quad (i = 1,2,3), \tag{9.160}$$

$$\frac{\partial p}{\partial t} - ha^2 \frac{\partial \rho}{\partial t} + \sum_{k=1}^{3} v_k \frac{\partial p}{\partial x_k} - ha^2 \sum_{k=1}^{3} v_k \frac{\partial \rho}{\partial x_k} = 0, \tag{9.161}$$

$$\begin{aligned}&\frac{\gamma^3 \rho}{c^2} \sum_{k=1}^{3} v_k \frac{\partial v_k}{\partial t} + \gamma \frac{\partial \rho}{\partial t} + \frac{\gamma^3 \rho}{c^2} \sum_{k,l=1}^{3} v_k v_l \frac{\partial v_l}{\partial x_k} \\ &\quad + \rho\gamma \sum_{k=1}^{3} \frac{\partial v_k}{\partial x_k} + \gamma \sum_{k=1}^{3} v_k \frac{\partial \rho}{\partial x_k} = 0,\end{aligned} \tag{9.162}$$

where

$$b = \frac{c^2 - v^2}{\varepsilon c^2 + p}, \quad h = \frac{\varepsilon c^2 + p}{\rho c^2}, \quad \gamma = \frac{1}{\sqrt{1 - \frac{v^2}{c^2}}}. \tag{9.163}$$

Denote

$$U = (v_1, v_2, v_3, p, \rho)^{\mathrm{T}}. \tag{9.164}$$

The system of equations (9.160)–(9.162) can be written as

$$A_0 \frac{\partial U}{\partial t} + \sum_{k=1}^{3} A_k \frac{\partial U}{\partial x_k} = 0, \tag{9.165}$$

where

$$A_0 = \begin{pmatrix} 1 & 0 & 0 & \frac{b}{c^2} v_1 & 0 \\ 0 & 1 & 0 & \frac{b}{c^2} v_2 & 0 \\ 0 & 0 & 1 & \frac{b}{c^2} v_3 & 0 \\ 0 & 0 & 0 & 1 & -ha^2 \\ \frac{\gamma^3 \rho v_1}{c^2} & \frac{\gamma^3 \rho v_2}{c^2} & \frac{\gamma^3 \rho v_3}{c^2} & 0 & \gamma \end{pmatrix},$$

$$A_1 = \begin{pmatrix} v_1 & 0 & 0 & b & 0 \\ 0 & v_1 & 0 & 0 & 0 \\ 0 & 0 & v_1 & 0 & 0 \\ 0 & 0 & 0 & v_1 & -ha^2 v_1 \\ \rho\gamma + \frac{\gamma^3 \rho v_1^2}{c^2} & \frac{\gamma^3 \rho v_1 v_2}{c^2} & \frac{\gamma^3 \rho v_1 v_3}{c^2} & 0 & \gamma v_1 \end{pmatrix},$$

$$A_2 = \begin{pmatrix} v_2 & 0 & 0 & 0 & 0 \\ 0 & v_2 & 0 & b & 0 \\ 0 & 0 & v_2 & 0 & 0 \\ 0 & 0 & 0 & v_2 & -ha^2 v_2 \\ \frac{\gamma^3 \rho v_2 v_1}{c^2} & \rho\gamma + \frac{\gamma^3 \rho v_2^2}{c^2} & \frac{\gamma^3 \rho v_2 v_3}{c^2} & 0 & \gamma v_2 \end{pmatrix},$$

and

$$A_3 = \begin{pmatrix} v_3 & 0 & 0 & 0 & 0 \\ 0 & v_3 & 0 & 0 & 0 \\ 0 & 0 & v_3 & b & 0 \\ 0 & 0 & 0 & v_3 & -ha^2 v_3 \\ \frac{\gamma^3 \rho v_3 v_1}{c^2} & \frac{\gamma^3 \rho v_3 v_2}{c^2} & \rho\gamma + \frac{\gamma^3 \rho v_3^2}{c^2} & 0 & \gamma v_3 \end{pmatrix}.$$

Direct calculation easily yields

$$\det A_0 = \gamma \left(1 - \frac{a^2 v^2}{c^4} \right). \tag{9.166}$$

Since a is the local speed of sound, we have $a < c$. Together with $v < c$, we immediately get

$$\det A_0 > 0. \tag{9.167}$$

For any given $\boldsymbol{\omega} = (\omega_1, \omega_2, \omega_3) \in \mathbb{R}^3$, $|\boldsymbol{\omega}| = 1$, we now check the generalized eigenvalues and generalized eigenvectors of the matrix

$$A(\boldsymbol{\omega}) = \sum_{k=1}^{3} \omega_k A_k;$$

i.e., we examine for which values of λ the determinant of the matrix

$$A(\boldsymbol{\omega}) - \lambda A_0 = \begin{pmatrix} \boldsymbol{\omega}\cdot\boldsymbol{v}-\lambda & 0 & 0 & g_1 & 0 \\ 0 & \boldsymbol{\omega}\cdot\boldsymbol{v}-\lambda & 0 & g_2 & 0 \\ 0 & 0 & \boldsymbol{\omega}\cdot\boldsymbol{v}-\lambda & g_3 & 0 \\ 0 & 0 & 0 & \boldsymbol{\omega}\cdot\boldsymbol{v}-\lambda & -ha^2(\boldsymbol{\omega}\cdot\boldsymbol{v}-\lambda) \\ \rho\gamma\omega_1+fv_1 & \rho\gamma\omega_2+fv_2 & \rho\gamma\omega_3+fv_3 & 0 & \gamma(\boldsymbol{\omega}\cdot\boldsymbol{v}-\lambda) \end{pmatrix} \tag{9.168}$$

is zero, and find the nontrivial solutions of the corresponding homogeneous linear algebraic system, where $f = \frac{\rho\gamma^3(\boldsymbol{\omega}\cdot\boldsymbol{v}-\lambda)}{c^2}$, $g_i = b(\omega_i - \frac{\lambda}{c^2}v_i)$ $(i = 1,2,3)$.

It is easy to see that

$$\begin{aligned}
&\det(A(\boldsymbol{\omega}) - \lambda A_0) \\
&= \gamma(\boldsymbol{\omega}\cdot\boldsymbol{v}-\lambda)^5 + ha^2 b(\boldsymbol{\omega}\cdot\boldsymbol{v}-\lambda) \\
&\quad\cdot \begin{vmatrix} \boldsymbol{\omega}\cdot\boldsymbol{v}-\lambda & 0 & 0 & \omega_1 - \frac{\lambda}{c^2}v_1 \\ 0 & \boldsymbol{\omega}\cdot\boldsymbol{v}-\lambda & 0 & \omega_2 - \frac{\lambda}{c^2}v_2 \\ 0 & 0 & \boldsymbol{\omega}\cdot\boldsymbol{v}-\lambda & \omega_3 - \frac{\lambda}{c^2}v_3 \\ \rho\gamma\omega_1+fv_1 & \rho\gamma\omega_2+fv_2 & \rho\gamma\omega_3+fv_3 & 0 \end{vmatrix} \\
&= \gamma(\boldsymbol{\omega}\cdot\boldsymbol{v}-\lambda)^5 + ha^2 b\gamma\rho(\boldsymbol{\omega}\cdot\boldsymbol{v}-\lambda) \\
&\quad\cdot \begin{vmatrix} \boldsymbol{\omega}\cdot\boldsymbol{v}-\lambda & 0 & 0 & \omega_1 - \frac{\lambda}{c^2}v_1 \\ 0 & \boldsymbol{\omega}\cdot\boldsymbol{v}-\lambda & 0 & \omega_2 - \frac{\lambda}{c^2}v_2 \\ 0 & 0 & \boldsymbol{\omega}\cdot\boldsymbol{v}-\lambda & \omega_3 - \frac{\lambda}{c^2}v_3 \\ \omega_1 & \omega_2 & \omega_3 & 0 \end{vmatrix} \\
&\quad + \frac{1}{c^2} ha^2 b\gamma^3\rho(\boldsymbol{\omega}\cdot\boldsymbol{v}-\lambda)^2 \\
&\quad\cdot \begin{vmatrix} \boldsymbol{\omega}\cdot\boldsymbol{v}-\lambda & 0 & 0 & \omega_1 - \frac{\lambda}{c^2}v_1 \\ 0 & \boldsymbol{\omega}\cdot\boldsymbol{v}-\lambda & 0 & \omega_2 - \frac{\lambda}{c^2}v_2 \\ 0 & 0 & \boldsymbol{\omega}\cdot\boldsymbol{v}-\lambda & \omega_3 - \frac{\lambda}{c^2}v_3 \\ v_1 & v_2 & v_3 & 0 \end{vmatrix}
\end{aligned}$$

$$
\begin{aligned}
&= \gamma(\boldsymbol{\omega}\cdot\boldsymbol{v}-\lambda)^5 \\
&\quad - ha^2 b\gamma\rho(\boldsymbol{\omega}\cdot\boldsymbol{v}-\lambda)^3\left(\omega_1\left(\omega_1-\frac{\lambda}{c^2}v_1\right)\right. \\
&\quad \left.+\omega_2\left(\omega_2-\frac{\lambda}{c^2}v_2\right)+\omega_3\left(\omega_3-\frac{\lambda}{c^2}v_3\right)\right) \\
&\quad -\frac{1}{c^2}ha^2 b\gamma^3\rho(\boldsymbol{\omega}\cdot\boldsymbol{v}-\lambda)^4\left(v_1\left(\omega_1-\frac{\lambda}{c^2}v_1\right)\right. \\
&\quad \left.+v_2\left(\omega_2-\frac{\lambda}{c^2}v_2\right)+v_3\left(\omega_3-\frac{\lambda}{c^2}v_3\right)\right) \\
&= \gamma(\boldsymbol{\omega}\cdot\boldsymbol{v}-\lambda)^3\left(\left(1-\frac{a^2}{c^4}v^2\right)\lambda^2-2\left(1-\frac{a^2}{c^2}\right)(\boldsymbol{\omega}\cdot\boldsymbol{v})\lambda\right. \\
&\quad \left.+\left(1-\frac{a^2}{c^2}\right)(\boldsymbol{\omega}\cdot\boldsymbol{v})^2-a^2\left(1-\frac{v^2}{c^2}\right)\right).
\end{aligned}
\tag{9.169}
$$

In the last step of the above computation, we used the definition (9.163) for b, h, and γ.

Obviously, $A(\boldsymbol{\omega})$ has a triple generalized eigenvalue

$$
\lambda_2 = \boldsymbol{\omega}\cdot\boldsymbol{v}, \tag{9.170}
$$

and the other two generalized eigenvalues are

$$
\begin{aligned}
\lambda_{1,3} = & \frac{\left(1-\frac{a^2}{c^2}\right)(\boldsymbol{\omega}\cdot\boldsymbol{v})}{1-\frac{a^2v^2}{c^4}} \\
& \mp \frac{a\left(1-\frac{v^2}{c^2}\right)^{\frac{1}{2}}\sqrt{1-\frac{a^2v^2}{c^4}-\frac{1}{c^2}\left(1-\frac{a^2}{c^2}\right)(\boldsymbol{\omega}\cdot\boldsymbol{v})^2}}{1-\frac{a^2v^2}{c^4}}.
\end{aligned}
\tag{9.171}
$$

From $|\boldsymbol{\omega}|=1$, we have $(\boldsymbol{\omega}\cdot\boldsymbol{v})^2 \leqslant v^2$; then

$$
0 \leqslant \frac{a^2v^2}{c^4}+\frac{1}{c^2}\left(1-\frac{a^2}{c^2}\right)(\boldsymbol{\omega}\cdot\boldsymbol{v})^2 \leqslant \frac{v^2}{c^2} < 1.
$$

So λ_1 and λ_3 given by (9.171) are two distinct real generalized eigenvalues.

Moreover, it can be verified that for any given $\boldsymbol{\omega}$ and $\boldsymbol{v}$, the generalized eigenvalues λ_1 and λ_3 are both not equal to $\lambda_2 = \boldsymbol{\omega}\cdot\boldsymbol{v}$. In fact, we can verify this without loss of generality for the case $\boldsymbol{\omega}\cdot\boldsymbol{v} > 0$. Since

$$
\frac{\left(1-\frac{a^2}{c^2}\right)(\boldsymbol{\omega}\cdot\boldsymbol{v})}{1-\frac{a^2v^2}{c^4}} < \boldsymbol{\omega}\cdot\boldsymbol{v},
$$

then $\lambda_1 \neq \lambda_2$. In order that $\lambda_3 = \lambda_2$, we must have

$$\left(1-\frac{a^2}{c^2}\right)(\boldsymbol{\omega}\cdot\boldsymbol{v})+a\left(1-\frac{v^2}{c^2}\right)^{\frac{1}{2}}\sqrt{1-\frac{a^2v^2}{c^4}-\frac{1}{c^2}\left(1-\frac{a^2}{c^2}\right)(\boldsymbol{\omega}\cdot\boldsymbol{v})^2}$$
$$=\left(1-\frac{a^2v^2}{c^4}\right)(\boldsymbol{\omega}\cdot\boldsymbol{v}).$$

But it is easy to verify that the above formula leads to

$$(\boldsymbol{\omega}\cdot\boldsymbol{v})^2 = c^2,$$

which cannot be true.

Thus, in order to show that the generalized eigenvectors of $A(\boldsymbol{\omega})$ constitute a complete system of vectors, we need only prove that there exist three linearly independent generalized eigenvectors corresponding to the triple generalized eigenvalue λ_2 of $A(\boldsymbol{\omega})$. From (9.168) we have

$$A(\boldsymbol{\omega})-\lambda_2 A_0 = \begin{pmatrix} 0 & 0 & 0 & b(\omega_1-\frac{\lambda}{c^2}v_1) & 0 \\ 0 & 0 & 0 & b(\omega_2-\frac{\lambda}{c^2}v_2) & 0 \\ 0 & 0 & 0 & b(\omega_3-\frac{\lambda}{c^2}v_3) & 0 \\ 0 & 0 & 0 & 0 & 0 \\ \rho\gamma\omega_1 & \rho\gamma\omega_2 & \rho\gamma\omega_3 & 0 & 0 \end{pmatrix}. \tag{9.172}$$

For any given $\boldsymbol{\omega} = (\omega_1, \omega_2, \omega_3)$ with $|\boldsymbol{\omega}| = 1$, there exist two linearly independent vectors $\boldsymbol{r} = (r_1, r_2, r_3)$ and $\boldsymbol{s} = (s_1, s_2, s_3)$ orthogonal to $\boldsymbol{\omega}$, i.e.,

$$\boldsymbol{\omega}\cdot\boldsymbol{r} = \boldsymbol{\omega}\cdot\boldsymbol{s} = 0.$$

It follows immediately from (9.172) that

$$(r_1, r_2, r_3, 0, 0)^{\mathrm{T}} \quad \text{and} \quad (s_1, s_2, s_3, 0, 0)^{\mathrm{T}}$$

are exactly two linearly independent generalized eigenvectors corresponding the generalized eigenvalue λ_2 of $A(\boldsymbol{\omega})$. In addition, it is obvious that

$$(0,0,0,0,1)^{\mathrm{T}}$$

is also a generalized eigenvector corresponding the generalized eigenvalue λ_2 of $A(\boldsymbol{\omega})$, which is linearly independent of the above two generalized eigenvectors. This indicates that the generalized eigenvectors of $A(\boldsymbol{\omega})$ form a complete system.

According to Definition 5.8 in Chapter 5 of Volume I, the system (9.160)–(9.162) of relativistic fluid dynamics is a *first-order quasi-linear hyperbolic system*.

Since any equivalent combination or reduction of the equations in a system of partial differential equations does not change the type of system, the system of conservation laws for relativistic fluid dynamics established in the last section is also hyperbolic. For example, instead of p and ρ, we can also take the thermodynamical variables ε and S as the unknown functions to discuss the system formed by the Euler equations (9.145), the conservation equation (9.124), and the entropy equation (9.143). We have the same result for the mathematical structure of this system (see Exercises 18 and 19 and section 9.6.3).

Hyperbolic conservation laws have more applications in relativistic fluid dynamics than in classical fluid dynamics. This is because the speed of any motion and action in relativistic fluid dynamics is supposed to be less than the speed of light c; then parabolic features like the infinite propagation speed presented by the Navier–Stokes equations in classical fluid dynamics cannot appear, even for fluids with dissipation effects of viscosity and heat conduction. Since the model of relativistic fluid dynamics with the above dissipation effects is still not well developed, and due to the limited space of this book, it is impossible for us to present an in-depth discussion of related problems here; the interested reader may refer to [2].

9.6.3 System of One-Dimensional Relativistic Fluid Dynamics

Suppose that the fluid flows along the $x = x_1$ direction, and any state variable takes the same value at any given cross section perpendicular to the x_1 axis. Now we can denote $v = v_1$, and $v_2 = v_3 = 0$, and if all variables depend only on t and $x = x_1$, then the system of relativistic fluid dynamics is reduced to a one-dimensional form. The one-dimensional form of the system of equations (9.124)–(9.125) and (9.131) is

$$\frac{\partial}{\partial t}\left(\frac{\varepsilon c^2+p}{c^2-v^2}-\frac{p}{c^2}\right)+\frac{\partial}{\partial x}\left(\frac{\varepsilon c^2+p}{c^2-v^2}v\right)=0, \tag{9.173}$$

$$\frac{\partial}{\partial t}\left(\frac{\varepsilon c^2+p}{c^2-v^2}v\right)+\frac{\partial}{\partial x}\left(\frac{\varepsilon c^2+p}{c^2-v^2}v^2+p\right)=0, \tag{9.174}$$

$$\frac{\partial}{\partial t}\left(\frac{\rho}{\sqrt{1-\frac{v^2}{c^2}}}\right)+\frac{\partial}{\partial x}\left(\frac{\rho v}{\sqrt{1-\frac{v^2}{c^2}}}\right)=0, \tag{9.175}$$

where $\varepsilon = \frac{\mu}{c^2}$ (see (9.154)).

Now we focus on some related properties of the system of one-dimensional relativistic fluid dynamics. Instead of the system (9.173)–(9.175), we consider the following equivalent system:

$$\frac{\partial v}{\partial t}+v\frac{\partial v}{\partial x}+\frac{c^2-v^2}{\varepsilon c^2+p}\left(\frac{a^2}{c^2}v\frac{\partial \varepsilon}{\partial t}+\frac{p_S}{c^2}v\frac{\partial S}{\partial t}+a^2\frac{\partial \varepsilon}{\partial x}+p_S\frac{\partial S}{\partial x}\right)=0, \tag{9.176}$$

$$\begin{aligned}&2\frac{(\varepsilon c^2+p)}{(c^2-v^2)^2}v\frac{\partial v}{\partial t}+\frac{c^4+a^2v^2}{c^2(c^2-v^2)}\frac{\partial \varepsilon}{\partial t}-\frac{1}{c^2}p_S\frac{\partial S}{\partial t}\\&+\frac{(\varepsilon c^2+p)(c^2+v^2)}{(c^2-v^2)^2}\frac{\partial v}{\partial x}+\frac{(c^2+a^2)v}{c^2-v^2}\frac{\partial \varepsilon}{\partial x}=0,\end{aligned} \tag{9.177}$$

$$\frac{\partial S}{\partial t}+v\frac{\partial S}{\partial x}=0, \tag{9.178}$$

where $a^2 = \frac{\partial p(\varepsilon,S)}{\partial \varepsilon}$ is the square of the local speed of sound, $p_S = \frac{\partial p(\varepsilon,S)}{\partial S}$. Equations (9.176) and (9.178) are the one-dimensional forms of the Euler equations (9.145) and the entropy equation (9.143), respectively, while equation (9.177) is obtained from equation (9.173) by expanding the partial derivatives in it and using (9.178).

Denote

$$U = (v, \varepsilon, S)^{\mathrm{T}}. \tag{9.179}$$

System (9.176)–(9.178) can be written as

$$A_0 \frac{\partial U}{\partial t} + A_1 \frac{\partial U}{\partial x} = 0, \tag{9.180}$$

where

$$A_0 = \begin{pmatrix} 1 & \dfrac{a^2(c^2 - v^2)v}{c^2(\varepsilon c^2 + p)} & \dfrac{(c^2 - v^2)p_S v}{c^2(\varepsilon c^2 + p)} \\ 2\dfrac{(\varepsilon c^2 + p)v}{(c^2 - v^2)^2} & \dfrac{c^4 + a^2 v^2}{c^2(c^2 - v^2)} & -\dfrac{1}{c^2} p_S \\ 0 & 0 & 1 \end{pmatrix}$$

and

$$A_1 = \begin{pmatrix} v & \dfrac{a^2(c^2 - v^2)}{\varepsilon c^2 + p} & \dfrac{(c^2 - v^2)p_S}{\varepsilon c^2 + p} \\ \dfrac{(\varepsilon c^2 + p)(c^2 + v^2)}{(c^2 - v^2)^2} & \dfrac{(c^2 + a^2)v}{c^2 - v^2} & 0 \\ 0 & 0 & v \end{pmatrix}.$$

It is easy to compute that

$$\det A_0 = \frac{c^4 - a^2 v^2}{c^2(c^2 - v^2)} > 0; \tag{9.181}$$

moreover,

$$A_1 - \lambda A_0 = \begin{pmatrix} v - \lambda & \frac{a^2(c^2 - v^2)(1 - \frac{v}{c^2}\lambda)}{\varepsilon c^2 + p} & \frac{(c^2 - v^2)p_S(1 - \frac{v}{c^2}\lambda)}{\varepsilon c^2 + p} \\ \frac{(\varepsilon c^2 + p)(c^2 + v^2 - 2v\lambda)}{(c^2 - v^2)^2} & \frac{(c^2 + a^2)v - (c^2 + \frac{a^2}{c^2}v^2)\lambda}{c^2 - v^2} & \frac{p_S \lambda}{c^2} \\ 0 & 0 & v - \lambda \end{pmatrix}, \tag{9.182}$$

and a direct computation yields

$$\det(A_1 - \lambda A_0) = \frac{v - \lambda}{c^2 - v^2}\left(\left(c^2 - \frac{a^2 v^2}{c^2}\right)\lambda^2 - 2(c^2 - a^2)v\lambda + c^2(v^2 - a^2)\right). \tag{9.183}$$

It follows immediately from (9.183) that the corresponding generalized eigenvalues are

$$\lambda_1 = \frac{v-a}{1-\frac{av}{c^2}}, \quad \lambda_2 = v, \quad \lambda_3 = \frac{v+a}{1+\frac{av}{c^2}}. \tag{9.184}$$

It is easy to verify that

$$\lambda_1 < \lambda_2 < \lambda_3, \tag{9.185}$$

so system (9.176)–(9.178) is strictly hyperbolic.

Before further discussion about the generalized eigenvectors corresponding to these generalized eigenvalues, we first give the following definition.

Definition 9.2. *For the first-order quasi-linear strictly hyperbolic system of partial differential equations*

$$A_0(U)\frac{\partial U}{\partial t} + A_1(U)\frac{\partial U}{\partial x} = 0, \tag{9.186}$$

where $U = (u_1,\dots,u_n)^T$*,* $A_0(U)$ *and* $A_1(U)$ *are* $n\times n$ *matrices, and* $\det A_0(U) \neq 0$*, denote by* $r_k(U)$ *the generalized eigenvector of the matrix* $A_1(U)$ *with respect to* $A_0(U)$ *corresponding to the generalized eigenvalue* $\lambda_k(U)$*. If*

$$\nabla\lambda_k(U)\cdot r_k(U) \neq 0 \tag{9.187}$$

on the domain under consideration, then we say that the kth generalized eigenvalue $\lambda_k(U)$ *of system* (9.186) *is* genuinely nonlinear*. If*

$$\nabla\lambda_k(U)\cdot r_k(U) \equiv 0 \tag{9.188}$$

on the domain under consideration, then we say that the kth generalized eigenvalue $\lambda_k(U)$ *is* linearly degenerate*. The gradients of* $\lambda_k(U)$ *in both* (9.187) *and* (9.188) *are taken with respect to the variable* U*.*

The notions of genuine nonlinearity and linear degeneracy for generalized eigenvalues play important roles in the study of the first-order quasi-linear hyperbolic system of partial differential equations (see [12]).

It is easy to see from (9.182) that the generalized eigenvector corresponding to $\lambda_2 = v$ can be taken as

$$\boldsymbol{r}_2 = (0, -p_S, a^2)^{\mathrm{T}}, \tag{9.189}$$

and it is clear that

$$\nabla\lambda_2 = (1,0,0)^{\mathrm{T}}. \tag{9.190}$$

Then

$$\nabla\lambda_2\cdot\boldsymbol{r}_2 \equiv 0; \tag{9.191}$$

i.e., the generalized eigenvalue λ_2 is linearly degenerate.

We can also infer from (9.182) that the generalized eigenvector corresponding to $\lambda_1 = \frac{v-a}{1-\frac{av}{c^2}}$ can be taken as

$$r_1 = \left(-a\left(1-\frac{v^2}{c^2}\right), \varepsilon + \frac{p}{c^2}, 0\right)^{\mathrm{T}}, \tag{9.192}$$

and

$$\nabla\lambda_1 = \left(\frac{1-\frac{a^2}{c^2}}{\left(1-\frac{av}{c^2}\right)^2}, -a_\varepsilon\frac{1-\frac{v^2}{c^2}}{\left(1-\frac{av}{c^2}\right)^2}, -a_S\frac{1-\frac{v^2}{c^2}}{\left(1-\frac{av}{c^2}\right)^2}\right)^{\mathrm{T}}, \tag{9.193}$$

where $a_\varepsilon = \frac{\partial a}{\partial \varepsilon}$, $a_S = \frac{\partial a}{\partial S}$. Then

$$\nabla\lambda_1 \cdot r_1 = -\frac{1-\frac{v^2}{c^2}}{\left(1-\frac{av}{c^2}\right)^2}\left(a\left(1-\frac{a^2}{c^2}\right) + a_\varepsilon\left(\varepsilon + \frac{p}{c^2}\right)\right). \tag{9.194}$$

It follows from (9.156) that

$$a_\varepsilon = \frac{1}{2a}\frac{\partial^2 p}{\partial \varepsilon^2}; \tag{9.195}$$

then we get

$$\nabla\lambda_1 \cdot r_1 = -\frac{1-\frac{v^2}{c^2}}{2ac^2\left(1-\frac{av}{c^2}\right)^2}\left(2a^2(c^2-a^2) + (\varepsilon c^2 + p)\frac{\partial^2 p}{\partial \varepsilon^2}\right). \tag{9.196}$$

Thus, the generalized eigenvalue λ_1 is genuinely nonlinear if and only if

$$\frac{\partial^2 p}{\partial \varepsilon^2} + \frac{2a^2(c^2-a^2)}{\varepsilon c^2 + p} \quad \text{is sign-preserved.} \tag{9.197}$$

Similarly, the generalized eigenvector corresponding to $\lambda_3 = \frac{v+a}{1+\frac{av}{c^2}}$ can be taken as

$$r_3 = \left(a\left(1-\frac{v^2}{c^2}\right), \varepsilon + \frac{p}{c^2}, 0\right)^{\mathrm{T}},$$

and (9.197) is also the necessary and sufficient condition for the generalized eigenvalues λ_3 being genuinely nonlinear.

The above-stated results are very similar to those in the case of classical ideal fluid dynamics (see Exercise 16).

Here we point out that the condition (9.197), under which the eigenvalues λ_1 and λ_3 for the system of one-dimensional relativistic fluid dynamics are genuinely nonlinear, is independent of v and depends only on the state equation $p = p(\varepsilon, S)$. In fact, as a structural condition, the genuine nonlinearity should not change under the Lorentz transformation, so the variable v should not be involved in the condition to determine the genuine nonlinearity.

For the hot extremely relativistic fluid, which is very significant (see section 9.5.3), the equation of state is $\mu = 3p$; i.e.,

$$p = \frac{c^2}{3}\varepsilon. \tag{9.198}$$

On this occasion, system (9.176)–(9.177) is already closed without the aid of the entropy equation (9.178). It is easy to prove that for this system of two equations, the necessary and sufficient condition for the genuine nonlinearity of the two generalized eigenvalues is still given by (9.197). But now

$$\frac{\mathrm{d}^2 p}{\mathrm{d}\varepsilon^2} \equiv 0 \quad \text{and} \quad a^2 = \frac{c^2}{3}, \tag{9.199}$$

so the two generalized eigenvalues of this system are certainly genuinely nonlinear (see Exercise 17).

9.7 System of Relativistic Magnetohydrodynamics

The main object of study of magnetohydrodynamics introduced in Chapter 3 of Volume I is plasma, and 99% of matter in the vast universe exists in the form of plasma. For the plasma in the universe, from both macroscopic and microscopic points of view, the relativistic effect should be taken into consideration. In this section we consider relativistic magnetohydrodynamics.

9.7.1 Lorentz Invariance of Maxwell's Equations

We first look at electric charges and currents. Experiments show that the charge of a charged particle remains unchanged when transformed from one inertial frame to another; i.e., the charge of a charged particle is independent of the motion state of this particle. Therefore, the charge should be invariant under the Lorentz transformation. Denote by ρ_0 the charge density of the fluid in the proper reference frame (i.e., the comoving frame), which is a four-scalar.

Now we check in a general inertial frame K. Since the proper frame moves with respect to the inertial frame K at the macroscopic velocity $\boldsymbol{v}$ of the fluid, from the discussion in section 9.3.3 we know that if we measure the length in the reference frame K, the length along the $\boldsymbol{v}$ direction is contracted to $\frac{1}{\gamma}$ times the proper length, where $\gamma = 1\Big/\sqrt{1-\frac{v^2}{c^2}}$ and $|v| = |\boldsymbol{v}|$, but the length along the direction vertical to the $\boldsymbol{v}$ direction remains unchanged. Thus, the volume element $\mathrm{d}V_0$ in the proper frame becomes the volume $\mathrm{d}V = \frac{1}{\gamma}\mathrm{d}V_0$ when measured in the reference frame K. Therefore, the charge density ρ in the reference frame K should be given by

$$\rho = \gamma\rho_0. \tag{9.200}$$

Once we have the charge density, we can give the current density in the reference frame K as

$$\boldsymbol{j} = \rho \boldsymbol{v}, \tag{9.201}$$

where $\boldsymbol{v}$ is the macroscopic velocity of the fluid observed in the reference frame K. Denote

$$(j^0, j^1, j^2, j^3) = (\rho c, \rho v_1, \rho v_2, \rho v_3). \tag{9.202}$$

Noting (9.200) and the definition (9.78) of the four-velocity, it is clear that

$$(j^0, j^1, j^2, j^3) = \rho_0(u^0, u^1, u^2, u^3). \tag{9.203}$$

This shows that (j^0, j^1, j^2, j^3) given by (9.202) is a (contravariant) four-vector called the *four-current density*. The continuity equation

$$\frac{\partial \rho}{\partial t} + \operatorname{div} \boldsymbol{j} = 0$$

(see (1.20) in Chapter 1 of Volume I), which describes the conservation of charges, now implies that the four-divergence of the four-current density is equal to zero:

$$\frac{\partial j^\alpha}{\partial x^\alpha} = 0. \tag{9.204}$$

In order to discuss the Lorentz invariance of Maxwell's equations, we need to rewrite them in a tensor form in the Minkowski four-space-time M. We first point out that the electric field and the magnetic field are not two independent but two related vectors. When transformed from one inertial frame to another under the Lorentz transformation, the electric field vector in the new frame depends not only on the electric field vector in the original frame but also on the magnetic field vector in the original frame. This is also the case for the magnetic field. As a matter of fact, this relation between the electric field and the magnetic field can be easily shown by noticing the following simple fact. In a reference frame, a charge at rest generates only an electrostatic field, while in another reference frame moving with respect to this reference frame, this charge becomes a charge in motion, so it generates a magnetic field in addition to an electric field. Therefore, the electric field vector and the magnetic field vector cannot be regarded as the spatial components of two independent four-dimensional vectors, respectively; instead, they should be put together for consideration. Since there are six components altogether, they can be regarded as a second-order skew-symmetric tensor in the Minkowski four-space-time.

To give this tensor, we start from the scalar potential and the vector potential in the electromagnetic field (see section 1.6.2 in Chapter 1 of Volume I). Under the Lorentz gauge, the scalar potential ϕ and the vector potential $\boldsymbol{A}$ of the electromagnetic field satisfy, respectively,

$$\frac{1}{c^2}\frac{\partial^2 \phi}{\partial t^2} - \Delta \phi = \frac{\rho}{\varepsilon_0}, \tag{9.205}$$

$$\frac{1}{c^2}\frac{\partial^2 \boldsymbol{A}}{\partial t^2} - \Delta \boldsymbol{A} = \mu_0 \boldsymbol{j} \tag{9.206}$$

and the Lorentz condition

$$\frac{1}{c^2}\frac{\partial \phi}{\partial t} + \operatorname{div} \boldsymbol{A} = 0 \tag{9.207}$$

(see (1.144)–(1.146) in Chapter 1 of Volume I), where ε_0 and μ_0 are the dielectric constant and the magnetic conductivity in a vacuum, which satisfy $\varepsilon_0\mu_0 = c^{-2}$. The right-hand side

$$\left(\frac{\rho}{c\varepsilon_0}, \mu_0 \boldsymbol{j}\right) = \mu_0(j^0, j^1, j^2, j^3)$$

of the system (9.205) (dividing both sides by c) and (9.206) is a (contravariant) four-vector, while the left-hand side is the wave operator, which is invariant under the Lorentz transformation. Therefore,

$$(A^0, A^1, A^2, A^3) = \left(\frac{1}{c}\phi, \boldsymbol{A}\right) \tag{9.208}$$

is a (contravariant) four-vector and the system (9.205)–(9.206) and the Lorentz condition (9.207) can be represented, respectively, in the forms

$$\Box A^\alpha = \mu_0 j^\alpha \quad (\alpha = 0, \ldots, 3) \tag{9.209}$$

and

$$\frac{\partial A^\beta}{\partial x^\beta} = 0, \tag{9.210}$$

where $\Box$ stands for the wave operator, $\Box = g^{\alpha\beta}\frac{\partial}{\partial x^\alpha}\frac{\partial}{\partial x^\beta}$, and

$$(g^{\alpha\beta}) = \begin{pmatrix} 1 & & & \\ & -1 & & \\ & & -1 & \\ & & & -1 \end{pmatrix}.$$

From the (contravariant) four-vector (A^α), we can define the second-order (contravariant) four-tensor

$$F^{\alpha\beta} = g^{\beta\delta}\frac{\partial A^\alpha}{\partial x^\delta} - g^{\alpha\gamma}\frac{\partial A^\beta}{\partial x^\gamma} \quad (\alpha, \beta = 0, \ldots, 3). \tag{9.211}$$

It is easy to show that

$$\begin{aligned}
F^{00} &= F^{11} = F^{22} = F^{33} = 0, \\
F^{01} &= -F^{10} = -\frac{1}{c}\left(\frac{\partial A_1}{\partial t} + \frac{\partial \phi}{\partial x_1}\right), \\
F^{02} &= -F^{20} = -\frac{1}{c}\left(\frac{\partial A_2}{\partial t} + \frac{\partial \phi}{\partial x_2}\right), \\
F^{03} &= -F^{30} = -\frac{1}{c}\left(\frac{\partial A_3}{\partial t} + \frac{\partial \phi}{\partial x_3}\right), \\
F^{12} &= -F^{21} = \frac{\partial A_2}{\partial x_1} - \frac{\partial A_1}{\partial x_2}, \\
F^{13} &= -F^{31} = \frac{\partial A_3}{\partial x_1} - \frac{\partial A_1}{\partial x_3}, \\
F^{23} &= -F^{32} = \frac{\partial A_3}{\partial x_2} - \frac{\partial A_2}{\partial x_3}.
\end{aligned} \tag{9.212}$$

Noting furthermore that the electric field intensity $\boldsymbol{E}$ and the magnetic induction intensity $\boldsymbol{B}$ can be given by the scalar potential ϕ and the vector potential $\boldsymbol{A}$ of the electromagnetic field (see (1.139) and (1.138) in Chapter 1 of Volume I),

$$\boldsymbol{E} = -\operatorname{grad}\phi - \frac{\partial \boldsymbol{A}}{\partial t}, \tag{9.213}$$

$$\boldsymbol{B} = \operatorname{rot}\boldsymbol{A}, \tag{9.214}$$

we have

$$(F^{\alpha\beta}) = \begin{pmatrix} 0 & \frac{1}{c}E_1 & \frac{1}{c}E_2 & \frac{1}{c}E_3 \\ -\frac{1}{c}E_1 & 0 & B_3 & -B_2 \\ -\frac{1}{c}E_2 & -B_3 & 0 & B_1 \\ -\frac{1}{c}E_3 & B_2 & -B_1 & 0 \end{pmatrix}. \tag{9.215}$$

This tensor is called the (contravariant) *field intensity tensor* of the electromagnetic field. It is a second-order skew-symmetric tensor.

Denote

$$({}^*F^{\alpha\beta}) = \begin{pmatrix} 0 & B_1 & B_2 & B_3 \\ -B_1 & 0 & -\frac{1}{c}E_3 & \frac{1}{c}E_2 \\ -B_2 & \frac{1}{c}E_3 & 0 & -\frac{1}{c}E_1 \\ -B_3 & -\frac{1}{c}E_2 & \frac{1}{c}E_1 & 0 \end{pmatrix}. \tag{9.216}$$

It can be shown that $({}^*F^{\alpha\beta})$ is also a second-order (contravariant) tensor (see Exercise 20) called the *dual tensor* of $(F^{\alpha\beta})$.

Once we have the field intensity tensor $(F^{\alpha\beta})$ and its dual tensor $({}^*F^{\alpha\beta})$, Maxwell's equations,

$$\operatorname{div}\boldsymbol{E} = \frac{\rho}{\varepsilon_0}, \tag{9.217}$$

$$\operatorname{rot}\boldsymbol{E} = -\frac{\partial \boldsymbol{B}}{\partial t}, \tag{9.218}$$

$$\operatorname{div}\boldsymbol{B} = 0, \tag{9.219}$$

$$\operatorname{rot}\boldsymbol{B} = \mu_0\left(\varepsilon_0\frac{\partial \boldsymbol{E}}{\partial t} + \boldsymbol{j}\right), \tag{9.220}$$

in a vacuum (see (1.54)–(1.57) in Chapter 1 of Volume I) can be written in simple tensor form. Equations (9.217) and (9.220) can be combined as

$$\frac{\partial F^{\alpha\beta}}{\partial x^\beta} = \mu_0 j^\alpha \quad (\alpha = 0,\ldots,3), \tag{9.221}$$

while equations (9.218) and (9.219) can be combined as

$$\frac{\partial^* F^{\alpha\beta}}{\partial x^\beta} = 0 \quad (\alpha = 0,\dots,3). \tag{9.222}$$

When checking the electromagnetic field in a medium, instead of $(F^{\alpha\beta})$, we introduce the field intensity tensor

$$(G^{\alpha\beta}) = \begin{pmatrix} 0 & D_1 & D_2 & D_3 \\ -D_1 & 0 & \frac{1}{c}H_3 & -\frac{1}{c}H_2 \\ -D_2 & -\frac{1}{c}H_3 & 0 & \frac{1}{c}H_1 \\ -D_3 & \frac{1}{c}H_2 & -\frac{1}{c}H_1 & 0 \end{pmatrix}, \tag{9.223}$$

where $\boldsymbol{D}$ is the electric flux density or the electric displacement vector, and $\boldsymbol{H}$ is the magnetic field intensity. Its dual tensor is

$$(^*G^{\alpha\beta}) = \begin{pmatrix} 0 & \frac{1}{c}H_1 & \frac{1}{c}H_2 & \frac{1}{c}H_3 \\ -\frac{1}{c}H_1 & 0 & -D_3 & D_2 \\ -\frac{1}{c}H_2 & D_3 & 0 & -D_1 \\ -\frac{1}{c}H_3 & -D_2 & D_1 & 0 \end{pmatrix}. \tag{9.224}$$

Thus, Maxwell's equations,

$$\operatorname{div} \boldsymbol{D} = \rho, \tag{9.225}$$

$$\operatorname{rot} \boldsymbol{E} = -\frac{\partial \boldsymbol{B}}{\partial t}, \tag{9.226}$$

$$\operatorname{div} \boldsymbol{B} = 0, \tag{9.227}$$

$$\operatorname{rot} \boldsymbol{H} = \frac{\partial \boldsymbol{D}}{\partial t} + \boldsymbol{j}, \tag{9.228}$$

in a medium (see (1.183)–(1.186) in Chapter 1 of Volume I) can be written as

$$\frac{\partial G^{\alpha\beta}}{\partial x^\beta} = \frac{1}{c} j^\alpha \quad (\alpha = 0,\dots,3) \tag{9.229}$$

and

$$\frac{\partial^* G^{\alpha\beta}}{\partial x^\beta} = 0 \quad (\alpha = 0,\dots,3). \tag{9.230}$$

Thanks to the tensor forms in the Minkowski four-space-time, it goes without saying that Maxwell's equations (9.221)–(9.222) and (9.229)–(9.230) are invariant under the Lorentz transformation.

9.7.2 Energy–Momentum Tensor of the Electromagnetic Field

Let $(\tau^{\alpha\beta})$ be the energy–momentum tensor of the electromagnetic field. Referring to the discussion of the energy–momentum tensor for the relativistic fluid in section 9.5, and noting $x^0 = ct$ and $x^i = x_i$ $(i = 1,2,3)$, then τ^{00} is the electromagnetic energy density, and $(\tau^{01}, \tau^{02}, \tau^{03})$ is the electromagnetic energy flux density vector (divided by c), while $(\tau^{10}, \tau^{20}, \tau^{30})^{\mathrm{T}}$ is the electromagnetic momentum density vector (multiplied by c), and (τ^{ij}) $(i, j = 1,2,3)$ is the electromagnetic momentum flux density tensor.

From section 1.7 in Chapter 1 of Volume I, we know that

the electromagnetic energy density is $\frac{1}{2}(\varepsilon E^2 + \mu H^2)$;

the electromagnetic energy flux density vector is $\boldsymbol{S} = \boldsymbol{E} \times \boldsymbol{H}$;

the electromagnetic momentum density vector is $\frac{1}{c^2}\boldsymbol{S} = \frac{1}{c^2}\boldsymbol{E} \times \boldsymbol{H}$;

the electromagnetic momentum flux density tensor is

$\frac{1}{2}(\varepsilon E^2 + \mu H^2)\boldsymbol{I} - \varepsilon \boldsymbol{E} \otimes \boldsymbol{E} - \mu \boldsymbol{H} \otimes \boldsymbol{H}$,

where ε and μ are, respectively, the dielectric constant and the magnetic conductivity of the medium in the electromagnetic field. Thus we have

$$\tau^{00} = \frac{1}{2}(\varepsilon E^2 + \mu H^2), \tag{9.231}$$

$$\tau^{0i} = \tau^{i0} = \frac{1}{c}(\boldsymbol{E} \times \boldsymbol{H})_i \quad (i = 1,2,3), \tag{9.232}$$

$$\tau^{ij} = -(\varepsilon E_i E_j + \mu H_i H_j) + \frac{1}{2}(\varepsilon E^2 + \mu H^2)\delta^{ij} \quad (i, j = 1,2,3), \tag{9.233}$$

where δ^{ij} is the Kronecker symbol.

The above formulas can be unified in the following form:

$$\tau^{\alpha\beta} = -c g_{\delta\gamma} F^{\alpha\delta} G^{\beta\gamma} + \frac{1}{2} g^{\alpha\beta}(\mu H^2 - \varepsilon E^2), \tag{9.234}$$

where $g_{\alpha\beta} = g^{\alpha\beta}$. It is not hard to directly verify (9.234) by the expressions (9.215) and (9.223) of $(F^{\alpha\beta})$ and $(G^{\alpha\beta})$. In addition, we need to show that $(\tau^{\alpha\beta})$ given by (9.234) is indeed a second-order (contravariant) tensor. This is guaranteed by the following lemma.

Lemma 9.1. $\mu H^2 - \varepsilon E^2$, *being invariant under the Lorentz transformation, is a scalar in the Minkowski four-space-time.*

The proof of this lemma is left to the reader (Exercise 21).

9.7.3 Energy–Momentum Tensor of the Ideal Magnetic Fluid

Combining (9.121) and (9.234), we know that the *energy–momentum tensor of the relativistic electromagnetic fluid* is

$$T^{\alpha\beta} = \frac{1}{c^2}(\widetilde{\mu} + p)u^\alpha u^\beta - pg^{\alpha\beta} + \tau^{\alpha\beta}. \tag{9.235}$$

Note here that in the above formula we denote by $\widetilde{\mu}$ the proper energy density of the fluid to avoid notational confusion, while in the expression of $\tau^{\alpha\beta}$ given by (9.234) we denote by μ the magnetic conductivity. From (9.235) we can obtain the general system of relativistic magnetohydrodynamics with regard to the dissipation (viscosity and heat conduction) effect, but we confine our discussion to the case of ideal magnetic fluids.

From (3.8) in Chapter 3 of Volume I, the classical conduction current density is

$$\sigma(\boldsymbol{E} + \mu\boldsymbol{v} \times \boldsymbol{H}), \tag{9.236}$$

where σ is the electric conductivity, μ is the magnetic conductivity, and $\boldsymbol{v}$ is the velocity of the fluid. From (9.200) and (9.201), the relativistic conduction current density should be

$$\boldsymbol{j} = \sigma\gamma(\boldsymbol{E} + \mu\boldsymbol{v} \times \boldsymbol{H}). \tag{9.237}$$

From (9.202), $\gamma(\boldsymbol{E} + \mu\boldsymbol{v} \times \boldsymbol{H})$ should be the spatial components of a (contravariant) four-vector, by which we define the spatial components of the electric field four-vector (e^α) (see (9.240)). Similarly, define $\gamma(\boldsymbol{H} - \varepsilon\boldsymbol{v} \times \boldsymbol{E})$ as the spatial components of the magnetic field four-vector (h^α) (see (9.241)).

Definition 9.3. *The (contravariant) four-vectors*

$$e^\alpha = u_\beta F^{\alpha\beta} \quad (\alpha = 0,\ldots,3) \tag{9.238}$$

and

$$h^\alpha = u_\beta {}^*G^{\alpha\beta} \quad (\alpha = 0,\ldots,3) \tag{9.239}$$

are called the electric field four-vector *and the* magnetic field four-vector, *respectively, where* (u_β), ($F^{\alpha\beta}$), *and* ($^*G^{\alpha\beta}$) *are given by* (9.133), (9.215), *and* (9.224), *respectively.*

From the expressions of ($F^{\alpha\beta}$) and ($^*G^{\alpha\beta}$) and the definitions of the electric field four-vector and the magnetic field four-vector, noting $\boldsymbol{D} = \varepsilon\boldsymbol{E}$ and $\boldsymbol{B} = \mu\boldsymbol{H}$, we immediately get

$$\begin{aligned} e^0 &= \frac{1}{c}(u^1E_1 + u^2E_2 + u^3E_3), \\ e^1 &= \gamma E_1 + u^2B_3 - u^3B_2, \\ e^2 &= \gamma E_2 + u^3B_1 - u^1B_3, \\ e^3 &= \gamma E_3 + u^1B_2 - u^2B_1 \end{aligned} \tag{9.240}$$

and

$$\begin{aligned} h^0 &= \frac{1}{c}(u^1H_1 + u^2H_2 + u^3H_3), \\ h^1 &= \gamma H_1 - u^2D_3 + u^3D_2, \\ h^2 &= \gamma H_2 - u^3D_1 + u^1D_3, \\ h^3 &= \gamma H_3 - u^1D_2 + u^2D_1. \end{aligned} \tag{9.241}$$

It is clear from (9.240) and (9.241) that, in the proper frame,

$$e^0 = 0, \quad e^i = E_i \quad (i = 1,2,3) \tag{9.242}$$

and

$$h^0 = 0, \quad h^i = H_i \quad (i = 1,2,3). \tag{9.243}$$

Now we discuss the magnetic fluid, i.e., the high-temperature plasma with very large electric conductivity σ (see section 3.1 in Chapter 3 of Volume I). On this occasion, we can approximately take $\mu = \mu_0$ as the magnetic conductivity in a vacuum. Furthermore, we assume that the magnetic fluid is ideal, i.e., there is no effect of viscosity and heat conduction, and $\sigma = +\infty$. Since the current $\boldsymbol{j}$ is always finite, from (9.237), for the ideal magnetic fluid, we have

$$\boldsymbol{E} = -\mu_0 \boldsymbol{v} \times \boldsymbol{H}. \tag{9.244}$$

Noting also (9.242), it is easy to show that $(e^\alpha) = 0$ in the proper frame. Since (e^α) is a four-vector, in any inertial frame we have

$$(e^\alpha) = 0. \tag{9.245}$$

Thus, the energy–momentum tensor $(\tau^{\alpha\beta})$ in the electromagnetic field can be expressed only by the magnetic field four-vector (h^α).

In the proper frame we have $\boldsymbol{E} = \boldsymbol{D} = 0$; using (9.242) and the expressions (9.215) and (9.223) of $(F^{\alpha\beta})$ and $(G^{\alpha\beta})$, it can be verified directly from (9.234) that

$$\begin{aligned} \tau^{\alpha\beta} &= -c g_{\delta\gamma} F^{\alpha\delta} G^{\beta\gamma} + \frac{1}{2} g^{\alpha\beta} \mu_0 H^2 \\ &= \begin{cases} -\mu_0 h^\alpha h^\beta & \text{if } \alpha \neq \beta, \\ -\mu_0 (\frac{1}{2}|h|^2 + (h^\alpha)^2) & \text{if } \alpha = \beta, \end{cases} \end{aligned} \tag{9.246}$$

where $|h|^2 = (h^0)^2 - (h^1)^2 - (h^2)^2 - (h^3)^2$. Noting that, in the proper frame,

$$(u^0, u^1, u^2, u^3) = (c, 0, 0, 0), \tag{9.247}$$

formula (9.246) can be rewritten as

$$\tau^{\alpha\beta} = -\mu_0 h^\alpha h^\beta + \mu_0 \left(\frac{1}{2} g^{\alpha\beta} - \frac{1}{c^2} u^\alpha u^\beta \right) |h|^2. \tag{9.248}$$

Since (h^0, h^1, h^2, h^3) is a four-vector, $|h|^2$ is a four-scalar, and then the right-hand side of (9.248) is a (contravariant) four-tensor. This four-tensor is the same as the (contravariant) four-tensor $(\tau^{\alpha\beta})$ in the proper frame, so it is the same as $(\tau^{\alpha\beta})$ in any given inertial reference frame, i.e., (9.248) is valid for any inertial reference frame. Equation (9.248) is exactly the expression of the electromagnetic energy–momentum tensor for the ideal magnetic fluid. Plugging it into (9.235) gives the expression for the *energy–momentum tensor of the ideal magnetic fluid*.

9.7.4 System of Ideal Relativistic Magnetohydrodynamics

From (9.235) and (9.248), the energy–momentum tensor of the ideal magnetic fluid can be written as

$$T^{\alpha\beta} = \frac{1}{c^2}(\widetilde{\mu} + p - \mu_0|h|^2)u^\alpha u^\beta - \left(p - \frac{1}{2}\mu_0|h|^2\right)g^{\alpha\beta} - \mu_0 h^\alpha h^\beta$$
$$(\alpha, \beta = 0, \ldots, 3), \tag{9.249}$$

where $\widetilde{\mu}$ is the proper internal energy density (not including the magnetic field energy) of the fluid and μ_0 is the magnetic conductivity in a vacuum.

From this, the conservation system of energy and momentum for the ideal magnetic fluid can be given as

$$\frac{\partial T^{\alpha\beta}}{\partial x^\beta} = 0 \quad (\alpha = 0, \ldots, 3), \tag{9.250}$$

i.e.,

$$\frac{\partial}{\partial x^\beta}\left(\frac{1}{c^2}(\widetilde{\mu} + p - \mu_0|h|^2)u^\alpha u^\beta - \left(p - \frac{1}{2}\mu_0|h|^2\right)g^{\alpha\beta} - \mu_0 h^\alpha h^\beta\right) = 0$$
$$(\alpha = 0, \ldots, 3). \tag{9.251}$$

For an ideal magnetic fluid, from (9.245) we know that $({}^*F^{\alpha\beta})$ depends only on (h^α), so Maxwell's equations (9.229)–(9.230) can be reduced to (9.230), and (9.229) is only used to determine the current (j^α). Furthermore, noting (9.242)–(9.243), it can be shown from the expression (9.216) of $({}^*F^{\alpha\beta})$ that, in the proper frame,

$${}^*F^{\alpha\beta} = \begin{cases} \mu_0 h^\beta & \text{for } \alpha = 0, \\ -\mu_0 h^\alpha & \text{for } \beta = 0, \\ 0 & \text{for } \alpha\beta \neq 0. \end{cases}$$

But now $(u^0, u^1, u^2, u^3) = (c, 0, 0, 0)$, so the above formula can be written as

$${}^*F^{\alpha\beta} = \frac{\mu_0}{c}(u^\alpha h^\beta - u^\beta h^\alpha). \tag{9.252}$$

Since the right-hand side of the above formula is a second-order (contravariant) tensor and is the same as the tensor $({}^*F^{\alpha\beta})$ in the proper frame, then (9.252) is true in any given inertial reference frame.

Hence, for an ideal magnetic fluid, Maxwell's equations can be taken in the following simple form:

$$\frac{\partial}{\partial x^\beta}(u^\alpha h^\beta - u^\beta h^\alpha) = 0 \quad (\alpha = 0, \ldots, 3). \tag{9.253}$$

In relativistic magnetohydrodynamics, in addition to the system (9.251) describing the conservation of momentum and energy, and Maxwell's equations (9.253), we need the continuity equation describing the conservation of particle numbers. This equation has the same form as that in the absence of the electromagnetic field discussed in section 9.5, which is given by (see (9.130))

$$\frac{\partial}{\partial x^\alpha}(\widetilde{\rho} u^\alpha) = 0. \tag{9.254}$$

To avoid confusion with the charge density, here we denote by $\widetilde{\rho}$ the proper mass density of the fluid.

Thus, we obtain the *system of relativistic ideal magnetohydrodynamics*, which is a system of nine equations, (9.251), (9.253), and (9.254).

Now we discuss the above system further.

First, we show that in the scope of classical solutions, the continuity equation (9.254) is equivalent to the entropy equation

$$\frac{\partial}{\partial x^\alpha}(\widetilde{\rho} S u^\alpha) = 0 \tag{9.255}$$

or

$$u^\alpha \frac{\partial S}{\partial x^\alpha} = 0, \tag{9.256}$$

where S is the entropy per unit mass of the fluid in the proper frame.

As a matter of fact, system (9.251) can be rewritten as

$$\begin{aligned}
&\frac{1}{c^2} u^\alpha \frac{\partial}{\partial x^\beta}[(\widetilde{\mu} + p - \mu_0 |h|^2) u^\beta] + \frac{1}{c^2}(\widetilde{\mu} + p - \mu_0 |h|^2) u^\beta \frac{\partial u^\alpha}{\partial x^\beta} \\
&- g^{\alpha\beta} \frac{\partial}{\partial x^\beta}\left(p - \frac{1}{2}\mu_0 |h|^2\right) - \mu_0 \frac{\partial h^\alpha}{\partial x^\beta} h^\beta - \mu_0 h^\alpha \frac{\partial h^\beta}{\partial x^\beta} = 0 \\
&\qquad (\alpha = 0, 1, 2, 3).
\end{aligned} \tag{9.257}$$

From the definition (9.239) of (h^α) and the skew-symmetry of $({}^*G^{\alpha\beta})$, it is easy to see that

$$u_\alpha h^\alpha = 0. \tag{9.258}$$

Multiplying both sides of (9.257) by u_α, then summing over α from 0 to 3, and noting that, due to (9.87) and (9.133), we have

$$u_\alpha u^\alpha = c^2, \quad u_\alpha \frac{\partial u^\alpha}{\partial x^\beta} = 0 \tag{9.259}$$

as well as (9.258), system (9.257) can be reduced to

$$\begin{aligned}
&\frac{\partial}{\partial x^\beta}((\widetilde{\mu} + p - \mu_0 |h|^2) u^\beta) - g^{\alpha\beta} u_\alpha \frac{\partial}{\partial x^\beta}\left(p - \frac{1}{2}\mu_0 |h|^2\right) \\
&- \mu_0 u_\alpha \frac{\partial h^\alpha}{\partial x^\beta} h^\beta = 0,
\end{aligned} \tag{9.260}$$

i.e.,

$$\begin{aligned}
&\frac{\partial}{\partial x^\beta}((\widetilde{\mu} + p) u^\beta) - u^\beta \frac{\partial p}{\partial x^\beta} \\
&- \mu_0 \left(\frac{1}{2} u^\beta \frac{\partial |h|^2}{\partial x^\beta} + |h|^2 \frac{\partial u^\beta}{\partial x^\beta} + u_\alpha h^\beta \frac{\partial h^\alpha}{\partial x^\beta}\right) = 0.
\end{aligned} \tag{9.261}$$

During the computation of (9.261) from (9.260), we used

$$u^\beta = g^{\alpha\beta} u_\alpha. \tag{9.262}$$

Maxwell's equations (9.253) can be written as

$$u^\alpha \frac{\partial h^\beta}{\partial x^\beta} + h^\beta \frac{\partial u^\alpha}{\partial x^\beta} - u^\beta \frac{\partial h^\alpha}{\partial x^\beta} - h^\alpha \frac{\partial u^\beta}{\partial x^\beta} = 0. \tag{9.263}$$

Multiplying both sides of the above formula by h_α, then summing over 0 to 3 with respect to α, and noting

$$h_\alpha u^\alpha = 0 \tag{9.264}$$

(it follows immediately from (9.258)), we get

$$\frac{1}{2} u^\beta \frac{\partial |h|^2}{\partial x^\beta} + |h|^2 \frac{\partial u^\beta}{\partial x^\beta} - h_\alpha h^\beta \frac{\partial u^\alpha}{\partial x^\beta} = 0, \tag{9.265}$$

where $h_0 = h^0$, $h_i = -h^i$ $(i = 1,2,3)$. It follows easily from (9.264) that

$$h_\alpha \frac{\partial u^\alpha}{\partial x^\beta} = -u^\alpha \frac{\partial h_\alpha}{\partial x^\beta} = -u_\alpha \frac{\partial h^\alpha}{\partial x^\beta}. \tag{9.266}$$

Then (9.265) can also be written as

$$\frac{1}{2} u^\beta \frac{\partial |h|^2}{\partial x^\beta} + |h|^2 \frac{\partial u^\beta}{\partial x^\beta} + u_\alpha h^\beta \frac{\partial h^\alpha}{\partial x^\beta} = 0. \tag{9.267}$$

Thus, (9.261) is reduced to

$$\frac{\partial}{\partial x^\beta}((\widetilde{\mu} + p) u^\beta) - u^\beta \frac{\partial p}{\partial x^\beta} = 0. \tag{9.268}$$

The above formula is exactly the same as (9.134) in section 9.5. Then, using the same arguments as in section 9.5, we conclude that if the continuity equation (9.254) holds, then thanks to (9.268), so does the conservation equation (9.256) of entropy, and vice versa.

Second, we point out that there are still some unsatisfactory aspects in the nine equations (9.251), (9.253), and (9.254) or (9.256) of the obtained system of ideal magnetohydrodynamics. In fact, the 0th equation (corresponding to $\alpha = 0$) in Maxwell's equations (9.253) is

$$\frac{\partial}{\partial x^i}(u^0 h^i - u^i h^0) = 0, \tag{9.269}$$

where, according to the convention, the repeated Latin letter indicates the summation over this index from 1 to 3. Equation (9.269) involves no partial derivatives with respect to x^0 (i.e., the partial derivatives with respect to t), so it is not an evolutionary equation. In addition, the problem of whether the solution of the above system satisfies (9.258) is still open. In order to avoid the trouble caused by this, we first prove the following result.

Lemma 9.2. *Equation* (9.269) *can be reduced to the requirement on the initial condition, i.e., if* (9.269) *is satisfied at $x^0 = 0$, then it is satisfied for any $x^0 \in \mathbb{R}$.*

Proof. The last three equations in Maxwell's equations (9.253) (corresponding to $\alpha = 1,2,3$) are

$$\frac{\partial}{\partial x^0}(u^i h^0 - u^0 h^i) + \frac{\partial}{\partial x^j}(u^i h^j - u^j h^i) = 0 \quad (i = 1,2,3). \tag{9.270}$$

Taking partial derivatives with respect to x^i in the above formula, and summing over i from 1 to 3, we get

$$\frac{\partial}{\partial x^0}\left(\frac{\partial}{\partial x^i}(u^i h^0 - u^0 h^i)\right) = 0. \tag{9.271}$$

This yields the conclusion of the lemma easily. □

Now we obtain the system of ideal magnetohydrodynamics composed of eight equations altogether, which are the equations (9.251) of momentum conservation and energy conservation, the three equations corresponding to $\alpha = 1,2,3$ in Maxwell's equations (9.253), and the continuity equation (9.254) or the entropy conservation equation (9.256). Noting furthermore that $u^0 = c\gamma$ and from (9.258), we have

$$h^0 = \frac{1}{c}(v_1 h^1 + v_2 h^2 + v_3 h^3); \tag{9.272}$$

the independent unknown functions in the above system can be taken as v_1, v_2, v_3, h^1, h^2, h^3, and two thermodynamical state variables, so the total is also eight. Thus, the above eight equations constitute a closed system. Compared with the system of nonrelativistic ideal magnetohydrodynamics in section 3.4 in Chapter 3 of Volume I, we have more reason to conclude that this system is hyperbolic, but a rigorous mathematical conclusion remains to be studied.

Exercises

1. Prove that the form of the wave equation

$$\frac{\partial^2 u}{\partial t^2} - c^2 \Delta u = 0$$

remains unchanged under the Lorentz transformation.

2. Prove the inverse triangle inequality (9.53).
3. Prove that in the Minkowski four-space-time M, a Lorentz transformation must be an orthogonal transformation.
4. Let A be the coefficient matrix of the linear transformation L in the Minkowski four-space-time M. Prove that L is an orthogonal transformation if and only if

$$A^{\mathrm{T}}\begin{pmatrix} 1 & & & \\ & -1 & & \\ & & -1 & \\ & & & -1 \end{pmatrix} A = \begin{pmatrix} 1 & & & \\ & -1 & & \\ & & -1 & \\ & & & -1 \end{pmatrix}.$$

5. A supersonic plane flies at a speed of $v = 3 \times 10^{-6}c$. How long does it take for this plane to make its clock one second slower than a clock on the earth?
6. Suppose that a rod, with a proper length l_0 and located parallel to the x axis in the inertial frame K, moves at a speed v along the positive direction of the x axis. Suppose further that the inertial frame $\overline{K}$ moves with respect to K at a speed V along the positive direction of the x axis. Find the length $\overline{l}$ of the rod measured in the inertial frame $\overline{K}$.
7. Verify the formula (9.58).
8. Suppose that a particle moves in the xy plane of the inertial frame K at a speed v along the direction with an angel θ to the x axis. Suppose furthermore that the inertial frame $\overline{K}$ with respect to K moves at a speed V along the positive direction of the x axis. Find the speed $\overline{v}$ of the particle with respect to $\overline{K}$ and the angle $\overline{\theta}$ between its direction and the $\overline{x}$ axis.
9. Suppose that the momentum of a particle is twice its nonrelativistic momentum $m_0 v$, where m_0 is the rest mass of the particle. What is the velocity v of this particle?
10. Let the velocity of an electron be $v = 0.99c$ and the rest mass $m_0 = 9.1 \times 10^{-31}$ kg. Compute the kinetic energy of this electron, and compare it with the kinetic energy computed in classical Newtonian mechanics.
11. Suppose that a static particle with rest mass m_0 collides with and then absorbs a photon with energy Q. Find the velocity v, the mass M, the rest mass M_0, and the kinetic energy E_k of the new particle.
12. Use the Euler equations (9.145) to prove that in steady flows,

$$\frac{\mu + p}{\rho\sqrt{1 - \dfrac{v^2}{c^2}}}$$

remains unchanged along the streamline, i.e.,

$$\boldsymbol{v} \cdot \nabla \left(\frac{\mu + p}{\rho\sqrt{1 - \dfrac{v^2}{c^2}}} \right) = 0.$$

13. Consider the motion of a charged particle in an electromagnetic field. Suppose that the rest mass of the particle is m_0, and its charge q does not change with the motion state of the particle. In classical Newtonian mechanics, the motion equation of the particle is

$$m_0 \frac{\mathrm{d}^2 \boldsymbol{x}}{\mathrm{d}t^2} = q(\boldsymbol{E} + \boldsymbol{v} \times \boldsymbol{B})$$

(where the expression of the Lorentz force can be found in (1.64) in Chapter 1 of Volume I). Prove that in relativistic mechanics, the above equation should be replaced by the following system of equations:

$$\frac{\mathrm{d}\boldsymbol{x}}{\mathrm{d}t} = \left(m_0^2 + \frac{p^2}{c^2}\right)^{-\frac{1}{2}} \boldsymbol{p}(= \boldsymbol{v}),$$

$$\frac{\mathrm{d}\boldsymbol{p}}{\mathrm{d}t} = q(\boldsymbol{E} + \boldsymbol{v} \times \boldsymbol{B}),$$

where $\boldsymbol{v}$ and $\boldsymbol{p}$ represent the velocity and the momentum of the particle, respectively.

14. In the kinetic theory of gases, the Vlasov–Maxwell system can be established by taking the momentum $\boldsymbol{p}$ as a variable instead of the velocity $\boldsymbol{v}$ and denoting by $f(t, \boldsymbol{x}, \boldsymbol{p})$ the distribution function of particles (see section 8.7 in Chapter 8). Derive the following relativistic version of this system:

$$\frac{\partial f}{\partial t} + \boldsymbol{v} \cdot \nabla_x f + q(\boldsymbol{E} + \boldsymbol{v} \times \boldsymbol{B}) \cdot \nabla_p f = 0,$$

$$\rho(t, \boldsymbol{x}) = q \int_\infty f(t, \boldsymbol{x}, \boldsymbol{p}) \mathrm{d}p,$$

$$\boldsymbol{j}(t, \boldsymbol{x}) = q \int_\infty \boldsymbol{v} f(t, \boldsymbol{x}, \boldsymbol{p}) \mathrm{d}p,$$

where $\boldsymbol{E}$, $\boldsymbol{B}$, ρ, and $\boldsymbol{j}$ satisfy the standard Maxwell's equations (see section 1.3 in Chapter 1 of Volume I).

15. Prove that the characteristic curve of the Vlasov equation (the first equation) in Exercise 14 is exactly given by the system in Exercise 13.

16. It is known that the eigenvalues of the classical system of one-dimensional ideal fluid dynamics, given by (2.90)–(2.92) in Chapter 2 of Volume I, are

$$\lambda_1 = u - c, \quad \lambda_2 = u, \quad \lambda_3 = u + c,$$

where c is the local speed of sound: $c^2 = \frac{\partial p(\rho,S)}{\partial \rho}$ (see (2.107) in Chapter 2 of Volume I). Prove that λ_2 is linearly degenerate, and λ_1 and λ_3 are genuinely nonlinear if and only if $\frac{\partial^2 p}{\partial \rho^2}$ is sign-preserved (the case that really makes sense in applications is $\frac{\partial^2 p}{\partial \rho^2} > 0$).

17. Suppose that $p = a^2 \varepsilon$, where $\varepsilon = \frac{\mu}{c^2}$ and a is a positive constant less than c. Write down the system of one-dimensional relativistic ideal fluid dynamics composed of two equations. Compute the generalized eigenvalues and the generalized eigenvectors of this system, and prove that its generalized eigenvalues are genuinely nonlinear.

18. Taking $(v_1, v_2, v_3, \varepsilon, S)$ as unknown functions, give the matrix form of the system of first-order partial differential equations composed of the Euler equations (9.145), the conservation equation (9.124), and the entropy equation (9.143).

19. For the system of first-order partial differential equations given in Exercise 18, compute its generalized eigenvalues and generalized eigenvectors, and discuss its mathematical structure.

20. Prove that for the tensor $(F^{\alpha\beta})$ given by (9.216), its dual tensor $({}^*F^{\alpha\beta})$ is also a second-order (contravariant) tensor.

21. Prove that $\mu H^2 - \varepsilon E^2$ is a four-scalar.
22. Use the expression of $(F^{\alpha\beta})$ and the definition of the second-order (contravariant) tensor to prove that, under the Lorentz transformation given by (9.31), the electric field intensity $\boldsymbol{E}$ and the magnetic induction intensity $\boldsymbol{B}$ are transformed according to the following laws:

$$\overline{\boldsymbol{E}} = \gamma \boldsymbol{E} - (\gamma - 1)\frac{1}{v^2}(\boldsymbol{v} \cdot \boldsymbol{E})\boldsymbol{v} + \gamma(\boldsymbol{v} \times \boldsymbol{B}),$$
$$\overline{\boldsymbol{B}} = \gamma \boldsymbol{B} - (\gamma - 1)\frac{1}{v^2}(\boldsymbol{v} \cdot \boldsymbol{B})\boldsymbol{v} - \gamma(\boldsymbol{v} \times \boldsymbol{E}).$$

23. Use the continuity equation (9.254) and Maxwell's equations (9.253) to prove that

$$u^\beta \frac{\partial}{\partial x^\beta}\left(\frac{h^\alpha}{\rho}\right) = \frac{h^\beta}{\rho}\frac{\partial u^\alpha}{\partial x^\beta} + \frac{u^\alpha}{\rho}\frac{\partial h^\beta}{\partial x^\beta} \quad (\alpha = 0, \ldots, 3).$$

This is the corresponding relativistic version of the "frozen" principle of magnetic field lines (see (3.65) in Chapter 3 of Volume I).

Bibliography

[1] Anile A. M. *Relativistic Fluids and Magneto-Fluids*. Cambridge Monographs on Mathematical Physics. Cambridge: Cambridge University Press, 1989.

[2] Anile A. M., Choquet-Bruhat Y., eds. *Relativistic Fluid Dynamics*, Noto, 1987. Lecture Notes in Mathematics 1385. Berlin, Heidelberg: Springer–Verlag, 1989.

[3] Cai B. *Special Relativity* (in Chinese). Beijing: Higher Education Press, 1991.

[4] Cwiklik R. *Albert Einstein and the Theory of Relativity*. New York: Barron's Educational Series, 1987.

[5] Einstein A. et al. *The Principle of Relativity: A Collection of Original Memoirs on the Special and General Theory of Relativity*, Second Edition. New York: Dover Publications, 1952.

[6] Godlewski E., Raviart P.-A. *Numerical Approximation of Hyperbolic Systems of Conservation Laws*. New York: Springer–Verlag, 1996.

[7] Gu C., Li T. et al. *Mathematical Physics Equations* (in Chinese). Shanghai: Shanghai Science and Technology Press, 1987.

[8] Hawking S. *A Brief History of Time: From the Big Bang to Black Holes*. New York: Bantam Books, 1988.

[9] Jackson J. D. *Classical Electrodynamics*, Third Edition. New York: John Wiley & Sons, 1998.

[10] Landau L. D., Lifshitz E. M. *The Classical Theory of Fields*, Fourth Edition. Oxford: Elsevier, 1975.

[11] Landau L. D., Lifshitz E. M. *Fluid Mechanics*, Second Edition. Oxford: Pergamon Press, 1987.

[12] Li T. *Global Classical Solutions for Quasilinear Hyperbolic Systems*. Paris: Masson/Chichester: J. Wiley, 1994.

[13] Lichnerowicz A. *Relativistic Hydrodynamics and Magnetohydrodynamics*. New York: Benjamin, 1967.

[14] Petrowsky I. G. *Lectures on Partial Differential Equations*. New York: Interscience Publishers, 1954.

[15] Penrose R. *Emperor's New Mind: Concerning Computers, Minds, and the Laws of Physics*. Oxford: Oxford University Press, 1989.

[16] Shi C. *Relativistic Fluid Dynamics* (in Chinese). Beijing: Science Press, 1992.

[17] Smoller J., Temple B. *Global existence of the relativistic Euler equations*. Commun. Math. Phys., 156 (1993), 67–99.

[18] Weinberg S. *Gravitation and Cosmology: Principles and Applications of the General Theory of Relativity*. New York: John Wiley & Sons, 1972.

[19] You G. *How Did Einstein Create the Theory of Relativity* (in Chinese). Changsha: Hunan Education Press, 1993.

Chapter 10

Quantum Mechanics

10.1 Establishment of Quantum Mechanics

Until the end of the 19th century, the theory of classical physics (Newtonian mechanics, electrodynamics, thermodynamics, and statistical physics) had developed to a close to perfect degree. There was in general a kind of optimism among physicists, thinking that all the major theoretical problems in physics had been resolved, and all that was left were only minor issues. Lord Kelvin, a famous British physicist, said in 1900, "The ship of science finally pulled into a tranquil bay after conquering a large number of 'underwater reefs' and 'severe storms'. There is nothing new to be discovered in physics now. All that remains is more and more precise measurement."

However, in this optimistic atmosphere, there hid a crisis of classical physics and bred a revolution in physics. This revolution gave birth to two fundamental theoretical systems of modern physics: relativity and quantum mechanics. The birth of relativity and the basic contents of (special) relativity were introduced in the previous chapter. In this chapter, we will introduce the basic contents of quantum mechanics and focus on the basic equations of quantum mechanics: the Schrödinger equation and the Dirac equations.

First, we briefly explain what brought about the birth of quantum mechanics at the end of the 19th century, since classical physics was almost perfect then. Some new physical phenomena occurred in scientific experiments at that time with the development of science and technology, and explaining these phenomena posed insurmountable obstacles for classical physics. These difficulties mainly occurred in problems such as the blackbody radiation, the photoelectric effect, and the atomic spectrum. The attempts made by scientists when they tried to explain these phenomena prompted the birth of quantum mechanics.

10.1.1 Blackbody Radiation and Planck's Quantum Theory

The needs of metallurgical high-temperature measurement techniques, astronomy, etc., pushed forward research on thermal radiation. *Thermal radiation* means the electromagnetic radiation issued by an object when its temperature is higher than absolute zero. By the end of the 19th century, people had recognized that thermal radiation and light

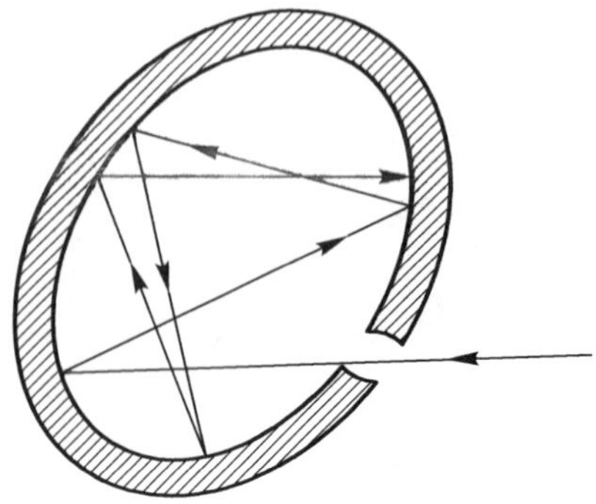

Figure 10.1

radiation are both electromagnetic waves. All objects emit thermal radiation to their surroundings, meanwhile absorbing thermal radiation from their surroundings. Suppose that an object exchanges heat with its surroundings only through thermal radiation; when this object and its surroundings are in the thermal equilibrium state, the rates of emission and absorption of thermal radiation are the same; otherwise thermal equilibrium would be destroyed. An important problem is to study the distribution of thermal radiation energy in different frequency ranges when the object is in thermal equilibrium. Since the spectrum of electromagnetic waves produced by thermal radiation is very complicated and related to the material type and surface state of the radiator, as an ideal state, physicists study an object that can absorb all thermal radiation, called a *blackbody*. For a blackbody, when thermal equilibrium is reached, the thermal radiation spectrum has nothing to do with either the material type or the surface states, but depends only on the temperature, which facilitates the study of the thermal radiation spectrum. Real blackbodies do not exist. However, if we open a small hole in a cavity (see Figure 10.1), as long as the hole area is very small compared with the surface area of this cavity, then this cavity can be regarded as a blackbody, and where the hole is located is the surface of the blackbody. In fact, the thermal radiation entering the cavity through the small hole escapes through the same hole with extremely small possibility, and will eventually be almost completely absorbed by the cavity wall.

Now we consider a certain blackbody that has definite temperature and reaches thermal equilibrium, and we study the distribution of its thermal radiation energy density E_ν with respect to frequency ν, where $E_\nu \mathrm{d}\nu$ stands for the radiation energy emitted by unit volume with frequency in the range $(\nu, \nu + \mathrm{d}\nu)$. From the experiment, the law of change of radiation energy density E_ν with respect to frequency ν is shown by the solid line in Figure 10.2. However, at the beginning of the 20th century, according to classical theories of electrodynamics and statistical physics, J. W. Rayleigh and J. H. Jeans obtained the blackbody radiation formula

$$E_\nu = \frac{8\pi}{c^3} kT\nu^2, \tag{10.1}$$

where c is the speed of light, T is the absolute temperature, and $k = 1.38 \times 10^{-16}$ J/K is the *Boltzmann constant*. Equation (10.1) is also known as the *Rayleigh–Jeans formula*, and its graph is shown by the dotted line in Figure 10.2. From Figure 10.2, the result given

Figure 10.2

by the Rayleigh–Jeans formula matches quite well with the experimental curve in the low-frequency part; however, in the high-frequency part, the differences between the two curves are very obvious. Moreover, from the Rayleigh–Jeans formula, E_ν monotonically tends to infinity as $\nu \to \infty$. This will lead to an absurd conclusion: the energy will continuously be radiated into the electromagnetic field, and the higher the frequency is, the larger the radiation energy is, hence energy will be radiated unlimitedly. This is known as the "UV catastrophe." This conclusion goes against common sense, which tells us that at a given temperature, the maximum of thermal radiation energy density always occurs at a limited frequency corresponding to a visible color of a thermal radiator, e.g., a hot poker is red, the sun is yellow and white, etc.

Almost at the same time as the above work was done, based on Wien's semiempirical formula (which earned W. Wien the 1911 Nobel Prize in Physics), Planck proposed a formula that is now named for him:

$$E_\nu = \frac{c_1 \nu^3}{e^{c_2\nu/T} - 1}, \tag{10.2}$$

where c_1, c_2 are positive constants. Not only is Planck's formula simple in form, but it also matches very well with the experimental data. It is believed that this is not accidental, and that this formula certainly contains a very important scientific principle that has not yet been revealed. After two months of effort, Planck proposed an adventurous and revolutionary "quantum" hypothesis: For radiation with a given frequency ν, objects can only absorb or emit energy that is an integer multiple of the smallest energy unit ε. In other words, the absorption or emission of electromagnetic radiation proceeds only in the form of a "quantum," and the energy of the "quantum" is

$$\varepsilon = h\nu, \tag{10.3}$$

where $h = 6.6256 \times 10^{-34}$ J·s is called the *Planck constant*. Using this assumption, Planck derived his blackbody radiation formula (10.2) in theory. The radiation of electromagnetic energy can proceed only in a discrete and discontinuous way; this idea of Planck's cannot been understood in classical physics. He himself also tried to find evidence for energy quantization in classical theories, but undoubtedly all ended in failure. Despite this, his bold hypothesis became the dawn of quantum theory. Planck won the 1918 Nobel Prize in Physics for this contribution.

10.1.2 Photoelectric Effect and Einstein's Light Quantum Theory

The phenomenon that a metal emits electrons when its surface is irradiated by light is called the *photoelectric effect*, and the emitted electrons are called *photoelectrons*. This phenomenon occurs because the electromagnetic wave transmits energy to electrons in the metal such that electrons escape from the metal surface. This phenomenon was discovered by H. Hertz in 1887.

Experiments show that in order to have the photoelectric effect, the frequency of irradiation light has a critical value ν_0: when the irradiation light frequency $\nu < \nu_0$, no matter how intense the light is, it cannot make electrons escape from the metal surface. In addition, the kinetic energy of the escaping photoelectron is irrelevant to the intensity of the irradiation light, but it is proportional to the light frequency ν. The intensity of the irradiation light affects only the intensity of the photocurrent, i.e., the number of photoelectrons escaping per unit time from unit area of the metal surface.

This experimental fact also cannot be explained by classical physics. According to classical electromagnetic theory, the energy of an electromagnetic wave is only determined by its intensity (the square of the amplitude; see section 1.4 in Chapter 1); the greater the intensity, the greater the energy of the electromagnetic wave. Therefore, no matter what the frequency is, as long as the metal is irradiated by sufficiently intense light, there should be a number of electrons escaping from the surface at high speed, but real experimental results do not show this.

To explain the photoelectric effect, Einstein proposed the *photon hypothesis* in 1905: light with frequency ν is made up of a group of particles with energy $h\nu$. This kind of particle is called the *photon* or the *light quantum*, and each photon has energy

$$E = h\nu, \tag{10.4}$$

where h is the Planck constant. When a photon hits the metal surface, an electron in the metal can obtain all of its energy.

Using Einstein's photon hypothesis, the photoelectric effect problem was smoothly solved. Suppose that an electron needs energy W to escape from the metal. Then only when

$$h\nu > W \tag{10.5}$$

are there electrons escaping from the metal. Suppose that the velocity of the escaping electron is v; then we have

$$\frac{1}{2}mv^2 = h\nu - W, \tag{10.6}$$

where m is the mass of the electron. When $\nu < \nu_0 = W/h$ (critical frequency), the electron cannot break the barriers of the metal surface to escape from the metal.

Light quantum theory was furthermore verified in 1922–1923 by the Compton–Wu Youxun effect on X-ray scattering.

Einstein's light quantum theory pointed out not only the discontinuity of the electromagnetic energy (made up of discrete light quanta) but also that light itself is made up of particles (although this is what Newton insisted on more than two centuries ago, it is not a simple returning to Newtonian corpuscular theory). But we know that Maxwell and Hertz once pointed out that light is an electromagnetic wave (see Chapter 1), and this statement has also been verified by many experiments. Therefore, light should be either a particle or a wave. Moreover, according to Einstein's photon hypothesis, we can also establish the relation between the photon momentum and the light wavelength. In fact, the motion velocity of a photon is the speed of light c; from Einstein's formula $E = mc^2$ (see (9.85) in Chapter 9) in the theory of relativity, we know that the photon momentum is $p = E/c$. Noting (10.4), we get the following relation between the photon momentum p and the wavelength λ $(= \frac{c}{\nu})$:

$$p = \frac{h}{\lambda}. \tag{10.7}$$

This formula establishes the relation between a particle and a wave, which seem to be two absolutely distinct concepts. However, the unity of these two concepts, particle and wave, is one of the basic features of quantum mechanics.

Young Einstein published four epochal papers in succession in Germany's most prestigious physics journal, *Annalen der Physik*, during 1905 and 1906. The first one is about the photoelectric effect, in which the concept of the light quantum was introduced; the fourth one is about the mathematical theory of molecular Brownian motion; the second and third ones which were already mentioned in Chapter 9 concern special relativity. It was just his first paper on the photoelectric effect that won him the 1921 Nobel Prize in Physics.

10.1.3 Atomic Spectrum and Bohr's Model of the Hydrogen Atom

Spectrum analysis originated from Newton, but later on the specific spectrum lines of different elements were used for component analysis of trace elements. In 1885 the Swiss physicist Balmer found that the wave number $\frac{1}{\lambda}$ of the visible light spectrum of the hydrogen atom satisfies the following laws:

$$\frac{1}{\lambda} = R_H\left(\frac{1}{2^2} - \frac{1}{n^2}\right), \quad n = 3, 4, \ldots, \tag{10.8}$$

where λ is the wavelength and R_H is called the *Rydberg constant*, whose value is

$$R_H = 1.096776 \times 10^7/\text{m}. \tag{10.9}$$

The spectrum line satisfying (10.8) is called the *Balmer series*.

Balmer formula (10.8) surprisingly matches the observations. However, classical Newtonian mechanics and Maxwell's electromagnetic theory cannot explain why the above spectrum does not distribute continuously but discretely.

The hydrogen atom is made up of one proton and one electron. According to Newtonian mechanics and Maxwell's electromagnetic theory, the electron cannot revolve around the proton stably forever, like the earth revolves around the sun. Because the motion with which the electron revolves around the proton is an accelerated one, according to classical electrodynamics, it will be accompanied by the emission of electromagnetic waves in the accelerated movement of charged particles. Therefore, the electron will lose energy because of continuous radiation, the radius of its orbit will continue shrinking, and eventually it will land on the proton (see Figure 10.3). Of course, this does not match the fact that the hydrogen atom does not emit radiation and is stable under normal conditions.

In addition, according to classical radiation theory, the frequency of the electromagnetic wave emitted by the electron has something to do with the period of the electron moving around the nucleus. Note that according to the above model, the frequency of

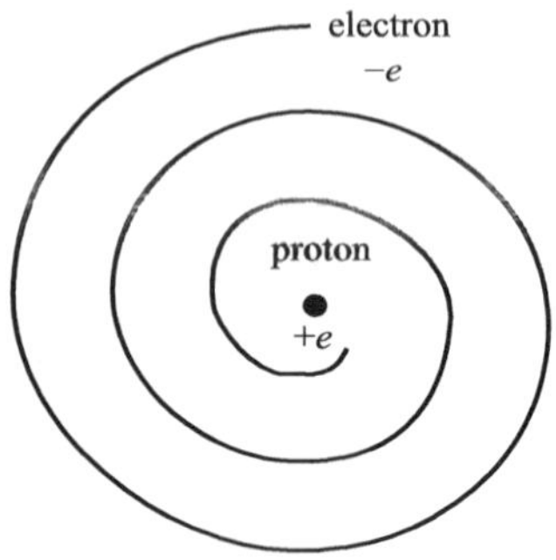

Figure 10.3

the electron revolving around the proton will be getting higher and higher continuously, so the frequency spectrum of the electromagnetic wave emitted by the electron should be continuous. This does not agree with the above-mentioned experimental results either.

In order to explain the stability of the hydrogen atom and the discreteness of the visible spectrum, the young Danish physicist N. Bohr proposed in 1913 the following *hydrogen atom model* by a clever use of the Planck relation.

Assumption 10.1. The electron can move only along a special set of circular orbits. The electron moving along this kind of special orbit is in a stable state (stationary state), neither absorbing nor emitting light (electromagnetic wave). The angular momentum of the electron moving on this kind of stable orbit can only be an integer multiple of $\hbar = h/2\pi$, i.e.,

$$mvr = n\hbar, \quad n = 1, 2, \ldots, \tag{10.10}$$

where m, v, and r represent the electron mass, velocity, and orbital radius, respectively, and h is the Planck constant introduced before.

Assumption 10.2. Only when the electron jumps from one stable orbit to another one do the emission and absorption of light occur. Suppose that the energies of the electron before and after the transition are E_m and E_n, respectively; then the frequency of emitting or absorbing light is

$$\nu = \frac{|E_m - E_n|}{h}. \tag{10.11}$$

Light is emitted when $E_m > E_n$ and absorbed when $E_m < E_n$.

Bohr successfully used these two assumptions to interpret the stability of the hydrogen atom and the discreteness of the hydrogen atomic spectrum, and the value of the Rydberg constant figured out by this matches quite well with the value given in spectroscopy. Bohr also predicted under these two assumptions that there exists another line series in the ultraviolet band. In the second year (1914) after this prediction, this line series was actually observed by C. V. Lyman. The ideas of the quantization of atomic energy and the quantum transition that Bohr first proposed in these assumptions are still correct. Bohr's successful theory won him the 1922 Nobel Prize in Physics.

Although Bohr's theory had been successful on many issues, its limitations were soon revealed. This theory cannot explain why these stable states of atoms exist. Meanwhile, although Bohr's assumptions can elegantly explain the spectrum of the hydrogen atom, the interpretation of the spectrum of complex atoms (even the helium atom) encountered great difficulties. In addition, Bohr's theory can only deal with simple periodic motion and is powerless to solve nonbounded state problems such as the scattering problem. Even so, this outstanding work of Bohr, which is now called "old quantum theory," played an important role in the establishment of quantum mechanics.

10.1.4 De Broglie Hypothesis of the Matter Wave

The understanding of the world by classical physics can be summarized as follows: there are two main objects in the universe—particles and fields (or waves). The motion of particles follows Newton's law, and electromagnetic fields obey Maxwell's equations. Einstein's light quantum theory showed that electromagnetic waves exhibit particle properties. Conversely, do particles exhibit wave properties?

In 1924, L. de Broglie, a young French noble and an insightful physicist, gave a positive answer to this issue in his doctoral dissertation. He thought that the coexistence of waves and particles was not limited to the case of light that Einstein had studied, but also applied to all particles. He pointed out that Einstein's light quantum formulas (10.4) and (10.7) are valid for any material particles, that is, a material particle with energy E and momentum p also has wave properties, and its frequency ν and wavelength λ are determined by energy E and momentum p, respectively, in the following formulas:

$$\nu = \frac{E}{h}, \quad \lambda = \frac{h}{p}. \tag{10.12}$$

De Broglie called the wave determined by the above formulas the *matter wave*.

Using the concept of the matter wave, de Broglie unified the theories on these two main objects (the two existence forms of matters) in classical physics. He also corresponded the stationary state of atoms in Bohr's theory to the standing wave, and then interpreted the quantization issue of the angular momentum of the electron. Suppose that the orbital radius of the electron is r; then the length of its orbit is $2\pi r$. After one rotation of the electron around the nucleus, the circumference of the circle is required to be an integer multiple of the wavelength λ so that standing waves can connect with each other smoothly, i.e.,

$$2\pi r = n\lambda, \quad n = 1,2,\dots. \tag{10.13}$$

From this we get $\lambda = 2\pi r/n$, and substituting this into the second formula of (10.12), we get $p = nh/2\pi r$, i.e.,

$$rp = n\hbar, \quad n = 1,2,\dots. \tag{10.14}$$

This is Bohr's quantization condition (10.10) for angular momentum.

In the third year (1927) after de Broglie proposed the matter wave theory, the wave properties of material particles were verified by the electron diffraction experiment of C. Davisson and L. Germer. This experiment also verified the correctness of the de Broglie relation $\lambda = h/p$.

Now that material particles bear wave properties, why do macroscopic objects not show their wave behaviors under normal conditions? Imagine that a bullet with mass 1 g is shot at the speed 100 m/s. By the second formula of (10.12) we can calculate that the corresponding wavelength of the matter wave is $\lambda = 6.6 \times 10^{-33}$ m. Since the wavelength is very short, under normal conditions, macroscopic objects do not show wave behavior.

De Broglie won the 1929 Nobel Prize in Physics because of his matter wave theory. So far, he is one of the very few young people, and is the earliest one, to have received the Nobel Prize in Physics through a Ph.D. dissertation.

10.1.5 Establishment of Quantum Mechanics

The theory of quantum mechanics was built up from 1925 to 1928.

In 1925, W. Heisenberg published a paper titled "Quantum Theoretical Re-interpretation of Kinematic and Mechanical Relations." On one hand, he inherited the reasonable core of early quantum theory, such as the quantization of atomic energy; on the other hand, he abandoned some contents without experimental basis in the early quantum theory, such as the concept of the electronic orbit. He emphasized that any physical theory should focus only on observable physical quantities, and based on these observable quantities each physical quantity can be endowed with a matrix (of infinite order); from this he created the theory of quantum mechanics that is now known as *matrix mechanics*, and for this he received the 1932 Nobel Prize in Physics. Bohr's ideas played an important role in the establishment of matrix mechanics.

In 1926, Austrian physicist E. Schrödinger further extended the concept of the de Broglie matter wave, deriving the motion equation satisfied by the matter wave under the quantum system—the *Schrödinger equation*—then established the theory of quantum mechanics that was later known as *wave mechanics*; thanks to this he received the 1933 Nobel Prize in Physics. Using the partial differential equations he established, Schrödinger successfully solved a series of significant problems, such as the hydrogen atom spectrum. After that Schrödinger also proved that, although the mathematical tools used to express matrix mechanics and wave mechanics are different, these two are equivalent (it should be said that this equivalence is very natural from the viewpoint of linear operator theory in Hilbert space). It was also in 1926 that M. Born gave the probability interpretation of the wave function in Schrödinger's wave mechanics, which earned him the 1954 Nobel Prize in Physics.

In 1927, W. Heisenberg proposed the famous *uncertainty principle*, which is now also known as the *uncertainty relation*. By then, the current form of nonrelativistic quantum mechanics was almost finished.

In 1928, P. A. M. Dirac, a distinguished British physicist, published a paper entitled "The Quantum Theory of Electrons," in which he introduced special relativity into quantum theory and laid the foundation for relativistic quantum mechanics, which earned him the 1933 Nobel Prize in Physics. Thus, the theoretical system of quantum mechanics was basically complete.

Quantum mechanics, as one of the two most important achievements in physics in the 20th century, has had a huge impact on modern physics and all of human civilization, and now all branches of modern physics and its related disciplines are based on quantum mechanics. The laws of quantum mechanics restrict the behaviors of transistors and integrated circuits, which are basic components of all kinds of electronic devices, such as televisions and computers; quantum mechanics is also the basis of modern chemistry and biology. The only areas of physical science to which quantum mechanics has not yet been applied are the gravitational field and the large-scale structure of the universe. But, unifying quantum theory or its more advanced form—quantum field theory—and Einstein's theory of general relativity into quantum gravity is just what the contemporary British physicist Stephen Hawking is pursuing.

10.2 Schrödinger Equation and Wave Function

The Schrödinger equation is a fundamental equation in (nonrelativistic) quantum mechanics, and it is as important as Newton's equation is to classical mechanics. Those fundamental equations (systems) in the previous nine chapters were all derived rigorously according to basic physical assumptions or laws on the premise that the unknown functions have specific physical meanings. However, in quantum mechanics, the Schrödinger equation itself can only be regarded as a basic assumption that cannot be proved by some other more fundamental assumptions. Even when Schrödinger figured out this equation, he himself did not know the specific meaning of the unknown function of this equation—the *wave function*. After Born gave the probability interpretation of the wave function, which is now accepted by most physicists, Schrödinger strongly opposed it. Even so, the correctness of the Schrödinger equation, and the quantum mechanics based on it, have so far been verified by a large number of scientific experiments and production practices.

10.2.1 Schrödinger Equation

We first discuss the system composed of only one free particle.

From the de Broglie relation (10.12), both the frequency and the wavelength of the matter wave related to the motion of this particle should be specific constants. Then it is reasonable to assume that this wave is a plane wave described by the following wave function:

$$\psi(t,\boldsymbol{x}) = A\,\mathrm{e}^{\mathrm{i}(\boldsymbol{k}\cdot\boldsymbol{x}-\omega t)}, \tag{10.15}$$

where ω and $\boldsymbol{k}$ are the angular frequency and the wave vector, respectively, and A is a complex number. The frequency of this plane wave is $\nu = \omega/2\pi$, and the wavelength is $\lambda = 2\pi/|\boldsymbol{k}|$ (see section 1.5.3 in Chapter 1). Thus, the de Broglie relation (10.12) can be written as

$$\omega = \frac{E}{\hbar}, \quad \boldsymbol{k} = \frac{\boldsymbol{p}}{\hbar}, \tag{10.16}$$

where $\hbar = h/2\pi$, and the second formula is actually the three-dimensional form of the second formula of (10.12). Hence, the plane wave (10.15) can also be written as

$$\psi(t,\boldsymbol{x}) = A\,\mathrm{e}^{\mathrm{i}(\boldsymbol{p}\cdot\boldsymbol{x}-Et)/\hbar}. \tag{10.17}$$

Taking partial derivatives with respect to t and $\boldsymbol{x}$, respectively, of the wave function given by (10.17), we obtain

$$\frac{\partial\psi}{\partial t} = -\mathrm{i}\frac{E}{\hbar}\psi, \tag{10.18}$$

$$\frac{\partial^2\psi}{\partial x_k^2} = -\frac{p_k^2}{\hbar^2}\psi \quad (k = 1,2,3), \tag{10.19}$$

where x_k and p_k are the components of $\boldsymbol{x}$ and $\boldsymbol{p}$, respectively. Suppose that the mass of the particle is m; then, in the classical sense, the energy E and the momentum $\boldsymbol{p}$ of the particle have the following relation:

$$E = \frac{|\boldsymbol{p}|^2}{2m}. \tag{10.20}$$

From (10.18)–(10.20), we immediately know that the *wave function* ψ of the free particle satisfies the equation

$$\mathrm{i}\hbar\frac{\partial\psi}{\partial t} = -\frac{\hbar^2}{2m}\Delta\psi. \tag{10.21}$$

For the general state of the free particle, i.e., the free particle is in a *quantum state* with neither specific momentum nor specific position at any give time, suppose that its wave function is a complex-valued function

$$\psi = \psi(t, \boldsymbol{x}). \tag{10.22}$$

Under certain conditions, this wave function can be represented by the superposition of plane waves, i.e.,

$$\psi(t, \boldsymbol{x}) = \frac{1}{(2\pi)^3}\int_{\mathbb{R}^3}\phi(\boldsymbol{p})\mathrm{e}^{\mathrm{i}(\boldsymbol{p}\cdot\boldsymbol{x}-Et)/\hbar}\,\mathrm{d}\boldsymbol{p}, \tag{10.23}$$

where E and $\boldsymbol{p}$ satisfy (10.20). Since the integrand in (10.23) is a plane wave satisfying equation (10.21), the wave function given by (10.22) should also satisfy equation (10.21).

It is worth mentioning that if we take the following transformation in the energy–momentum relation (10.20) of the particle,

$$E \to \mathrm{i}\hbar\frac{\partial}{\partial t}, \quad \boldsymbol{p} \to \widehat{\boldsymbol{p}} = -\mathrm{i}\hbar\nabla, \tag{10.24}$$

and use the obtained operators on the wave function ψ, we then get (10.21).

Now we discuss the particle moving in the potential field $V(\boldsymbol{x})$. At this moment, in the classical sense we have

$$E = \frac{1}{2m}|\boldsymbol{p}|^2 + V(\boldsymbol{x}). \tag{10.25}$$

This is exactly the *Hamiltonian* in classical mechanics. Operating the transformation given by (10.24) on the right-hand side of (10.25), we obtain the operator

$$\widehat{H} = -\frac{\hbar^2}{2m}\Delta + V(\boldsymbol{x}). \tag{10.26}$$

$\widehat{H}$ is called the *Hamiltonian operator*. Operating the transformation given by (10.24) on both sides of (10.25) and using the obtained operators on the wave function ψ, we get

$$\mathrm{i}\hbar\frac{\partial\psi}{\partial t} = \widehat{H}\psi, \tag{10.27}$$

where $\widehat{H}$ is given by (10.26). This equation is called the *Schrödinger equation*. The potential field V in the Schrödinger equation may depend on time t.

The above discussion is about the equation satisfied by the wave function of one particle. The wave function of a many-particle system composed of, e.g., N particles, has the following form:

$$\psi = \psi(t, \boldsymbol{x}^1, \ldots, \boldsymbol{x}^N), \tag{10.28}$$

where $\boldsymbol{x}^k = (x_1^k, x_2^k, x_3^k)$ stands for the space coordinates of the kth particle ($k = 1, \ldots, N$). Suppose that the external potential field received by the kth particle is $U_k(\boldsymbol{x}^k)$, and the interactive potential among particles is $V(\boldsymbol{x}^1, \ldots, \boldsymbol{x}^N)$; then the wave function ψ satisfies the Schrödinger equation

$$\mathrm{i}\hbar \frac{\partial \psi}{\partial t} = \sum_{k=1}^{N} \left(-\frac{\hbar^2}{2m_k} \Delta_k \psi + U_k(\boldsymbol{x}^k)\psi \right) + V(\boldsymbol{x}^1, \ldots, \boldsymbol{x}^N)\psi, \tag{10.29}$$

where m_k is the mass of the kth particle and Δ_k stands for the Laplace operator with respect to the variable $\boldsymbol{x}^k$, i.e.,

$$\Delta_k = \left(\frac{\partial}{\partial x_1^k}\right)^2 + \left(\frac{\partial}{\partial x_2^k}\right)^2 + \left(\frac{\partial}{\partial x_3^k}\right)^2.$$

For instance, for an atom with N electrons, consider the system composed of its electrons. For this system, the potential of the electric field formed by the nucleus (with charge Ne) is the external potential field, and

$$U_k(\boldsymbol{x}^k) = -\frac{Ne^2}{|\boldsymbol{x}^k|} \quad (k = 1, \ldots, N). \tag{10.30}$$

Here e is the charge of the electron, and the origin of the coordinate system is taken on the nucleus. The interactive potential among electrons is

$$V(\boldsymbol{x}^1, \ldots, \boldsymbol{x}^N) = \sum_{k<l}^{N} \frac{e^2}{|\boldsymbol{x}^k - \boldsymbol{x}^l|}. \tag{10.31}$$

10.2.2 Significance of the Wave Function

What does the wave function $\psi(t, \boldsymbol{x})$ actually mean in physics? What kind of wave is the matter wave? These essential problems were not clear at the beginning. Even the understanding of de Broglie and Schrödinger, who established matter wave theory and wave mechanics, was greatly influenced by classical physics. Schrödinger thought that the wave function itself represented a specific distribution of matter. According to this opinion, the matter particle is regarded as a continuously distributed matter wave packet in three-dimensional space; the size of the wave packet is just that of the matter particle, and the group velocity of the wave packet is just the motion velocity of the matter particle.

To show whether this interpretation of the wave function is reasonable, we first introduce the concept of a wave packet. For simplicity, we take the one-dimensional case as an example.

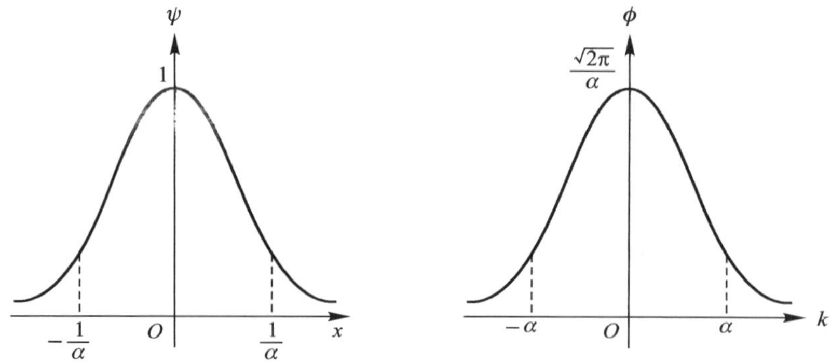

Figure 10.4

The so-called wave packet means the wave determined by a wave function whose strength is obviously not equal to zero only in a space domain that is not large. For example, for the wave given by the Gaussian function

$$\psi(x) = e^{-\frac{1}{2}\alpha^2 x^2}, \tag{10.32}$$

where $\alpha > 0$ is a constant, its strength focuses mainly in the domain $|x| < \frac{1}{\alpha}$ (see Figure 10.4).

The wave packet $\psi(x)$ can be regarded as the superposition of plane waves with different wave numbers k, i.e.,

$$\psi(x) = \frac{1}{2\pi}\int_{-\infty}^{\infty} \phi(k) e^{ikx}\, dx, \tag{10.33}$$

where $\phi(k)$ is the Fourier transform of $\psi(x)$:

$$\phi(k) = \int_{-\infty}^{\infty} \psi(x) e^{-ikx}\, dx. \tag{10.34}$$

The Fourier transform of the Gaussian wave packet (10.32) is

$$\phi(k) = \frac{\sqrt{2\pi}}{\alpha} e^{-\frac{k^2}{2\alpha^2}}. \tag{10.35}$$

It is still a Gaussian wave packet, whose strength focuses mainly in the range of $|k| < \alpha$ (see Figure 10.4).

Now we consider the motion of the wave packet, i.e., its time evolution. Suppose that the wave packet at $t = 0$ is given by (10.33). As $t > 0$ we have

$$\psi(t,x) = \frac{1}{2\pi}\int_{-\infty}^{\infty} \phi(k) e^{i(kx-\omega(k)t)}\, dk. \tag{10.36}$$

Here $\omega = \omega(k)$ is taken such that $\psi(t,x)$ is a solution of the equation under consideration, called the *dispersion relation*. If

$$\omega(k) = c_0 k, \tag{10.37}$$

where c_0 is a constant, then $\psi(t,x)$ given by (10.36) is a traveling wave propagating at velocity c_0: $\psi(t,x) = F(x - c_0 t)$. In this situation, although the center of the wave packet is moving, the shape of the wave packet remains unchanged. If (10.37) does not hold, i.e.,

$$\omega'(k) \not\equiv \text{constant}, \tag{10.38}$$

then plane waves with different wave numbers (wavelengths) propagate at different velocities

$$v_p = \frac{\omega(k)}{k}. \tag{10.39}$$

v_p is called the *phase velocity*. This means that a wave packet superposed by simple harmonic waves with different frequencies will gradually disperse during propagation since different frequencies lead to different phase velocities. This phenomenon is called *dispersion*, and the corresponding wave is called the *dispersive wave*.

$$v_g = \omega'(k) \tag{10.40}$$

is called the *group velocity* of the wave packet. For a dispersive wave, the group velocity v_g is different from the phase velocity v_p of each partial wave.

We take the more general Gaussian wave packet centered at k_0 as an example for further discussion. Suppose that

$$\phi(k) = \frac{\sqrt{2\pi}}{\alpha} \mathrm{e}^{-\frac{(k-k_0)^2}{2\alpha^2}}; \tag{10.41}$$

then

$$\psi(t,x) = \frac{1}{\sqrt{2\pi}\,\alpha} \int_{-\infty}^{\infty} \mathrm{e}^{-\frac{(k-k_0)^2}{2\alpha^2} + \mathrm{i}(kx - \omega(k)t)} \mathrm{d}k. \tag{10.42}$$

Since the strength of the Gauss function $\phi(k)$ is distributed mainly in the neighborhood of $k = k_0$, making Taylor's expansion of $\omega(k)$ near $k = k_0$, we have

$$\omega(k) \approx \omega_0 + v_g(k - k_0) + \frac{\gamma}{2}(k - k_0)^2, \tag{10.43}$$

where $\omega_0 = \omega(k_0)$, $v_g = \omega'(k_0)$ is the group velocity corresponding to the wave number k_0, and $\gamma = \omega''(k_0)$. Plugging (10.43) into (10.42), we get

$$\psi(t,x) \approx \frac{1}{\sqrt{2\pi}\,\alpha} \mathrm{e}^{\mathrm{i}(k_0 x - \omega_0 t)} \int_{-\infty}^{\infty} \exp\Big(\mathrm{i}(k - k_0)(x - v_g t) - \Big(\frac{\mathrm{i}}{2}\gamma t + \frac{1}{2\alpha^2}\Big)(k - k_0)^2\Big) \mathrm{d}k. \tag{10.44}$$

But

$$\mathrm{i}(k - k_0)(x - v_g t) - \left(\frac{\mathrm{i}}{2}\gamma t + \frac{1}{2\alpha^2}\right)(k - k_0)^2 = -z^2 - \frac{\alpha^2 (x - v_g t)^2}{2(1 + \mathrm{i}\gamma\alpha^2 t)},$$

where

$$z = \sqrt{\frac{1+\mathrm{i}\gamma\alpha^2 t}{2\alpha^2}}(k-k_0) - \frac{\mathrm{i}}{2}\sqrt{\frac{2\alpha^2}{1+\mathrm{i}\gamma\alpha^2 t}}(x - v_g t).$$

Substituting the above formulas into (10.44) and noting that $\int_{-\infty}^{\infty} \mathrm{e}^{-z^2}\,\mathrm{d}z = \sqrt{\pi}$, we obtain

$$\begin{aligned}\psi(t,x) &\approx \frac{1}{\sqrt{2\pi}\alpha}\sqrt{\frac{2\alpha^2}{1+\mathrm{i}\gamma\alpha^2 t}}\exp\Big(-\frac{\alpha^2(x-v_g t)^2}{2(1+\mathrm{i}\gamma\alpha^2 t)} + \mathrm{i}(k_0 x - \omega_0 t)\Big)\int_{-\infty}^{\infty} \mathrm{e}^{-z^2}\,\mathrm{d}z \\ &= \sqrt{\frac{1}{1+\mathrm{i}\gamma\alpha^2 t}}\,\mathrm{e}^{-\frac{\alpha^2(x-v_g t)^2}{2(1+\mathrm{i}\gamma\alpha^2 t)} + \mathrm{i}(k_0 x - \omega_0 t)}. \end{aligned} \tag{10.45}$$

The strength distribution of this wave function is

$$|\psi(t,x)|^2 = \frac{1}{\sqrt{1+\gamma^2\alpha^4 t^2}}\,\mathrm{e}^{-\frac{\alpha^2(x-v_g t)^2}{1+\gamma^2\alpha^4 t^2}}. \tag{10.46}$$

It is still a Gauss function with respect to x, which is a wave packet centered at

$$x = v_g t. \tag{10.47}$$

From the above formula we can see that the wave packet moves at the velocity v_g, which is why $v_g = \omega'(k)$ is called the group velocity. From the above discussion on the width of the Gaussian wave packet, the width of this wave packet is

$$\Delta x \approx \frac{1}{\sqrt{2}\alpha}\sqrt{1+\gamma^2\alpha^4 t^2}. \tag{10.48}$$

At $t = 0$, $\Delta x \approx \frac{1}{\sqrt{2}\alpha}$. But if $\gamma \neq 0$, the width Δx of the wave packet will be widened with increasing time t, and the bigger the time t is, the more obvious the width increment is. From (10.48) we can also see that the smaller the initial velocity $1/\alpha$ of the wave packet is (namely, the bigger the α is), the faster the diffusion of the wave packet is.

For the wave function of one free particle (i.e., a solution of the Schrödinger equation (10.21)), it is easy to obtain

$$\omega(k) = \frac{\hbar}{2m}k^2 \tag{10.49}$$

and

$$\gamma = \omega''(k) = \frac{\hbar}{m} \neq 0. \tag{10.50}$$

Therefore, the matter wave packet diffuses inevitably or, more figuratively speaking, the particle (e.g., electron) will get "fatter and fatter" as time goes by. This does not match with the experimental results, so this interpretation of the wave function is unreasonable.

In 1926, when using the Schrödinger equation to deal with the scattering problem, in order to explain the angular distribution of scattering particles, Born considered that what the wave function in the Schrödinger equation describes does not represent the wave motion of any specific physical quantity like classical waves, but the *probability wave* depicting the spatial probability distribution of the particle. Precisely speaking, the wave strength is

$$|\psi(t,\boldsymbol{x})|^2 = \psi(t,\boldsymbol{x})\cdot\overline{\psi}(t,\boldsymbol{x}), \tag{10.51}$$

where $\overline{\psi}(t,\boldsymbol{x})$ stands for the complex conjugate of $\psi(t,\boldsymbol{x})$. This describes the spatial probability distribution of the particle; that is, the probability that this particle appears in the element $\mathrm{d}x$ around $\boldsymbol{x}$ at the moment t is $|\psi(t,\boldsymbol{x})|^2\mathrm{d}x$. The concept of the probability wave proposed by Born unified the particle properties and the wave properties of microscopic objects. The so-called particle properties mean that the object bears particle-like properties such as having fixed mass or charge, which has no necessary relation with the classical notion that "the particle has specific orbit"; however, wave properties imply the superposable property of waves, which may not represent the wave motion of any specific physical quantity.

A system composed of N particles has wave function $\psi = \psi(t,\boldsymbol{x}^1,\ldots,\boldsymbol{x}^N)$, and

$$|\psi(t,\boldsymbol{x}^1,\ldots,\boldsymbol{x}^N)|^2\mathrm{d}x^1\cdots\mathrm{d}x^N \tag{10.52}$$

represents the probability that particle 1 appears in the element $\mathrm{d}x^1$ around $\boldsymbol{x}^1,\ldots,$ and particle N appears in the element $\mathrm{d}x^N$ around $\boldsymbol{x}^N$.

For the phenomenon that material particles neither generate nor annihilate, the number of particles does not change with time and is conserved. The probability of finding a certain particle in the whole space is evidently equal to 1. Then, according to Born's interpretation of the wave function, we should have

$$\frac{\mathrm{d}}{\mathrm{d}t}\int_{\mathbb{R}^3}|\psi(t,\boldsymbol{x})|^2\mathrm{d}x = 0. \tag{10.53}$$

Now we prove (10.53). Applying the complex conjugate on both sides of the Schrödinger equation

$$\mathrm{i}\hbar\frac{\partial\psi}{\partial t} = -\frac{\hbar^2}{2m}\Delta\psi + V(\boldsymbol{x})\psi, \tag{10.54}$$

and noting that the external potential field $V(\boldsymbol{x})$ is always assumed to be a real-valued function, we have

$$-\mathrm{i}\hbar\frac{\partial\overline{\psi}}{\partial t} = -\frac{\hbar^2}{2m}\Delta\overline{\psi} + V(\boldsymbol{x})\overline{\psi}. \tag{10.55}$$

Multiplying both sides of equation (10.54) by $\overline{\psi}$, and both sides of (10.55) by ψ, and subtracting the resulting formulas, we get

$$\mathrm{i}\hbar\frac{\partial}{\partial t}(|\psi|^2) = -\frac{\hbar^2}{2m}(\overline{\psi}\Delta\psi - \psi\Delta\overline{\psi}). \tag{10.56}$$

It is easy to verify directly that

$$\overline{\psi}\Delta\psi - \psi\Delta\overline{\psi} = \mathrm{div}(\overline{\psi}\nabla\psi - \psi\nabla\overline{\psi}).$$

Then (10.56) can also be written as the following form of conservation law:

$$\frac{\partial \rho}{\partial t} + \operatorname{div} \boldsymbol{j} = 0, \tag{10.57}$$

where

$$\rho(t,\boldsymbol{x}) = |\psi(t,\boldsymbol{x})|^2, \tag{10.58}$$

$$\boldsymbol{j}(t,\boldsymbol{x}) = -\frac{\mathrm{i}\hbar}{2m}(\overline{\psi}\nabla\psi - \psi\nabla\overline{\psi}). \tag{10.59}$$

Formally, (10.57) is exactly the same as the continuity equation in fluid dynamics and electrodynamics (see (2.10) in Chapter 2 and (1.50) in Chapter 1), so $\boldsymbol{j}$ is called the *probability flux density vector*.

Under certain conditions, e.g., $\psi(t,\boldsymbol{x})$ and its spatial partial derivatives tend to zero as $|\boldsymbol{x}| \to \infty$ at a sufficiently rapid speed, by Green's formula, (10.53) follows immediately from (10.57).

10.2.3 Mathematical Structure of the Schrödinger Equation

For simplicity, we discuss the case of a free particle (namely, $V(\boldsymbol{x}) \equiv 0$). The Schrödinger equation (10.21) can be written as

$$\frac{\partial \psi}{\partial t} - \mathrm{i}a^2\Delta\psi = 0, \tag{10.60}$$

where $a^2 = \hbar/2m$. We first consider this equation, which is very similar in form to the heat equation.

Suppose that

$$\psi(t,\boldsymbol{x}) = u_1(t,\boldsymbol{x}) + \mathrm{i}u_2(t,\boldsymbol{x}), \tag{10.61}$$

where the real-valued functions u_1 and u_2 are the real part and imaginary part of ψ, respectively. Plugging the above formula into equation (10.60), separating its real part and its imaginary part, we obtain

$$\frac{\partial u_1}{\partial t} + a^2\Delta u_2 = 0, \tag{10.62}$$

$$\frac{\partial u_2}{\partial t} - a^2\Delta u_1 = 0. \tag{10.63}$$

Denoting $U = (u_1, u_2)^{\mathrm{T}}$, the above system can be written in the following matrix form:

$$\frac{\partial U}{\partial t} = \sum_{k,l=1}^{3} A_{kl}\frac{\partial^2 U}{\partial x_k \partial x_l}, \tag{10.64}$$

where

$$A_{11} = A_{22} = A_{33} = \begin{pmatrix} 0 & -a^2 \\ a^2 & 0 \end{pmatrix}$$

and

$$A_{kl} = 0 \quad (k \neq l).$$

For any given $\boldsymbol{\xi} \in \mathbb{R}^3$, $|\boldsymbol{\xi}| = 1$, we have

$$\sum_{k,l=1}^{3} \xi_k \xi_l A_{kl} = \begin{pmatrix} 0 & -a^2 \\ a^2 & 0 \end{pmatrix}. \tag{10.65}$$

For the system of partial differential equations of the form (10.64), if for any given $\boldsymbol{\xi} \in \mathbb{R}^3$, $|\boldsymbol{\xi}| = 1$, the eigenvalue λ of the matrix on the left-hand side of (10.65) always satisfies $\operatorname{Re}\lambda > 0$, then (10.64) is called the *parabolic system in the sense of Petrowsky*. But for the system of Schrödinger equations (10.62)–(10.63), the eigenvalues of the corresponding matrix (10.65) are $\lambda = \pm \mathrm{i}a^2$; then $\operatorname{Re}\lambda = 0$, and thus system (10.62)–(10.63) is not a parabolic system in the sense of Petrowsky. But for system (10.62)–(10.63), matrix $\sum_{k,l=1}^{3} \xi_k \xi_l A_{kl}$ is an invertible positive semidefinite matrix; sometimes this type of system is called a *degenerate parabolic system*.

To show furthermore that the Schrödinger equation (10.60) is not of parabolic type, we discuss the inversion problem with respect to time. Parabolic equations usually describe dissipation phenomena like heat conduction, diffusion, and so on. For this kind of phenomenon, the entropy of the system increases with time, so the process is invertible. This means that this kind of equation cannot be solved along the direction that t decreases; for instance, the former state (when t is negative) cannot be determined by the state at $t = 0$. However, what is the case for the Schrödinger equation?

Applying the complex conjugate on both sides of equation (10.60), we get

$$\frac{\partial \overline{\psi}}{\partial t} + \mathrm{i}a^2 \Delta \overline{\psi} = 0. \tag{10.66}$$

Furthermore, taking the inversion transformation with respect to time, i.e., setting $t' = -t$, we have

$$-\frac{\partial \overline{\psi}(-t', \boldsymbol{x})}{\partial t'} + \mathrm{i}a^2 \Delta \overline{\psi}(-t', \boldsymbol{x}) = 0.$$

Rewriting t' as t in the above formula, we obtain

$$\frac{\partial \overline{\psi}(-t, \boldsymbol{x})}{\partial t} - \mathrm{i}a^2 \Delta \overline{\psi}(-t, \boldsymbol{x}) = 0. \tag{10.67}$$

Equation (10.67) satisfied by $\overline{\psi}(-t, \boldsymbol{x})$ is exactly the same in form as equation (10.60) satisfied by $\psi(t, \boldsymbol{x})$. This shows that if $\psi(t, \boldsymbol{x})$ is a solution of the Schrödinger equation (10.60), then so is the corresponding time inversion state $\overline{\psi}(-t, \boldsymbol{x})$. That is, the solvability of the Schrödinger equation along two directions of time t is the same. This is just the time inversion invariance of the Schrödinger equation. This feature of the Schrödinger equation is similar to the wave equation. This also reflects that, when solved by the method of semigroup, the solution operators of its well-posed problems form a group instead of only a semigroup (see [8]).

Now we investigate the fundamental solution of the Schrödinger equation (10.60). For simplicity, we only discuss the one-dimensional case:

$$\frac{\partial \psi}{\partial t} - \mathrm{i}a^2 \frac{\partial^2 \psi}{\partial x^2} = 0. \tag{10.68}$$

It is known that the fundamental solution of the heat equation

$$\frac{\partial u}{\partial t} - a^2 \frac{\partial^2 u}{\partial x^2} = 0 \tag{10.69}$$

is

$$E(t,x) = H(t)\frac{1}{2a\sqrt{\pi t}}\mathrm{e}^{-\frac{x^2}{4a^2 t}}, \tag{10.70}$$

where $H(t)$ is the Heaviside function (see, e.g., [2]). Formally, replacing a in (10.70) with $\sqrt{\mathrm{i}}a = \frac{\sqrt{2}}{2}(1+\mathrm{i})a$, we get

$$E(t,x) = H(t)\frac{1-\mathrm{i}}{2a\sqrt{2\pi t}}\mathrm{e}^{\mathrm{i}\frac{x^2}{4a^2 t}}. \tag{10.71}$$

It can be verified that $E(t,x)$ given by the above formula is just the fundamental solution of the Schrödinger equation (10.68), i.e., in the sense of distribution we have

$$\frac{\partial E}{\partial t} - \mathrm{i}a^2 \frac{\partial^2 E}{\partial x^2} = \delta(t,x), \tag{10.72}$$

where $\delta(t,x)$ is the Dirac δ function. To show this, it suffices to verify that for any given $\phi \in C_0^\infty(\mathbb{R}^2)$, it holds that

$$\int_{\mathbb{R}^2} E(t,x)\left(-\frac{\partial \phi}{\partial t} - \mathrm{i}a^2 \frac{\partial^2 \phi}{\partial x^2}\right)\mathrm{d}t\mathrm{d}x = \phi(0,0). \tag{10.73}$$

The proof is left to the reader as an exercise.

Definition 10.1. *Suppose that P is a linear partial differential operator with respect to $\boldsymbol{x} = (x_1,\ldots,x_n)$. If for any given open set $\Omega \subset \mathbb{R}^n$ and any given distribution $u \in \mathcal{D}'(\Omega)$, $Pu \in C^\infty(\Omega)$ can always lead to $u \in C^\infty(\Omega)$, then we say that P is* subelliptic. *Here $\mathcal{D}'(\Omega)$ is the space of distribution, which is composed of all linear continuous functionals on $C_0^\infty(\Omega)$ (see, e.g.,* [1]*).*

For the fundamental solution (10.70) of the heat equation, it is easy to prove that $E \in C^\infty$ in $\mathbb{R}^2 \backslash \{0\}$; thus we know that the heat operator is subelliptic (see [11]). But this is not the case for the fundamental solution $E(t,x)$ of the Schrödinger equation given by (10.71). It is not a C^∞ function in any given neighborhood of $(0,x^0)$, and then the Schrödinger operator $\mathrm{i}\hbar\frac{\partial}{\partial t} + \frac{\hbar}{2m}\frac{\partial^2}{\partial x^2}$ is not subelliptic. This is another important difference between the heat equation and the Schrödinger equation.

The Schrödinger equation is similar to the wave equation regarding the invariance of time inversion, but we can assert that it is essentially different from the wave equation just from the nonsymmetry of the Schrödinger equation with respect to time variable t and space variable x (first-order partial derivative with respect to t and second-order with respect to x). We can furthermore see this difference from the discussion on the Cauchy problem of the Schrödinger equation.

For the Schrödinger equation, we can propose the following Cauchy problem: find a solution of the Schrödinger equation (10.27) in $(0,\infty)\times\mathbb{R}^3$ such that it satisfies at $t=0$ the initial condition

$$\psi(0,\boldsymbol{x})=\psi_0(\boldsymbol{x}), \tag{10.74}$$

where $\psi_0(\boldsymbol{x})$ is a given function.

We still take the one-dimensional Schrödinger equation (10.68) as an example. Using the fundamental solution (10.71), the solution of the Cauchy problem (10.68) and (10.74) (the corresponding one-dimensional case) can be given by the following formula:

$$\psi(t,x)=\frac{1-\mathrm{i}}{2a\sqrt{2\pi t}}\int_{-\infty}^{\infty}\psi_0(\xi)\mathrm{e}^{\mathrm{i}\frac{(x-\xi)^2}{4a^2t}}\,\mathrm{d}\xi. \tag{10.75}$$

In fact, the above formula can be obtained by replacing a with $\sqrt{i}a$ in the Poisson formula of the solution to the Cauchy problem of heat equation (10.69). We do not discuss issues like what kind of requirement should be proposed on the initial function $\psi_0(x)$, in what sense the solution given by (10.75) satisfies the initial condition, and so on, here.

From (10.75) we can see that the value of the solution $\psi(t,x)$ of the Schrödinger equation at any given point (t,x) depends on the value of the initial function $\psi_0(x)$ on the whole x axis. Therefore, the perturbation of the initial function $\psi_0(x)$ around any given point propagates at infinite speed. This is similar to the heat equation, and essentially different from the wave equation, which has a finite propagation speed for perturbation.

The Schrödinger equation is not a hyperbolic equation, but since it describes the evolution of the probability wave of a particle, it is also called a wave equation by some authors. However, the wave described by its solution is not a hyperbolic wave but a *dispersive wave*.

Finally, we make some remarks on how to raise well-posed problems to the Schrödinger equation. In addition to the Cauchy problem mentioned before, we can also propose the initial-boundary value problem. Since the Schrödinger equation has specific physical meaning, what kind of boundary conditions we should assign depends on the specific problems. For some classical boundary conditions, such as Dirichlet-type or Neumann-type boundary conditions, the corresponding problems are all well-posed.

10.2.4 Steady-State Schrödinger Equation

In the Schrödinger equation (10.27), the potential field $V(\boldsymbol{x})$ depends only on space variable $\boldsymbol{x}$ and is independent of time variable t, so in many cases its solution can be expressed by the following form of separation of variables:

$$\psi(t,\boldsymbol{x})=f(t)u(\boldsymbol{x}). \tag{10.76}$$

Plugging the above formula into equation (10.27), we get

$$\mathrm{i}\hbar f'(t)u(\boldsymbol{x})=f(t)\widehat{H}u(\boldsymbol{x}).$$

Separating variables, we have

$$\mathrm{i}\hbar\frac{f'(t)}{f(t)}=\frac{\widehat{H}u(\boldsymbol{x})}{u(\boldsymbol{x})}. \tag{10.77}$$

The two sides of the above formula are functions of t and $\boldsymbol{x}$, respectively, so they have to be the same constant. Setting this constant to be E, we obtain

$$\mathrm{i}\hbar f'(t) = Ef(t) \tag{10.78}$$

and

$$\widehat{H}u(\boldsymbol{x}) = Eu(\boldsymbol{x}). \tag{10.79}$$

Up to a constant factor, the solution of (10.78) is

$$f(t) = \mathrm{e}^{-\mathrm{i}Et/\hbar}. \tag{10.80}$$

Equation (10.79) does not involve the time variable t, so it is called the *steady-state Schrödinger equation.*

Solving the steady-state Schrödinger equation means finding (under certain boundary conditions) the eigenvalues and eigenfunctions of the Hamiltonian operator $\widehat{H}$. Since $\widehat{H}$ is the energy operator of the system, its eigenvalues are also called the *energy eigenvalues* of the system, and the corresponding eigenfunctions are called the *energy eigenfunctions* of the system. The states of the particle described by the energy eigenfunctions are called the *energy eigenstates.*

Suppose that $\psi_E(\boldsymbol{x})$ is the energy eigenfunction of $\widehat{H}$ corresponding to the energy eigenvalue E; then the solution of the Schrödinger equation (10.27) with the initial function $\psi_E(\boldsymbol{x})$ is

$$\psi(t,\boldsymbol{x}) = \psi_E(\boldsymbol{x})\mathrm{e}^{-\mathrm{i}Et/\hbar}. \tag{10.81}$$

The state described by wave function (10.81) is a state with definite energy called the *stationary state*. Since

$$|\psi(t,\boldsymbol{x})|^2 = |\psi_E(\boldsymbol{x})|^2, \tag{10.82}$$

the spatial probability density $|\psi(t,\boldsymbol{x})|^2$ of a particle in steady state does not change with time.

Comparing (10.79) with (10.19), and noticing the definition of operator $\widehat{H}$ (see (10.26)) and (10.20), we can see that for the case of a free particle ($V \equiv 0$), the energy eigenvalue E is exactly the energy of the particle. This is also the case in general.

Example 10.1 (the motion of free electrons in metal). The motion of free electrons in metal is a basic problem in solid-state physics. But the microscopic structure of actual metal is fairly complicated in general, so it is very difficult in both physics and mathematics to derive rigorously the motion of its electrons according to the principles of quantum mechanics. For simplicity, we have to simplify the model of metals.

Here we discuss only the one-dimensional case. Metal is a crystal whose atoms are regularly aligned, and it is periodic in space structure. In this periodic structure, except at the very moment that the electrons collide with each other or when the electrons are very close to the surface of the metal, the forces received by the electrons are periodic, and so is the corresponding potential field. Therefore, the issue of the motion of electrons in a periodic potential field is very important. But for simplicity, here we adopt a simpler model: we ignore not only the collision of electrons with each other but also the periodic change of potential energy, and we assume that the electrons inside the metal are completely free (the potential energy is a constant and the electrons are force free) and are restricted only by a suddenly increased and very large potential energy "wall" when hitting the boundary.

Suppose that the metal lies on the interval $0 < x < L$, and that, without loss of generality, the potential energy vanishes in this interval. Then the Schrödinger equation and the corresponding boundary condition satisfied by the wave function of the electron are

$$i\hbar\frac{\partial\psi}{\partial t} = -\frac{\hbar^2}{2m}\frac{\partial^2\psi}{\partial x^2}, \quad 0 < x < L, \tag{10.83}$$

$$\psi(t,0) = \psi(t,L) = 0. \tag{10.84}$$

Here, the boundary condition takes form (10.84) because electrons are prevented by the very large potential wall at $x = 0$ and $x = L$ from escaping the metal, and thus the probability that the electrons will appear at the end face of the metal is zero. Of course, this is only an approximation, since from the viewpoint of quantum mechanics, there always exists a probability that the electrons will escape the metal, no matter how high the potential wall is (see, e.g., Example 10.2).

We first figure out the stationary solution

$$\psi(t,x) = e^{-iEt/\hbar}u(x) \tag{10.85}$$

of problem (10.83)–(10.84), where u satisfies the following steady-state Schrödinger equation and corresponding boundary condition:

$$-\frac{\hbar^2}{2m}\frac{d^2u}{dx^2} = Eu, \quad 0 < x < L, \tag{10.86}$$

$$u(0) = u(L) = 0. \tag{10.87}$$

This is exactly the well-known eigenvalue problem in mathematical physical equations (see, e.g., [2]). The eigenvalues are

$$E_n = \frac{\pi^2\hbar^2}{2mL^2}n^2 \quad (n = 1,2,\ldots), \tag{10.88}$$

and the corresponding eigenfunctions are

$$u_n(x) = a_n \sin\frac{n\pi}{L}x \quad (n = 1,2,\ldots), \tag{10.89}$$

where a_n are any given nonzero constants. Corresponding to eigenfunctions (10.89), we can obtain the stationary solutions

$$\psi_n(t,x) = a_n e^{-iE_nt/\hbar}\sin\frac{n\pi}{L} \quad (n = 1,2,\ldots), \tag{10.90}$$

where E_n stands for the total energy of the electrons.

We arrive at an important conclusion from the above discussion: the energy of a free electron moving in a metal does not have continuous values like in classical theory but is discrete, namely, quantized, and forms different "energy levels." The state for $n = 1$ is called the *ground state*, which corresponds to the lowest value $E_1 = \pi^2\hbar^2/2mL^2$ of energy.

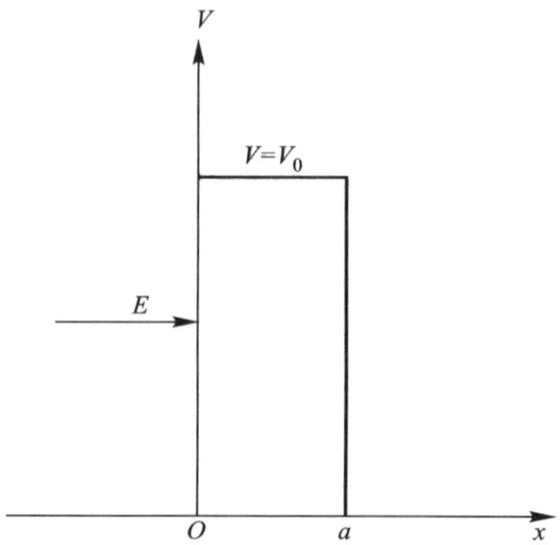

Figure 10.5

From a suitable linear superposition of stationary wave functions, we can give the wave function that describes the general state, namely, the quantum state, as follows:

$$\psi(t,x) = \sum_{n=1}^{\infty} a_n \mathrm{e}^{-\mathrm{i}E_n t/\hbar} \sin\frac{n\pi}{L}. \tag{10.91}$$

■

Example 10.2 (one-dimensional scattering problem). Given the rectangular potential barrier

$$V(x) = \begin{cases} V_0, & 0 < x < a, \\ 0, & x < 0 \text{ or } x > a, \end{cases} \tag{10.92}$$

where $V_0 > 0$ is a constant, suppose that a bunch of particles with energy E is shot at this potential barrier from the left along the positive direction of the x axis (see Figure 10.5).

From the viewpoint of classical mechanics, if $E > V_0$, particles will pass through the potential barrier, while if $E < V_0$, particles cannot enter the potential barrier and will be reflected back. But from the viewpoint of quantum mechanics, particles bear wave properties, so in any case, part of the probability wave will pass through the potential barrier and part of the probability wave will be reflected back.

Since the potential barrier $V(x)$ is independent of t, we can discuss the corresponding steady-state Schrödinger equation

$$-\frac{\hbar^2}{2m}\frac{\mathrm{d}^2 u}{\mathrm{d}x^2} + V(x)u = Eu. \tag{10.93}$$

Outside the potential barrier (i.e., $x < 0$ or $x > a$), $V \equiv 0$, and equation (10.93) is reduced to

$$\frac{\mathrm{d}^2 u}{\mathrm{d}x^2} + \frac{2mE}{\hbar^2}u = 0, \quad x < 0 \text{ or } x > a. \tag{10.94}$$

This is a second-order linear ordinary differential equation with constant coefficients, which has two linearly independent solutions:

$$u = \mathrm{e}^{-\mathrm{i}kx} \text{ and } \mathrm{e}^{\mathrm{i}kx}, \tag{10.95}$$

where

$$k = \frac{\sqrt{2mE}}{\hbar}. \tag{10.96}$$

We note that the wave function is now

$$\psi = \mathrm{e}^{-\mathrm{i}Et/\hbar} u = \mathrm{e}^{-\mathrm{i}(kx+Et/\hbar)} \text{ and } \mathrm{e}^{\mathrm{i}(kx-Et/\hbar)},$$

with $\mathrm{e}^{-\mathrm{i}kx}$ standing for the wave propagating to the left and $\mathrm{e}^{\mathrm{i}kx}$ the wave propagating to the right. For the problem under consideration, in domain $x < 0$, there should be both an incident wave propagating to the right and a reflecting wave propagating to the left, while in domain $x > a$, there should only be a transmissive wave propagating to the right. Thus, we have

$$u(x) = \begin{cases} \mathrm{e}^{\mathrm{i}kx} + R\,\mathrm{e}^{-\mathrm{i}kx}, & x < 0, \\ S\,\mathrm{e}^{\mathrm{i}kx}, & x > a. \end{cases} \tag{10.97}$$

Here, for simplicity, the coefficient (amplitude) of the incident wave $\mathrm{e}^{\mathrm{i}kx}$ is taken to be 1, and R and S are constants to be determined.

Inside the potential barrier (i.e., $0 < x < a$), equation (10.93) is reduced to

$$\frac{\mathrm{d}^2u}{\mathrm{d}x^2} - \frac{2m(V_0 - E)}{\hbar^2}u = 0, \quad 0 < x < a. \tag{10.98}$$

This is still a second-order linear ordinary differential equation with constant coefficients. To solve this equation, we divide our discussion into two cases.

1. *Case of* $E < V_0$. In this case, the general solution of equation (10.98) is

$$u = C_1\,\mathrm{e}^{-k'x} + C_2\,\mathrm{e}^{k'x}, \quad 0 < x < a, \tag{10.99}$$

where C_1 and C_2 are arbitrary constants and

$$k' = \frac{\sqrt{2m(V_0 - E)}}{\hbar}. \tag{10.100}$$

The continuity of the wave function $u = u(x)$ given by (10.97) and (10.99), and its first-order derivative at $x = 0$ (we cannot expect a solution with higher regularity since the coefficient $V(x)$ in (10.93) has discontinuity), leads to

$$1 + R = C_1 + C_2$$

and

$$\mathrm{i}k(1 - R) = k'(C_2 - C_1).$$

Then it can be solved that

$$C_1 = \frac{1}{2}\left[\left(1 - \frac{\mathrm{i}k}{k'}\right) + R\left(1 + \frac{\mathrm{i}k}{k'}\right)\right], \tag{10.101}$$

$$C_2 = \frac{1}{2}\left[\left(1 + \frac{\mathrm{i}k}{k'}\right) + R\left(1 - \frac{\mathrm{i}k}{k'}\right)\right]. \tag{10.102}$$

Moreover, from the continuity of the wave function $u = u(x)$ given by (10.97) and (10.99), and its first-order derivative at $x = a$, we have

$$S\,\mathrm{e}^{\mathrm{i}ka} = C_1\,\mathrm{e}^{-k'a} + C_2\,\mathrm{e}^{k'a}$$

and

$$\mathrm{i}kS\,\mathrm{e}^{\mathrm{i}ka} = -C_1k'\,\mathrm{e}^{-k'a} + C_2k'\,\mathrm{e}^{k'a}.$$

Then it can be solved that

$$C_1 = \frac{S}{2}\left(1 - \frac{\mathrm{i}k}{k'}\right)\mathrm{e}^{\mathrm{i}(k+k')a}, \tag{10.103}$$

$$C_2 = \frac{S}{2}\left(1 + \frac{\mathrm{i}k}{k'}\right)\mathrm{e}^{\mathrm{i}(k-k')a}. \tag{10.104}$$

Thus, by solving for R, S, C_1, and C_2 from (10.101)–(10.104) , we get the required wave function $u = u(x)$ from (10.97) and (10.99) . But what we are interested in is only reflection and transmission after the probability wave hits the potential barrier, so we need only compute R and S and discuss their physical explanation.

From (10.59), it is easy to see that the probability flux density in the one-dimensional case is

$$j = -\frac{\mathrm{i}\hbar}{2m}\left(\bar{u}\frac{\mathrm{d}u}{\mathrm{d}x} - u\frac{\mathrm{d}\bar{u}}{\mathrm{d}x}\right). \tag{10.105}$$

Taking the wave function in (10.105) to be the incident wave $u = \mathrm{e}^{\mathrm{i}kx}$, we get the incident particle flux density

$$j_i = -\frac{\mathrm{i}\hbar}{2m}\left(\mathrm{e}^{-\mathrm{i}kx}\frac{\mathrm{d}}{\mathrm{d}x}\mathrm{e}^{\mathrm{i}kx} - \mathrm{e}^{\mathrm{i}kx}\frac{\mathrm{d}}{\mathrm{d}x}\mathrm{e}^{-\mathrm{i}kx}\right) = \frac{\hbar k}{m}. \tag{10.106}$$

Taking the wave function in (10.105) to be the reflecting wave $u = R\,\mathrm{e}^{-\mathrm{i}kx}$, we get the reflecting particle flux density

$$|j_r| = \frac{\hbar k}{m}|R|^2. \tag{10.107}$$

Similarly, taking $u = S\,\mathrm{e}^{\mathrm{i}kx}$ in (10.105), we get the transmissive particle flux density

$$|j_t| = \frac{\hbar k}{m}|S|^2. \tag{10.108}$$

Thus, we obtain

$$\text{reflection coefficient} = \left|\frac{j_r}{j_i}\right| = |R|^2 \tag{10.109}$$

and

$$\text{transmission coefficient} = \left|\frac{j_t}{j_i}\right| = |S|^2. \tag{10.110}$$

Now we determine the reflection coefficient and the transmission coefficient. Eliminating C_1 and C_2 from (10.101)–(10.104), we get

$$\left(1-\frac{\mathrm{i}k}{k'}\right)+R\left(1+\frac{\mathrm{i}k}{k'}\right)=S\left(1-\frac{\mathrm{i}k}{k'}\right)\mathrm{e}^{\mathrm{i}(k+k')a} \tag{10.111}$$

and

$$\left(1+\frac{\mathrm{i}k}{k'}\right)+R\left(1-\frac{\mathrm{i}k}{k'}\right)=S\left(1+\frac{\mathrm{i}k}{k'}\right)\mathrm{e}^{\mathrm{i}(k-k')a}. \tag{10.112}$$

Eliminating R from the above formulas, we have

$$S\mathrm{e}^{\mathrm{i}ka} = -\frac{2\mathrm{i}k/k'}{\left(1-\left(\frac{k}{k'}\right)^2\right)\mathrm{sh}(k'a)-2\mathrm{i}\frac{k}{k'}\mathrm{ch}(k'a)}. \tag{10.113}$$

Then we immediately obtain the transmission coefficient

$$|S|^2 = \frac{4k^2k'^2}{(k^2-k'^2)^2\,\mathrm{sh}^2(k'a)+4k^2k'^2\,\mathrm{ch}^2(k'a)}. \tag{10.114}$$

Noting

$$\mathrm{ch}^2(k'a)-\mathrm{sh}^2(k'a)=1$$

and the definitions (10.96) and (10.100) of k and k', (10.114) can be rewritten as

$$\begin{aligned} |S|^2 &= \frac{4k^2k'^2}{(k^2+k'^2)^2\,\mathrm{sh}^2(k'a)+4k^2k'^2} \\ &= \left(1+\frac{1}{\frac{4E}{V_0}\left(1-\frac{E}{V_0}\right)}\mathrm{sh}^2(k'a)\right)^{-1}. \end{aligned} \tag{10.115}$$

Similarly, eliminating S from (10.111) and (10.112), we obtain the reflection coefficient

$$|R|^2 = \frac{(k^2+k'^2)^2\,\mathrm{sh}^2(k'a)}{(k^2+k'^2)^2\,\mathrm{sh}^2(k'a)+4k^2k'^2}. \tag{10.116}$$

From (10.115) and (10.116) we can immediately see that the reflection coefficient and the transmission coefficient have the following relation:

$$|R|^2+|S|^2=1. \tag{10.117}$$

This indicates the conservation of probability (or the conservation of particle number).

It is shown from (10.115) that the transmission coefficient $|S|^2 \neq 0$. This implies that even if the potential barrier is higher than the kinetic energy of particles, there still exists a probability for particles to pass through the potential barrier. This is called the *tunnel effect*, which is caused by the wave properties of particles.

2. *Case of* $E > V_0$. In this case, we need only replace k' with $\mathrm{i}k_1$ in (10.114), where

$$k_1 = \frac{\sqrt{2m(E-V_0)}}{\hbar}, \tag{10.118}$$

and note that

$$\mathrm{sh}(\mathrm{i}k_1 a) = \mathrm{i}\sin(k_1 a), \quad \mathrm{ch}(\mathrm{i}k_1 a) = \cos(k_1 a).$$

Then we obtain the transmission coefficient

$$\begin{aligned} |S|^2 &= \frac{4k^2k_1^2}{(k^2-k_1^2)^2\sin^2(k_1 a) + 4k^2k_1^2} \\ &= \frac{1}{1+\dfrac{1}{4}\left(\dfrac{k}{k_1}-\dfrac{k_1}{k}\right)^2\sin^2(k_1 a)}, \end{aligned} \tag{10.119}$$

where k and k_1 are given by (10.96) and (10.118), respectively. The reflection coefficient $|R|^2$ can be determined by the transmission coefficient $|S|^2$ and the conservation relation (10.117). ■

10.3 Introduction to the Fundamental Principles of Quantum Mechanics

For the sake of completeness, in this section we will briefly introduce the fundamental principles of quantum mechanics and their inferences. Strictly speaking, these fundamental principles (including the Schrödinger equation, which is regarded as the basic equation of quantum mechanics) are all assumptions based on experimental facts.

10.3.1 Basic Assumptions for Principles of Quantum Mechanics

1. The state of a particle in a certain system is described by a complex function—the wave function ψ of time and space coordinates. The probability of finding a particle in the spatial volume element $\mathrm{d}x$ is $|\psi|^2\mathrm{d}x = \overline{\psi}\psi\mathrm{d}x$.

If the wave function ψ satisfies

$$\int |\psi|^2\mathrm{d}x = 1, \tag{10.120}$$

we say that it is *normalized* or *uniformized*.

The wave function is not directly observable, and what we are concerned with are so-called observable quantities, such as particle position, momentum, and energy, corresponding to a certain state (described by the wave function). In order to obtain these mechanical quantities from the wave function describing the state of the particles, we need the operators that correspond to mechanical quantities.

2. Each mechanical quantity A (not limited to classical mechanical quantities, such as spin) corresponds to a linear Hermitian operator $\widehat{A}$ in a Hilbert space. The role of this operator is to transform one wave function (state) into another wave function (state). If the wave function ψ describing the system state is the eigenfunction of the operator $\widehat{A}$, that is,

$$\widehat{A}\psi = a\psi, \tag{10.121}$$

then the value obtained by measuring the mechanical quantity A corresponding to the state ψ is just the eigenvalue a corresponding to the eigenfunction ψ .

Example 10.3. For particle position $\boldsymbol{x} = (x_1, x_2, x_3)$, the result of the corresponding operators $\widehat{x_k}$ $(k = 1,2,3)$ acting upon the wave function ψ is $x_k\psi$, i.e., multiplying ψ by x_k. ∎

Example 10.4. For momentum $\boldsymbol{p} = (p_1, p_2, p_3)$, the corresponding operators are $\widehat{p_k} = -\mathrm{i}\hbar\frac{\partial}{\partial x_k}$ $(k = 1,2,3)$. Sometimes we denote $\widehat{\boldsymbol{p}} = -\mathrm{i}\hbar\nabla$, but it is a vector operator. ∎

Example 10.5. The classical angular momentum of particle is

$$\boldsymbol{l} = \boldsymbol{x} \times \boldsymbol{p}. \tag{10.122}$$

Its components are

$$\begin{aligned} l_1 &= x_2p_3 - x_3p_2, \\ l_2 &= x_3p_1 - x_1p_3, \\ l_3 &= x_1p_2 - x_2p_1. \end{aligned}$$

The corresponding operators are, respectively,

$$\begin{cases} \widehat{l_1} = -\mathrm{i}\hbar\left(x_2\dfrac{\partial}{\partial x_3} - x_3\dfrac{\partial}{\partial x_2}\right), \\ \widehat{l_2} = -\mathrm{i}\hbar\left(x_3\dfrac{\partial}{\partial x_1} - x_1\dfrac{\partial}{\partial x_3}\right), \\ \widehat{l_3} = -\mathrm{i}\hbar\left(x_1\dfrac{\partial}{\partial x_2} - x_2\dfrac{\partial}{\partial x_1}\right). \end{cases} \tag{10.123}$$

Equations (10.123) can also be written as

$$\widehat{\boldsymbol{l}} = \boldsymbol{x} \times \widehat{\boldsymbol{p}}. \tag{10.124}$$

∎

Example 10.6. Suppose that the energy of a particle in the potential field $V(\boldsymbol{x})$ is given by the Hamiltonian

$$H = \frac{1}{2m}|\boldsymbol{p}|^2 + V(\boldsymbol{x}).$$

The corresponding operator is just the Hamiltonian operator

$$\widehat{H} = -\frac{\hbar^2}{2m}\Delta + V(\boldsymbol{x}). \quad ∎$$

3. If the wave function ψ describing the state of a particle is not the eigenfunction of the operator $\widehat{A}$, then the result of one measurement of the mechanical quantity A remains an eigenvalue of the operator $\widehat{A}$, and after this measurement, the previous state changes into the eigenfunction of the operator $\widehat{A}$ corresponding to this eigenvalue.

In other words, when a system is in a quantum state ψ (noneigenstate), measuring the mechanical quantity A may give rise to all possible results (corresponding to different eigenstates), and each possible result has a definite probability of occurring. If we measure repeatedly a large number of identical systems whose states are all described by ψ (each system is measured at most once), the average of the measurement results is

$$\langle A \rangle = \int \overline{\psi} \widehat{A} \psi \mathrm{d}x, \tag{10.125}$$

where we assume that ψ is normalized. The fluctuation ΔA of measurements around the average is determined by

$$(\Delta A)^2 = \langle (A - \langle A \rangle)^2 \rangle = \int \overline{\psi} (\widehat{A} - \langle A \rangle)^2 \psi \mathrm{d}x. \tag{10.126}$$

4. The evolution of the wave function describing the state of a particle system (with respect to time t) satisfies the Schrödinger equation

$$\mathrm{i}\hbar \frac{\partial \psi}{\partial t} = \widehat{H} \psi, \tag{10.127}$$

where $\widehat{H}$ is the Hamiltonian operator.

10.3.2 Commutation Relations of Operators

Let $\widehat{A}$ and $\widehat{B}$ be two linear operators in a Hilbert space. In general, their product (understood as a continuous action of operators) does not obey the commutative law. The commutator between $\widehat{A}$ and $\widehat{B}$,

$$[\widehat{A}, \widehat{B}] = \widehat{A}\widehat{B} - \widehat{B}\widehat{A}, \tag{10.128}$$

is often referred to as the *commutation formula* in quantum mechanics. We have the following commutation relations.

(a) Basic commutation formula:

$$[\widehat{x}_k, \widehat{p}_j] = \mathrm{i}\hbar \delta_{kj}. \tag{10.129}$$

In fact,

$$x_1 \widehat{p}_1 \psi = -\mathrm{i}\hbar x_1 \frac{\partial \psi}{\partial x_1},$$

$$\widehat{p}_1 (x_1 \psi) = -\mathrm{i}\hbar \frac{\partial}{\partial x_1} (x_1 \psi);$$

thus

$$[\widehat{x}_1, \widehat{p}_1] = x_1 \widehat{p}_1 \psi - \widehat{p}_1 (x_1 \psi) = \mathrm{i}\hbar \psi.$$

The other formulas can be proved similarly.

(b) Commutation formula for angular momentum:

$$[\widehat{l_k},\widehat{l_k}]=0 \quad (k=1,2,3), \tag{10.130}$$

$$[\widehat{l_1},\widehat{l_2}]=\mathrm{i}\hbar\widehat{l_3}, \quad [\widehat{l_2},\widehat{l_3}]=\mathrm{i}\hbar\widehat{l_1}, \quad [\widehat{l_3},\widehat{l_1}]=\mathrm{i}\hbar\widehat{l_2}. \tag{10.131}$$

We only prove the first formula in (10.131). According to the definition (10.123) of the components of $\widehat{\boldsymbol{l}}$, we have

$$\begin{aligned}\widehat{l_1}\widehat{l_2}\psi &= -\mathrm{i}\hbar\left(x_2\frac{\partial}{\partial x_3}-x_3\frac{\partial}{\partial x_2}\right)(-\mathrm{i}\hbar)\left(x_3\frac{\partial\psi}{\partial x_1}-x_1\frac{\partial\psi}{\partial x_3}\right)\\ &= -\hbar^2\left(x_2\frac{\partial}{\partial x_3}\left(x_3\frac{\partial\psi}{\partial x_1}\right)-x_1x_2\frac{\partial^2\psi}{\partial x_3^2}-x_3^2\frac{\partial^2\psi}{\partial x_1\partial x_2}+x_1x_3\frac{\partial^2\psi}{\partial x_2\partial x_3}\right)\\ &= -\hbar^2\left(x_2x_3\frac{\partial^2\psi}{\partial x_1\partial x_3}-x_1x_2\frac{\partial^2\psi}{\partial x_3^2}-x_3^2\frac{\partial^2\psi}{\partial x_1\partial x_2}+x_1x_3\frac{\partial^2\psi}{\partial x_2\partial x_3}\right)-\hbar^2x_2\frac{\partial\psi}{\partial x_1}.\end{aligned}$$

Similarly,

$$\widehat{l_2}\widehat{l_1}\psi = -\hbar^2\left(x_2x_3\frac{\partial^2\psi}{\partial x_1\partial x_3}-x_3^2\frac{\partial^2\psi}{\partial x_1\partial x_2}-x_1x_2\frac{\partial^2\psi}{\partial x_3^2}+x_1x_3\frac{\partial^2\psi}{\partial x_2\partial x_3}\right)-\hbar^2x_1\frac{\partial\psi}{\partial x_2}.$$

Subtracting the above two formulas, we have

$$[\widehat{l_1},\widehat{l_2}]\psi=\hbar^2\left(x_1\frac{\partial\psi}{\partial x_2}-x_2\frac{\partial\psi}{\partial x_1}\right)=\mathrm{i}\hbar\widehat{l_3}\psi;$$

then the first formula in (10.131) is proved. The other formulas can be proved similarly (and are left as an exercise).

10.3.3 Uncertainty Relation

First, we discuss the problem of the simultaneous measurement of two mechanical quantities. This is not a problem in classical physics: in principle, one can obtain definite values when measuring simultaneously several mechanical quantities under any state. However, this is not the case in quantum mechanics. According to the above-stated assumption 2, if we measure simultaneously two mechanical quantities under a certain state and require that they have definite values, then this state must be the common eigenstate of the operators corresponding to these two mechanical quantities.

Now we investigate this issue in more detail. For a state ψ, measuring the uncertainty of mechanical quantities A and B is expressed by their fluctuations ΔA and ΔB, respectively. From the definition (10.126) of fluctuation and the Hermitian properties of the operator (complex symmetry), we have

$$\begin{aligned}(\Delta A)^2 &= \int |(\widehat{A}-\langle A\rangle)\psi|^2\mathrm{d}x,\\ (\Delta B)^2 &= \int |(\widehat{B}-\langle B\rangle)\psi|^2\mathrm{d}x.\end{aligned}$$

Using the Hermitian properties of the operator and the Schwarz inequality, we obtain

$$\begin{aligned}&\left|\int \overline{\psi}(\widehat{A}-\langle A\rangle)(\widehat{B}-\langle B\rangle)\psi \mathrm{d}x\right|^2\\&=\left|\int \overline{(\widehat{A}-\langle A\rangle)\psi}\cdot(\widehat{B}-\langle B\rangle)\psi \mathrm{d}x\right|^2\\&\leqslant \int |(\widehat{A}-\langle A\rangle)\psi|^2 \mathrm{d}x\cdot\int |(\widehat{B}-\langle B\rangle)\psi|^2 \mathrm{d}x\\&=(\Delta A\cdot\Delta B)^2.\end{aligned} \tag{10.132}$$

It is not difficult to verify directly that

$$(\widehat{A}-\langle A\rangle)(\widehat{B}-\langle B\rangle)=\widehat{G}+\mathrm{i}\widehat{C}, \tag{10.133}$$

where

$$\widehat{G}=\frac{1}{2}((\widehat{A}-\langle A\rangle)(\widehat{B}-\langle B\rangle)+(\widehat{B}-\langle B\rangle)(\widehat{A}-\langle A\rangle)), \tag{10.134}$$

$$\widehat{C}=-\frac{\mathrm{i}}{2}[\widehat{A},\widehat{B}], \tag{10.135}$$

and both $\widehat{G}$ and $\widehat{C}$ are Hermitian operators. Thus, $\int\overline{\psi}\widehat{G}\psi\mathrm{d}x$ and $\int\overline{\psi}\widehat{C}\psi\mathrm{d}x$ are both real, and from (10.133) we easily obtain

$$\left|\int \overline{\psi}(\widehat{A}-\langle A\rangle)(\widehat{B}-\langle B\rangle)\psi \mathrm{d}x\right|^2=\left|\int\overline{\psi}\widehat{G}\psi\mathrm{d}x\right|^2+\left|\int\overline{\psi}\widehat{C}\psi\mathrm{d}x\right|^2.$$

By (10.132) and noticing (10.135), we have

$$(\Delta A\cdot\Delta B)^2\geqslant\left|\int\overline{\psi}\widehat{C}\psi\mathrm{d}x\right|^2=\frac{1}{4}\left|\int\overline{\psi}[\widehat{A},\widehat{B}]\psi\mathrm{d}x\right|^2,$$

i.e.,

$$|\Delta A\cdot\Delta B|\geqslant\frac{1}{2}\langle[\widehat{A},\widehat{B}]\rangle. \tag{10.136}$$

This is just the *uncertainty relation*, which the fluctuations of any two given mechanical quantities must satisfy in terms of quantum states.

Example 10.7. Suppose that $\widehat{A}=\widehat{x}_1$ and $\widehat{B}=\widehat{p}_1$. By the commutation relation (10.129), we have

$$[\widehat{x}_1,\widehat{p}_1]=\mathrm{i}\hbar.$$

Using the uncertainty relation (10.136), we obtain

$$|\Delta x_1\cdot\Delta p_1|\geqslant\hbar/2. \tag{10.137}$$

This shows that the exact values of position and velocity of a particle cannot be measured simultaneously. ∎

From (10.136) we can see that if the operators $\widehat{A}$ and $\widehat{B}$ corresponding to mechanical quantities A and B are not commutative (i.e., $[\widehat{A},\widehat{B}] \neq 0$), then generally speaking ΔA and ΔB cannot be zero simultaneously, that is, A and B cannot have definite values simultaneously, and $\widehat{A}$ and $\widehat{B}$ cannot have common eigenfunctions simultaneously. Therefore, the condition that two mechanical quantities A and B have definite observed values simultaneously is

$$[\widehat{A},\widehat{B}] = 0. \tag{10.138}$$

10.4 Relativistic Quantum Mechanics and the Dirac Equation

The Schrödinger equation is a fundamental equation in nonrelativistic quantum mechanics. It is very successful in describing most phenomena of atoms and molecules, including many phenomena in low-energy nuclear physics. This is because the motion velocity of particles is much less than the speed of light in these phenomena, and the relativistic effect is not remarkable, so the Schrödinger equation is a very good approximation. However, when it comes to high-energy fields, the nonrelativistic Schrödinger equation is powerless in many cases, and the relativistic effect must be taken into account for the equation satisfied by the wave function.

10.4.1 Klein–Gordon Equation

The Schrödinger equation

$$\mathrm{i}\hbar\frac{\partial\psi}{\partial t} = \left(-\frac{\hbar^2}{2m}\Delta + V\right)\psi$$

does not have equity in form for time and space variables (it involves the first-order partial derivative with respect to time t, but the second-order partial derivatives with respect to space variables), so this equation is not invariant under the Lorentz transformation.

At about the same time the Schrödinger equation was proposed, the relativistic wave equation was put forward. From the energy–momentum formula

$$E^2 - c^2|\boldsymbol{p}|^2 = E_0^2 \tag{10.139}$$

in special relativity, where E and $\boldsymbol{p}$ are the total energy and momentum of the particle, respectively, $E_0 = m_0c^2$ is the rest energy of the particle, c is the speed of light, and m_0 is the rest mass of the particle (see (9.89) and (9.90) in Chapter 9), we have

$$E^2 = c^2|\boldsymbol{p}|^2 + m_0^2c^4. \tag{10.140}$$

In (10.140), applying the formal transformation

$$E \to \mathrm{i}\hbar\frac{\partial}{\partial t}, \quad \boldsymbol{p} \to -\mathrm{i}\hbar\nabla$$

given by (10.24) and acting the resulting operator on the wave function ψ, we obtain

$$-\hbar^2\frac{\partial^2\psi}{\partial t^2} = -c^2\hbar^2\Delta\psi + m_0^2c^4\psi. \tag{10.141}$$

This equation is called the *Klein–Gordon equation.* Noting that equation (10.141) can be written as

$$\frac{1}{c^2}\frac{\partial^2\psi}{\partial t^2} - \Delta\psi + \frac{m_0^2 c^2}{\hbar^2}\psi = 0, \tag{10.142}$$

from the Lorentz invariance (see section 9.2 in Chapter 9) of the wave equation

$$\frac{1}{c^2}\frac{\partial^2\psi}{\partial t^2} - \Delta\psi = 0,$$

we immediately get the Lorentz invariance of the Klein–Gordon equation.

Although the Klein–Gordon equation satisfies the Lorentz invariance, it gets into trouble when used to explain the wave function because of the presence of the second-order t derivative of the wave function. It is no longer reasonable to define the probability density of the particle by

$$\rho = |\psi(t,\boldsymbol{x})|^2,$$

because now it does not hold that

$$\frac{\mathrm{d}}{\mathrm{d}t}\int_{\mathbb{R}^3} \rho \mathrm{d}x = 0.$$

Various attempts had been made to revise the definition of ρ such that the above formula holds, and all have failed. Thus, the Klein–Gordon equation is not appropriate to be adopted as the wave equation describing the state of free particles, and this equation was laid aside for as long as seven years. But in 1934, W. Pauli and V. Weisskopf gave this equation a new interpretation: it is not the wave equation of an individual particle, but the equation of a field (just like Maxwell's equations are equations of the electromagnetic field). Pauli and Weisskopf also quantized this equation, drawing scientists' attention again.

10.4.2 Spin of the Electron

For the quantum system composed of particles, the previous description using only one component of the wave function is a little too rough. Experiments have shown that there exists "spin" for most discovered particles; there is null spin only for a small number of particles called scalar quanta, such as the nucleus of the π meson and some atoms. For particles without spin, only one component of the wave function is enough, but this is not the case for particles with spin.

In history, the spin of the electron was introduced in 1925 to explain the fine (multiple) structure of spectrum lines, especially the double-line structure of alkali metals. For instance, when a very intense yellow line in the spectrum of the Na atom is observed by high-resolution spectrograph, it is found to be composed of two spectrum lines very close to each other, whose wavelengths are 5890 Å and 5896 Å, respectively, with only a 6 Å (1 Å $= 10^{-10}$ m) difference. To explain this phenomenon, the hypothesis of the electron's spin was proposed: electrons spin in addition to moving around the nucleus, and the spin angular momentum takes only two values when projected to any direction in space.

In the initial hypothesis of the electron's spin, the motion of electrons is regarded as motion like the earth's orbit around the sun: on one hand, electrons move around the nucleus with corresponding orbital angular momentum; on the other hand, electrons spin with corresponding spin angular momentum. However, it is not correct to regard the electron's spin as mechanical autorotation. In fact, the spin angular momentum of an electron cannot be expressed by the angular momentum $\boldsymbol{x} \times \boldsymbol{p}$ of ordinary mechanical movement; it needs to be discussed starting from the commutation relations of angular momentum.

Let $\boldsymbol{S} = (S_1, S_2, S_3)$ represent the spin angular momentum of an electron. From the commutation relation (10.131) of the angular momentum operator, we have

$$\begin{cases} \widehat{S}_1\widehat{S}_2 - \widehat{S}_2\widehat{S}_1 = \mathrm{i}\hbar\widehat{S}_3, \\ \widehat{S}_2\widehat{S}_3 - \widehat{S}_3\widehat{S}_2 = \mathrm{i}\hbar\widehat{S}_1, \\ \widehat{S}_3\widehat{S}_1 - \widehat{S}_1\widehat{S}_3 = \mathrm{i}\hbar\widehat{S}_2, \end{cases} \tag{10.143}$$

where $\widehat{S}_j$ stands for the operator corresponding to S_j. From (10.143) we can prove that the angular momentum component $\widehat{S}_3$ (or $\widehat{S}_1$ or $\widehat{S}_2$) has altogether $2s+1$ eigenvalues,

$$\lambda = m_s\hbar \quad (m_s = -s, -s+1, \ldots, s-1, s)$$

(see [3]). But experiments show that there are only two possible values of the components of the electron's spin angular momentum in any direction; therefore, "spin" is sometimes regarded as the most quantum mechanical quantity of all physical quantities. Thus, we should have $2s+1 = 2$, i.e., $s = \frac{1}{2}$, and then $m_s = \pm\frac{1}{2}$. That is, the components of angular momentum can only take values $\pm\hbar/2$ in any direction, which is exactly half the minimum quantized angular momentum (see (10.10)) given by Bohr to electrons. This is the case not only for electrons but also for protons and neutrons, and so on. This kind of particle, with spin an odd multiple of $\hbar/2$, is called the *fermion*.

Thus, an electron is not a particle with just three degrees of freedom, it also has the spin degree of freedom. Therefore, to completely describe the state of an electron, its spin state must be taken into account; precisely speaking, we have to consider the two possible values of the spin projection in a certain direction (usually taken as the direction of the x_3 axis). Thus, the wave function depends not only on $\boldsymbol{x}$ but also on the projection component of spin (usually taken as S_3):

$$\psi(\boldsymbol{x}, S_3) = \begin{pmatrix} \psi(\boldsymbol{x}, \hbar/2) \\ \psi(\boldsymbol{x}, -\hbar/2) \end{pmatrix}. \tag{10.144}$$

This wave function is called the *spinor wave function*, and its physical explanation is as follows: $|\psi(\boldsymbol{x}, \hbar/2)|^2$ represents the probability density of the spinning-up (i.e., $S_3 = \hbar/2$) electron located at $\boldsymbol{x}$, while $|\psi(\boldsymbol{x}, -\hbar/2)|^2$ represents the probability density of the spinning-down (i.e., $S_3 = -\hbar/2$) electron located at $\boldsymbol{x}$.

Let

$$\alpha = \begin{pmatrix} 1 \\ 0 \end{pmatrix}, \quad \beta = \begin{pmatrix} 0 \\ 1 \end{pmatrix}. \tag{10.145}$$

The wave function given by (10.144) can be expressed as

$$\psi(\boldsymbol{x}, S_3) = \psi(\boldsymbol{x}, \hbar/2)\alpha + \psi(\boldsymbol{x}, -\hbar/2)\beta. \tag{10.146}$$

Therefore, the spinor wave function can be expressed by 2×1 matrices α and β, and, correspondingly, the spin operator has to be expressed by a second-order square matrix such that the product of an operator matrix and a wave function is still a wave function. For simplicity, instead of the spin angular momentum operator $\widehat{S}$, we introduce the *Pauli operator* σ by the following formula:

$$\widehat{S} = \frac{1}{2}\hbar\sigma. \tag{10.147}$$

Suppose that the components of σ are σ_1, σ_2, and σ_3, respectively; from the commutation relation (10.143) of $\widehat{S}$ we have

$$\begin{cases} \sigma_1\sigma_2 - \sigma_2\sigma_1 = 2\mathrm{i}\sigma_3, \\ \sigma_2\sigma_3 - \sigma_3\sigma_2 = 2\mathrm{i}\sigma_1, \\ \sigma_3\sigma_1 - \sigma_1\sigma_3 = 2\mathrm{i}\sigma_2. \end{cases} \tag{10.148}$$

Since the component of $\widehat{S}$ in any direction can only take values (i.e., its eigenvalues) $\pm\hbar/2$, the component of σ in any direction can only take values ± 1. Therefore, if we denote components of σ by second-order square matrices, then the eigenvalues of these matrices are all ± 1; thus

$$\sigma_1^2 = \sigma_2^2 = \sigma_3^2 = I, \tag{10.149}$$

where I is the second-order unit matrix.

Multiplying the second formula of (10.148) by σ_2 from the left and from the right, respectively, and using (10.149), we get

$$\sigma_3 - \sigma_2\sigma_3\sigma_2 = 2\mathrm{i}\sigma_2\sigma_1,$$
$$\sigma_2\sigma_3\sigma_2 - \sigma_3 = 2\mathrm{i}\sigma_1\sigma_2.$$

Adding the above two formulas together, we have

$$\sigma_1\sigma_2 + \sigma_2\sigma_1 = 0.$$

Similarly, we can prove that σ_1, σ_2, and σ_3 satisfy the following anticommutation relation:

$$\sigma_j\sigma_k + \sigma_k\sigma_j = 0, \quad \forall j \neq k. \tag{10.150}$$

Using also (10.148), we obtain the following algebraic relations satisfied by σ_1, σ_2, and σ_3:

$$\begin{cases} \sigma_1\sigma_2 = -\sigma_2\sigma_1 = \mathrm{i}\sigma_3, \\ \sigma_2\sigma_3 = -\sigma_3\sigma_2 = \mathrm{i}\sigma_1, \\ \sigma_3\sigma_1 = -\sigma_1\sigma_3 = \mathrm{i}\sigma_2. \end{cases} \tag{10.151}$$

Now we figure out the specific matrix expressions of σ_1, σ_2, and σ_3. We first look at σ_3. Let

$$\sigma_3 = \begin{pmatrix} a_{11} & a_{12} \\ a_{21} & a_{22} \end{pmatrix}.$$

Since α and β given by (10.145) are the eigenvectors of σ_3 corresponding to eigenvalues 1 and -1, respectively, we have

$$\begin{pmatrix} a_{11} & a_{12} \\ a_{21} & a_{22} \end{pmatrix}\begin{pmatrix} 1 \\ 0 \end{pmatrix} = \begin{pmatrix} 1 \\ 0 \end{pmatrix}$$

and

$$\begin{pmatrix} a_{11} & a_{12} \\ a_{21} & a_{22} \end{pmatrix}\begin{pmatrix} 0 \\ 1 \end{pmatrix} = -\begin{pmatrix} 0 \\ 1 \end{pmatrix}.$$

From these, it follows immediately that

$$\sigma_3 = \begin{pmatrix} 1 & 0 \\ 0 & -1 \end{pmatrix}. \tag{10.152}$$

Suppose furthermore that

$$\sigma_1 = \begin{pmatrix} b_{11} & b_{12} \\ b_{21} & b_{22} \end{pmatrix}.$$

As the operator corresponding to a mechanical quantity, σ_1 must be Hermitian, and its corresponding matrix is a Hermitian matrix: $\sigma_1^* = \sigma_1$, where σ_1^* is the conjugate transpose of σ_1. This requires that $b_{21} = \overline{b}_{12}$, so

$$\sigma_1 = \begin{pmatrix} b_{11} & b_{12} \\ \overline{b}_{12} & b_{22} \end{pmatrix}.$$

From the third formula of the anticommutation relation (10.151) and (10.152), we have

$$\begin{pmatrix} b_{11} & b_{12} \\ \overline{b}_{12} & b_{22} \end{pmatrix}\begin{pmatrix} 1 & 0 \\ 0 & -1 \end{pmatrix} = -\begin{pmatrix} 1 & 0 \\ 0 & -1 \end{pmatrix}\begin{pmatrix} b_{11} & b_{12} \\ \overline{b}_{12} & b_{22} \end{pmatrix};$$

then we get

$$b_{11} = b_{22} = 0.$$

Moreover, using $\sigma_1^2 = I$ (see (10.149)), we also have

$$|b_{12}|^2 = 1;$$

then $b_{12} = \mathrm{e}^{\mathrm{i}\theta}$. Usually we take $\theta = 0$; thus we have

$$\sigma_1 = \begin{pmatrix} 0 & 1 \\ 1 & 0 \end{pmatrix}. \tag{10.153}$$

After getting σ_1 and σ_3, using again the third formula of (10.151), we immediately get

$$\sigma_2 = \begin{pmatrix} 0 & -\mathrm{i} \\ \mathrm{i} & 0 \end{pmatrix}. \tag{10.154}$$

The second-order matrices σ_1, σ_2, and σ_3 given by (10.153), (10.154), and (10.152) are known as *Pauli matrices*. Pauli matrices not only are important representations of spin operators but are also extensively used in other applications.

10.4.3 Dirac Equations

In order to overcome the difficulties encountered by the Klein–Gordon equation, P. A. M. Dirac put forward in 1928 the relativistic wave equation of electrons.

Inspired by Pauli's method of describing electron motion by using two components of the wave function in nonrelativistic quantum mechanics, Dirac proposed that the wave function should be of vector form:

$$\psi = \begin{pmatrix} \psi_1(t,\boldsymbol{x}) \\ \psi_2(t,\boldsymbol{x}) \\ \vdots \\ \psi_n(t,\boldsymbol{x}) \end{pmatrix}, \tag{10.155}$$

where ψ_k $(k = 1,\ldots,n)$ are all complex-valued functions, and the probability density of the electron in space is defined by

$$\rho(t,\boldsymbol{x}) = \sum_{k=1}^{n} |\psi_k(t,\boldsymbol{x})|^2. \tag{10.156}$$

The above definition of probability density obviously satisfies nonnegativity, i.e.,

$$\rho(t,\boldsymbol{x}) \geqslant 0. \tag{10.157}$$

For the wave function given by (10.155), we have the following requirements:

1. To ensure the conservation of total probability,

$$\frac{\mathrm{d}}{\mathrm{d}t}\int_{\mathbb{R}^3} \rho(t,\boldsymbol{x})\mathrm{d}x = 0. \tag{10.158}$$

2. The equation of the wave function is Lorentz invariant.

3. Each component $\psi_k(t,\boldsymbol{x})$ of the wave function satisfies the Klein–Gordon equation (10.142).

Requirement 1 is very natural. 2 is a requirement for relativistic physical laws. Requirement 3 is inspired by Maxwell's equations in an electromagnetic field. In fact, Maxwell's system satisfied by the free electromagnetic wave is a system of first-order partial differential equations (next we will see that, under requirements 1 and 2, the equations satisfied by the wave function $\psi(t,\boldsymbol{x})$ should also constitute a system of first-order partial differential equations), but each component of the electric field and magnetic field satisfies the wave equation (see (1.111) and (1.112) in Chapter 1). Comparing the equations of the wave function with Maxwell's equations, Dirac proposed requirement 3 for the wave function.

Now we try to figure out what kinds of equations the wave function ψ should satisfy under requirements 1, 2, and 3.

In order that (10.158) holds, i.e.,

$$\frac{\mathrm{d}}{\mathrm{d}t}\int_{\mathbb{R}^3} \rho(t,\boldsymbol{x})\mathrm{d}x = \sum_{k=1}^{n}\int_{\mathbb{R}^3}\left(\frac{\partial \psi_k}{\partial t}\overline{\psi}_k + \psi_k\frac{\partial \overline{\psi}_k}{\partial t}\right)\mathrm{d}x = 0, \tag{10.159}$$

when ψ_k $(k = 1,\dots,n)$ is given, $\frac{\partial \psi_k}{\partial t}$ $(k = 1,\dots,n)$ cannot be given arbitrarily. Thus, the partial derivative with respect to time t involved in the equations satisfied by the wave function can only be a first-order partial derivative. Meanwhile, in relativistic space-time, space coordinates have equal status with the time coordinate, so if required to be Lorentz invariant, these equations should also involve the first-order partial derivatives of the wave function with respect to space variable $\boldsymbol{x}$. Therefore, regarding the form of the Klein–Gordon equation (10.142), Dirac suggested the following form of wave equation satisfied by the wave function:

$$\frac{1}{c}\frac{\partial \psi}{\partial t} + \sum_{j=1}^{3} A_j \frac{\partial \psi}{\partial x_j} + \frac{\mathrm{i} m_0 c}{\hbar} B\psi = 0, \tag{10.160}$$

where A_j $(j = 1,2,3)$ and B are all dimensionless nth-order square matrices. Considering the homogeneity of time and space, A_j and B should also be independent of t and $\boldsymbol{x}$, and their elements are complex constants.

System (10.160) can also be written as the following form of Hamiltonian system:

$$\mathrm{i}\hbar \frac{\partial \psi}{\partial t} = \widehat{H}\psi, \tag{10.161}$$

where the Hamiltonian operator is

$$\begin{aligned} \widehat{H} &= -\mathrm{i}\hbar c \sum_{j=1}^{3} A_j \frac{\partial}{\partial x_j} + m_0 c^2 B \\ &= c\sum_{j=1}^{3} A_j \widehat{p}_j + m_0 c^2 B; \end{aligned} \tag{10.162}$$

here we used $\boldsymbol{p} = -\mathrm{i}\hbar\nabla$. The corresponding Hamiltonian is

$$H = c\sum_{j=1}^{3} A_j p_j + m_0 c^2 B. \tag{10.163}$$

The Hamiltonian operator $\widehat{H}$ is also required to be Hermitian such that the probability conservation law (10.158) holds, i.e., A_j $(j = 1,2,3)$ and B are all Hermitian matrices:

$$A_j^* = A_j \quad (j = 1,2,3), \qquad B^* = B, \tag{10.164}$$

where A_j^* and B^* stand for the conjugate transposes of A_j and B, respectively, i.e., $A_j^* = (\overline{A}_j)^{\mathrm{T}}$, and so on.

In fact, under the assumption that A_j $(j = 1,2,3)$ and B are all Hermitian matrices, taking the conjugate transpose of both sides of system (10.160), we obtain

$$\frac{1}{c}\frac{\partial \psi^*}{\partial t} + \sum_{j=1}^{3} \frac{\partial \psi^*}{\partial x_j} A_j - \frac{\mathrm{i} m_0 c}{\hbar} \psi^* B = 0. \tag{10.165}$$

Multiplying system (10.160) by ψ^* from the left and system (10.165) by ψ from the right, respectively, and adding together the resulting formulas, we obtain

$$\frac{1}{c}\frac{\partial}{\partial t}(\psi^*\psi)+\sum_{j=1}^{3}\frac{\partial}{\partial x_j}(\psi^* A_j\psi)=0, \tag{10.166}$$

i.e.,

$$\frac{\partial\rho}{\partial t}+\operatorname{div}\boldsymbol{j}=0, \tag{10.167}$$

where $\rho=\psi^*\psi$ is the probability density given by (10.156) and

$$\boldsymbol{j}=c(\psi^* A_1\psi,\psi^* A_2\psi,\psi^* A_3\psi)^{\mathrm{T}} \tag{10.168}$$

is called the *probability flux density vector*. By Green's formula, the probability conservation law (10.158) follows immediately from (10.167), provided that $\psi(t,\boldsymbol{x})$ goes to zero as $|\boldsymbol{x}|\to\infty$ at a sufficiently rapid velocity.

According to requirement 3, each component of the wave function $\psi(t,\boldsymbol{x})$ should satisfy the Klein–Gordon equation (10.142), i.e.,

$$\frac{1}{c^2}\frac{\partial^2\psi}{\partial t^2}-\Delta\psi+\frac{m_0^2c^2}{\hbar^2}\psi=0. \tag{10.169}$$

What can we conclude from this requirement? Applying the operator

$$\frac{1}{c}\frac{\partial}{\partial t}-\sum_{j=1}^{3}A_j\frac{\partial}{\partial x_j}-\frac{\mathrm{i}m_0c}{\hbar}B$$

to both sides of equation (10.160), we have

$$\left(\frac{1}{c}\frac{\partial}{\partial t}-\sum_{j=1}^{3}A_j\frac{\partial}{\partial x_j}-\frac{\mathrm{i}m_0c}{\hbar}B\right)\left(\frac{1}{c}\frac{\partial}{\partial t}+\sum_{j=1}^{3}A_j\frac{\partial}{\partial x_j}+\frac{\mathrm{i}m_0c}{\hbar}B\right)\psi=0,$$

i.e.,

$$\begin{aligned}&\frac{1}{c^2}\frac{\partial^2\psi}{\partial t^2}-\frac{1}{2}\sum_{j,k=1}^{3}(A_jA_k+A_kA_j)\frac{\partial^2\psi}{\partial x_j\partial x_k}\\&-\frac{\mathrm{i}m_0c}{\hbar}\sum_{j=1}^{3}(A_jB+BA_j)\frac{\partial\psi}{\partial x_j}+\frac{m_0^2c^2}{\hbar^2}B^2\psi=0.\end{aligned} \tag{10.170}$$

Comparing the coefficient matrices of systems (10.170) and (10.169), we obtain, for $j,k=1,2,3$,

$$\begin{aligned}&\frac{1}{2}(A_jA_k+A_kA_j)=\begin{cases}I, & j=k,\\ 0, & j\neq k;\end{cases}\\&A_jB+BA_j=0;\\&B^2=I,\end{aligned}$$

where I stands for the nth-order unit matrix. In other words, these coefficient matrices should satisfy the following properties:

(1) $A_1^2 = A_2^2 = A_3^2 = B^2 = I$.

(2) Any two of A_1, A_2, A_3, and B are anticommutative, i.e., the exchange of their product order changes only the sign.

So far, the Hermitian properties that the coefficient matrices of system (10.160) should satisfy and the above two requirements have summarized all the algebraic properties of A_1, A_2, A_3, and B. Now we determine the expressions of these matrices.

We first show that the order of these matrices satisfying the above conditions, namely, the number of components of the Dirac wave function, can only be even. In fact, from the anticommutativity of A_j and B we have

$$A_j B = -B A_j.$$

From this we get

$$\det A_j \cdot \det B = (-1)^n \det B \cdot \det A_j.$$

From (1) we know that $\det A_j$ and $\det B$ are all nonzero. So $(-1)^n = 1$, i.e., n is even.

Second, we show that $n \geqslant 4$ must hold. For this we prove the following lemma first.

Lemma 10.1. *Suppose that the second-order matrices A_1, A_2, and A_3 are anticommutative to each other, and $A_1^2 = A_2^2 = A_3^2 = I$; then A_1, A_2, A_3, and I are complete in the real vector space composed of all the second-order Hermitian matrices; that is, any given second-order Hermitian matrix can be expressed by a real linear combination of A_1, A_2, A_3, and I.*

Proof. Noticing that the real vector space composed of all the second-order Hermitian matrices is equivalent to $\mathbb{R}^4$, it suffices to prove that A_1, A_2, A_3, and I are linearly independent. Suppose that there exist real numbers λ_0, λ_1, λ_2, and λ_3 such that

$$\lambda_0 I + \lambda_1 A_1 + \lambda_2 A_2 + \lambda_3 A_3 = 0; \tag{10.171}$$

we want to prove $\lambda_0 = \lambda_1 = \lambda_2 = \lambda_3 = 0$.

Multiplying the above formula by A_3 from the left and from the right, respectively, we get

$$\lambda_0 A_3 + \lambda_1 A_3 A_1 + \lambda_2 A_3 A_2 + \lambda_3 A_3^2 = 0$$

and

$$\lambda_0 A_3 + \lambda_1 A_1 A_3 + \lambda_2 A_2 A_3 + \lambda_3 A_3^2 = 0.$$

Adding the above formulas together and noting the anticommutativity of A_3 with A_1 and A_2 as well as $A_3^2 = I$, we immediately have

$$\lambda_0 A_3 + \lambda_3 I = 0.$$

But from the above-mentioned properties (1) and (2) satisfied by A_3, we can easily see that A_3 cannot be either a zero matrix or the unit matrix multiplied by a real number; then from the above formula we have $\lambda_0 = \lambda_3 = 0$. Thus, (10.171) is reduced to

$$\lambda_1 A_1 + \lambda_2 A_2 = 0.$$

Multiplying the above formula by A_2 from the left and from the right, respectively, and adding the resulting formulas together, from the anticommutativity of A_1 and A_2 we also have $\lambda_2 = 0$; then we get $\lambda_1 = 0$ from the above formula. The lemma is proved. □

If $n = 2$, according to Lemma 10.1, B can certainly be expressed by a linear combination of A_1, A_2, A_3, and I. But by the same method we used to prove Lemma 10.1, we can also prove that A_1, A_2, A_3, B, and I are linearly independent (left as an exercise). This leads to a contradiction; then $n \geqslant 4$.

Take $n = 4$. Now A_j ($j = 1,2,3$) and B satisfying the above requirements (1) and (2) are not unique and may have different representations. The most commonly used one is called the *Pauli–Dirac representation*. In this representation, B is taken as a diagonal matrix, and regarding $B^2 = I$, we take

$$B = \begin{pmatrix} I & 0 \\ 0 & -I \end{pmatrix}, \tag{10.172}$$

where I is the second-order unit matrix. Moreover, suppose that

$$A_j = \begin{pmatrix} \alpha_j & \beta_j \\ \gamma_j & \delta_j \end{pmatrix} \quad (j = 1,2,3), \tag{10.173}$$

where α_j, β_j, γ_j, and δ_j are all second-order matrices. Using the anticommutativity of A_j and B and noting the expression (10.172) of B, we obtain

$$\begin{pmatrix} \alpha_j & -\beta_j \\ \gamma_j & -\delta_j \end{pmatrix} = \begin{pmatrix} -\alpha_j & -\beta_j \\ \gamma_j & \delta_j \end{pmatrix}.$$

So $\alpha_j = \delta_j = 0$, and then (10.173) is reduced to

$$A_j = \begin{pmatrix} 0 & \beta_j \\ \gamma_j & 0 \end{pmatrix} \quad (j = 1,2,3).$$

Since A_j is a Hermitian matrix, we should have $\gamma_j = \beta_j^*$, so

$$A_j = \begin{pmatrix} 0 & \beta_j \\ \beta_j^* & 0 \end{pmatrix} \quad (j = 1,2,3).$$

Together with $A_j^2 = I$, we also have

$$\beta_j \beta_j^* = \beta_j^* \beta_j = I.$$

Without loss of generality, we take β_j to be a Hermitian matrix, namely, $\beta_j^* = \beta_j$; then we have

$$A_j = \begin{pmatrix} 0 & \beta_j \\ \beta_j & 0 \end{pmatrix} \quad (j = 1,2,3), \tag{10.174}$$

where

$$\beta_j^2 = I \quad (j = 1,2,3). \tag{10.175}$$

Moreover, using the anticommutation relation, $A_jA_k = -A_kA_j$ $(k \neq j)$, from (10.174) we get

$$\begin{pmatrix} \beta_j\beta_k & 0 \\ 0 & \beta_j\beta_k \end{pmatrix} = -\begin{pmatrix} \beta_k\beta_j & 0 \\ 0 & \beta_k\beta_j \end{pmatrix}, \quad k \neq j.$$

This shows that β_1, β_2, and β_3 also satisfy the following anticommutation relations with each other:

$$\beta_j\beta_k = -\beta_k\beta_j, \quad k \neq j. \tag{10.176}$$

From the discussion of the spin angular momentum of electron in the second paragraph of this section, the Pauli matrices

$$\sigma_1 = \begin{pmatrix} 0 & 1 \\ 1 & 0 \end{pmatrix}, \quad \sigma_2 = \begin{pmatrix} 0 & -\mathrm{i} \\ \mathrm{i} & 0 \end{pmatrix}, \quad \sigma_3 = \begin{pmatrix} 1 & 0 \\ 0 & -1 \end{pmatrix}$$

meet the requirements of (10.175), and (10.176) on β_j $(j = 1,2,3)$, and according to Lemma 10.1, the above three Pauli matrices along with the second-order unit matrix are complete in the real vector space composed of all the second-order Hermitian matrices. So we can take

$$\beta_j = \sigma_j \quad (j = 1,2,3).$$

Then from (10.174) we have

$$A_j = \begin{pmatrix} 0 & \sigma_j \\ \sigma_j & 0 \end{pmatrix} \quad (j = 1,2,3). \tag{10.177}$$

From (10.177) and (10.172), using the expressions of Pauli matrices, we obtain

$$A_1 = \begin{pmatrix} 0 & 0 & 0 & 1 \\ 0 & 0 & 1 & 0 \\ 0 & 1 & 0 & 0 \\ 1 & 0 & 0 & 0 \end{pmatrix},$$

$$A_2 = \begin{pmatrix} 0 & 0 & 0 & -\mathrm{i} \\ 0 & 0 & \mathrm{i} & 0 \\ 0 & -\mathrm{i} & 0 & 0 \\ \mathrm{i} & 0 & 0 & 0 \end{pmatrix},$$

$$A_3 = \begin{pmatrix} 0 & 0 & 1 & 0 \\ 0 & 0 & 0 & -1 \\ 1 & 0 & 0 & 0 \\ 0 & -1 & 0 & 0 \end{pmatrix},$$

$$B = \begin{pmatrix} 1 & 0 & 0 & 0 \\ 0 & 1 & 0 & 0 \\ 0 & 0 & -1 & 0 \\ 0 & 0 & 0 & -1 \end{pmatrix}.$$

These are the expressions of A_j $(j = 1,2,3)$ and B in the Pauli–Dirac representation.

Substituting the above expressions into system (10.160), we know that the system satisfied by the wave function in the Pauli–Dirac representation is

$$\begin{aligned}
&\frac{1}{c}\frac{\partial}{\partial t}\begin{pmatrix}\psi_1\\ \psi_2\\ \psi_3\\ \psi_4\end{pmatrix}+\begin{pmatrix}0&0&0&1\\0&0&1&0\\0&1&0&0\\1&0&0&0\end{pmatrix}\frac{\partial}{\partial x_1}\begin{pmatrix}\psi_1\\ \psi_2\\ \psi_3\\ \psi_4\end{pmatrix}\\
&+\begin{pmatrix}0&0&0&-\mathrm{i}\\0&0&\mathrm{i}&0\\0&-\mathrm{i}&0&0\\\mathrm{i}&0&0&0\end{pmatrix}\frac{\partial}{\partial x_2}\begin{pmatrix}\psi_1\\ \psi_2\\ \psi_3\\ \psi_4\end{pmatrix}\\
&+\begin{pmatrix}0&0&1&0\\0&0&0&-1\\1&0&0&0\\0&-1&0&0\end{pmatrix}\frac{\partial}{\partial x_3}\begin{pmatrix}\psi_1\\ \psi_2\\ \psi_3\\ \psi_4\end{pmatrix}\\
&+\frac{\mathrm{i}m_0c}{\hbar}\begin{pmatrix}1&0&0&0\\0&1&0&0\\0&0&-1&0\\0&0&0&-1\end{pmatrix}\begin{pmatrix}\psi_1\\ \psi_2\\ \psi_3\\ \psi_4\end{pmatrix}=0.
\end{aligned}\tag{10.178}$$

This is called the *Dirac system*, which is a fundamental system in relativistic quantum mechanics.

10.4.4 Plane Wave Solutions of the Free Electron

We discuss the plane wave solutions of the Dirac system (10.178); that is, we find solutions of the form

$$\psi(t,\boldsymbol{x})=\phi(\boldsymbol{p})\mathrm{e}^{\mathrm{i}(\boldsymbol{p}\cdot\boldsymbol{x}-Et)/\hbar}\tag{10.179}$$

(see (10.23)). Plugging the above formula into system (10.160), we know that $\phi=(\phi_1,\phi_2,\phi_3,\phi_4)^{\mathrm{T}}$ satisfies

$$\left(\sum_{j=1}^{3}cp_jA_j+m_0c^2B\right)\phi=E\phi.\tag{10.180}$$

Noting the matrix expressions (10.177) and (10.172) of A_j $(j=1,2,3)$ and B, it is easy to see that the above formula can be written as

$$(E-m_0c^2)\begin{pmatrix}\phi_1\\ \phi_2\end{pmatrix}-c(\sigma\cdot\boldsymbol{p})\begin{pmatrix}\phi_3\\ \phi_4\end{pmatrix}=0\tag{10.181}$$

and

$$-c(\sigma\cdot\boldsymbol{p})\begin{pmatrix}\phi_1\\ \phi_2\end{pmatrix}+(E+m_0c^2)\begin{pmatrix}\phi_3\\ \phi_4\end{pmatrix}=0,\tag{10.182}$$

where

$$\sigma\cdot\boldsymbol{p}=\sum_{j=1}^{3}p_j\sigma_j.\tag{10.183}$$

The linear algebraic system (10.181)–(10.182) has nontrivial solutions if and only if

$$\det\begin{pmatrix} (E-m_0c^2)I & -c\sigma\cdot\boldsymbol{p} \\ -c\sigma\cdot\boldsymbol{p} & (E+m_0c^2)I \end{pmatrix}=0,$$

which is equivalent to

$$\det((E^2-m_0^2c^4)I-c^2(\sigma\cdot\boldsymbol{p})^2)=0.$$

But from (10.175)–(10.176) satisfied by the Pauli matrices, we have $(\sigma\cdot\boldsymbol{p})^2=|\boldsymbol{p}|^2I$; then the above condition is reduced to

$$E^2-m_0^2c^4-c^2|\boldsymbol{p}|^2=0. \tag{10.184}$$

From this equation we can solve that

$$E_\pm=\pm c\sqrt{|\boldsymbol{p}|^2+m_0^2c^2}. \tag{10.185}$$

These two solutions are called the *positive energy solution* and the *negative energy solution*, respectively. Next we will see that, corresponding to the positive and negative energy solutions, linear algebraic system (10.181)–(10.182) admits two linearly independent solutions, respectively.

From system (10.181)–(10.182) we can solve that

$$\begin{pmatrix} \phi_3 \\ \phi_4 \end{pmatrix}=\frac{c}{E+m_0c^2}(\sigma\cdot\boldsymbol{p})\begin{pmatrix} \phi_1 \\ \phi_2 \end{pmatrix} \tag{10.186}$$

or

$$\begin{pmatrix} \phi_1 \\ \phi_2 \end{pmatrix}=\frac{c}{E-m_0c^2}(\sigma\cdot\boldsymbol{p})\begin{pmatrix} \phi_3 \\ \phi_4 \end{pmatrix}. \tag{10.187}$$

Here matrix

$$\sigma\cdot\boldsymbol{p}=\begin{pmatrix} p_3 & p_1-\mathrm{i}\,p_2 \\ p_1+\mathrm{i}\,p_2 & -p_3 \end{pmatrix} \tag{10.188}$$

has two eigenvalues,

$$\lambda_\pm=\pm|\boldsymbol{p}|, \tag{10.189}$$

and the two eigenvectors corresponding to these two eigenvalues can be taken as, respectively,

$$\begin{pmatrix} p_1-\mathrm{i}\,p_2 \\ |\boldsymbol{p}|-p_3 \end{pmatrix} \quad\text{and}\quad \begin{pmatrix} p_1-\mathrm{i}\,p_2 \\ -|\boldsymbol{p}|-p_3 \end{pmatrix}. \tag{10.190}$$

In terms of $E=E_+$, taking $(\phi_1,\phi_2)^{\mathrm{T}}$ in (10.186) to be the above two eigenvectors, we obtain the following two linearly independent solutions of system (10.181)–(10.182):

$$\phi=\begin{pmatrix} p_1-\mathrm{i}\,p_2 \\ |\boldsymbol{p}|-p_3 \\ \dfrac{c|\boldsymbol{p}|(p_1-\mathrm{i}\,p_2)}{E_++m_0c^2} \\ \dfrac{c|\boldsymbol{p}|(|\boldsymbol{p}|-p_3)}{E_++m_0c^2} \end{pmatrix} \quad\text{and}\quad \begin{pmatrix} p_1-\mathrm{i}\,p_2 \\ -(|\boldsymbol{p}|+p_3) \\ \dfrac{-c|\boldsymbol{p}|(p_1-\mathrm{i}\,p_2)}{E_++m_0c^2} \\ \dfrac{c|\boldsymbol{p}|(|\boldsymbol{p}|+p_3)}{E_++m_0c^2} \end{pmatrix}. \tag{10.191}$$

Plugging these two solutions into (10.179), we get the following two plane wave solutions of Dirac system (10.178) corresponding to positive energy $E = E_+$:

$$\psi = \begin{pmatrix} p_1 - \mathrm{i}\, p_2 \\ |\boldsymbol{p}| - p_3 \\ \dfrac{c|\boldsymbol{p}|(p_1 - \mathrm{i}\, p_2)}{E_+ + m_0 c^2} \\ \dfrac{c|\boldsymbol{p}|(|\boldsymbol{p}| - p_3)}{E_+ + m_0 c^2} \end{pmatrix} \mathrm{e}^{\mathrm{i}(\boldsymbol{p}\cdot\boldsymbol{x} - E_+ t)/\hbar} \tag{10.192}$$

and

$$\psi = \begin{pmatrix} p_1 - \mathrm{i}\, p_2 \\ -(|\boldsymbol{p}| + p_3) \\ \dfrac{-c|\boldsymbol{p}|(p_1 - \mathrm{i}\, p_2)}{E_+ + m_0 c^2} \\ \dfrac{c|\boldsymbol{p}|(|\boldsymbol{p}| + p_3)}{E_+ + m_0 c^2} \end{pmatrix} \mathrm{e}^{\mathrm{i}(\boldsymbol{p}\cdot\boldsymbol{x} - E_+ t)/\hbar}. \tag{10.193}$$

Similarly, in terms of $E = E_-$, taking $(\phi_3, \phi_4)^{\mathrm{T}}$ in (10.187) to be the two eigenvectors given by (10.190), respectively, we obtain the following two linearly independent solutions of system (10.181)–(10.182):

$$\phi = \begin{pmatrix} \dfrac{-c|\boldsymbol{p}|(p_1 - \mathrm{i}\, p_2)}{|E_-| + m_0 c^2} \\ \dfrac{-c|\boldsymbol{p}|(|\boldsymbol{p}| - p_3)}{|E_-| + m_0 c^2} \\ p_1 - \mathrm{i}\, p_2 \\ |\boldsymbol{p}| - p_3 \end{pmatrix} \quad \text{and} \quad \begin{pmatrix} \dfrac{c|\boldsymbol{p}|(p_1 - \mathrm{i}\, p_2)}{|E_-| + m_0 c^2} \\ \dfrac{-c|\boldsymbol{p}|(|\boldsymbol{p}| + p_3)}{|E_-| + m_0 c^2} \\ p_1 - \mathrm{i}\, p_2 \\ -(|\boldsymbol{p}| + p_3) \end{pmatrix}; \tag{10.194}$$

then the two plane wave solutions of the Dirac system (10.178) corresponding to negative energy $E = E_-$ are

$$\psi = \begin{pmatrix} \dfrac{-c|\boldsymbol{p}|(p_1 - \mathrm{i}\, p_2)}{|E_-| + m_0 c^2} \\ \dfrac{-c|\boldsymbol{p}|(|\boldsymbol{p}| - p_3)}{|E_-| + m_0 c^2} \\ p_1 - \mathrm{i}\, p_2 \\ |\boldsymbol{p}| - p_3 \end{pmatrix} \mathrm{e}^{\mathrm{i}(\boldsymbol{p}\cdot\boldsymbol{x} - E_- t)/\hbar} \tag{10.195}$$

and

$$\psi = \begin{pmatrix} \dfrac{c|\boldsymbol{p}|(p_1 - \mathrm{i}\, p_2)}{|E_-| + m_0 c^2} \\ \dfrac{-c|\boldsymbol{p}|(|\boldsymbol{p}| + p_3)}{|E_-| + m_0 c^2} \\ p_1 - \mathrm{i}\, p_2 \\ -(|\boldsymbol{p}| + p_3) \end{pmatrix} \mathrm{e}^{\mathrm{i}(\boldsymbol{p}\cdot\boldsymbol{x} - E_- t)/\hbar}. \tag{10.196}$$

If we take the direction of the electron momentum to be that of the x_3 axis, then $p_1 = p_2 = 0$ and $p_3 = p$. Noting that

$$\lim_{p_1,p_2\to 0} \frac{|\boldsymbol{p}| - p_3}{p_1 - \mathrm{i}\, p_2} = 0,$$

choosing suitable scalar factors in the above four plane wave solutions of the Dirac system (10.178), and passing to the limit as $p_1, p_2 \to 0$, we obtain, respectively,

$$\psi = \begin{pmatrix} 1 \\ 0 \\ \dfrac{cp}{E_+ + m_0c^2} \\ 0 \end{pmatrix} \mathrm{e}^{\mathrm{i}(\boldsymbol{p}\cdot\boldsymbol{x} - E_+ t)/\hbar}, \tag{10.197}$$

$$\psi = \begin{pmatrix} 0 \\ 1 \\ 0 \\ -\dfrac{cp}{E_+ + m_0c^2} \end{pmatrix} \mathrm{e}^{\mathrm{i}(\boldsymbol{p}\cdot\boldsymbol{x} - E_+ t)/\hbar}, \tag{10.198}$$

$$\psi = \begin{pmatrix} -\dfrac{cp}{|E_-| + m_0c^2} \\ 0 \\ 1 \\ 0 \end{pmatrix} \mathrm{e}^{\mathrm{i}(\boldsymbol{p}\cdot\boldsymbol{x} - E_- t)/\hbar}, \tag{10.199}$$

and

$$\psi = \begin{pmatrix} 0 \\ \dfrac{cp}{|E_-| + m_0c^2} \\ 0 \\ 1 \end{pmatrix} \mathrm{e}^{\mathrm{i}(\boldsymbol{p}\cdot\boldsymbol{x} - E_- t)/\hbar}, \tag{10.200}$$

where $\boldsymbol{p}\cdot\boldsymbol{x} = px_3$.

In terms of positive energy $E = E_+$, the Dirac system (10.178) has two linearly independent plane wave solutions (10.198) and (10.199), which correspond to two spin states of the electron. This implies that each spin state of the electron is associated with a wave function. To further show this fact, we look at the case

$$p \ll m_0 c.$$

In this situation, for solutions (10.198) and (10.199) corresponding to positive energy, ψ_3 and ψ_4 are both very small, called small components, while ψ_1 and ψ_2 are called large components. Ignoring small component parts, their large component parts are

$$\begin{pmatrix} 1 \\ 0 \end{pmatrix} \mathrm{e}^{\mathrm{i}(\boldsymbol{p}\cdot\boldsymbol{x} - E_+ t)/\hbar} \tag{10.201}$$

and

$$\begin{pmatrix} 0 \\ 1 \end{pmatrix} e^{i(\boldsymbol{p}\cdot\boldsymbol{x}-E_+t)/\hbar}. \tag{10.202}$$

This just indicates that the spin states of the electron are only two eigenstates.

In the plane wave solutions we obtained, there are solutions corresponding not only to positive energy but also to negative energy. Negative energy of a free particle does not make sense in classical physics. However, in quantum mechanics, the eigenfunction system will not be complete if the eigenstates corresponding to negative energy are excluded, and from the expansion of the wave function with respect to the eigenfunction system we know that there does exist a probability of transferring to negative energy states. Dirac thought that the negative energy state represented a different particle from the electron, namely, the *positive electron* state. Three years after Dirac made this prediction, C. D. Anderson discovered positive electrons in an experiment in 1932.

10.4.5 Mathematical Structure of the Dirac System

In the Dirac system

$$\frac{1}{c}\frac{\partial\psi}{\partial t} + \sum_{j=1}^{3} A_j \frac{\partial\psi}{\partial x_j} + \frac{i m_0 c}{\hbar} B\psi = 0, \tag{10.203}$$

ψ is a complex-valued unknown vector function, the coefficient matrix $\frac{1}{c}I$ of $\frac{\partial\psi}{\partial t}$ is a real positive definite symmetric matrix, and the coefficient matrices A_j ($j = 1,2,3$) are all Hermitian matrices. This type of system of first-order partial differential equations is called a *first-order Hermitian hyperbolic system of partial differential equations*.

A first-order Hermitian hyperbolic system has almost the same properties as a first-order symmetric hyperbolic system, and well-posed problems are defined in a similar way. We do not go into further detail here.

We need to point out that since the Dirac system is a first-order Hermitian hyperbolic system, according to the properties of the first-order symmetric hyperbolic system (see section 1.5 in Chapter 1), its solution has a finite domain of dependence, i.e., the perturbation has a finite propagation speed. This is an important difference between the Dirac system and the nonrelativistic Schrödinger equation. In addition, based on the theory of special relativity (see section 9.3 in Chapter 9), we can also assert that for the Dirac system, the propagation speed of perturbation cannot exceed the light speed c. Now we prove rigorously this assertion in mathematics.

Consider the Cauchy problem of the Dirac system (10.203) satisfying the initial condition

$$\psi(0,\boldsymbol{x}) = \psi^0(\boldsymbol{x}). \tag{10.204}$$

Suppose that $(t^0, \boldsymbol{x}^0)$ ($t^0 > 0$) is any given point in four-space-time, and $\Gamma(t^0)$ is the cone-shaped domain enclosed by the light cone surface

$$|\boldsymbol{x} - \boldsymbol{x}^0| = c(t^0 - t)$$

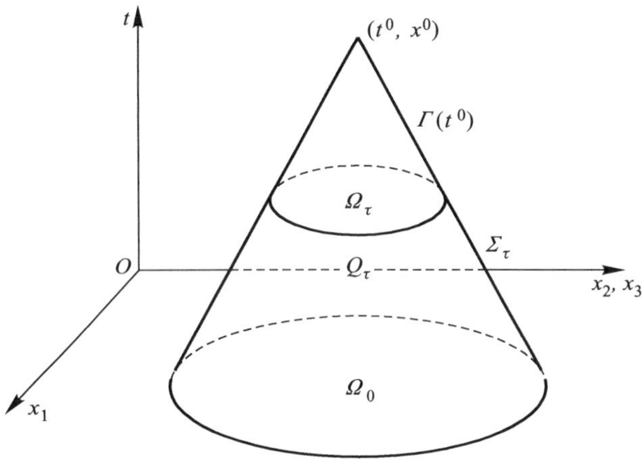

Figure 10.6

passing through this point and the initial plane $t = 0$ (see Figure 10.6):

$$\Gamma(t^0) : |\boldsymbol{x}-\boldsymbol{x}^0| \leqslant c(t^0-t) \quad (0 \leqslant t \leqslant t^0). \tag{10.205}$$

For any given moment $\tau \in [0,t^0]$, denote by Ω_τ the section of the superplane $t=\tau$ crossed with the cone-shaped domain $\Gamma(t^0)$:

$$\Omega_\tau : |\boldsymbol{x}-\boldsymbol{x}^0| \leqslant c(t^0-t), \quad t=\tau. \tag{10.206}$$

We have the following theorem.

Theorem 10.1. *Suppose that $\psi(t,\boldsymbol{x})$ is a solution to the Cauchy problem of the Dirac system* (10.203) *satisfying the initial condition* (10.204)*; then*

$$\int_{\Omega_\tau} \rho(\tau,\boldsymbol{x})\mathrm{d}x \leqslant \int_{\Omega_0} \rho(0,\boldsymbol{x})\mathrm{d}x \quad (0 \leqslant \tau \leqslant t^0), \tag{10.207}$$

where $\rho(t,\boldsymbol{x})$ is given by (10.156).

Proof. Denote by Q_τ the part of the cone $\Gamma(t^0)$ between the initial plane $t=0$ and $t=\tau$, where Σ_τ is its lateral boundary, i.e.,

$$Q_\tau : |\boldsymbol{x}-\boldsymbol{x}^0| \leqslant c(t^0-t) \quad (0 \leqslant t \leqslant \tau), \tag{10.208}$$
$$\Sigma_\tau : |\boldsymbol{x}-\boldsymbol{x}^0| = c(t^0-t) \quad (0 \leqslant t \leqslant \tau). \tag{10.209}$$

Integrating the conservation equation (10.167) in Q_τ and using Green's formula, we obtain

$$\int_{\Omega_\tau} \rho(\tau,\boldsymbol{x})\mathrm{d}x - \int_{\Omega_0} \rho(0,\boldsymbol{x})\mathrm{d}x + \int_{\Sigma_\tau} \left(\rho n_0 + \sum_{k=1}^{3} j_k n_k\right) \mathrm{d}S = 0, \tag{10.210}$$

where (n_0, n_1, n_2, n_3) stands for the outward unit normal vector on Σ_τ. Using equation (10.209) of Σ_τ, we can easily see that

$$(n_0, n_1, n_2, n_3) = \frac{1}{\sqrt{1+c^2}} \left(c, \frac{x_1 - x_1^0}{|\boldsymbol{x} - \boldsymbol{x}^0|}, \frac{x_2 - x_2^0}{|\boldsymbol{x} - \boldsymbol{x}^0|}, \frac{x_3 - x_3^0}{|\boldsymbol{x} - \boldsymbol{x}^0|} \right).$$

Noting also the definitions (10.156) and (10.168) of ρ and $\boldsymbol{j}$, we have

$$\begin{aligned} \rho n_0 + \sum_{k=1}^{n} j_k n_k &= c\psi^* \left(\frac{n_0}{c} I + \sum_{k=1}^{3} n_k A_k \right) \psi \\ &= \frac{c}{\sqrt{1+c^2}} \psi^* \left(I + \sum_{k=1}^{3} a_k A_k \right) \psi, \end{aligned} \tag{10.211}$$

where

$$a_k = \frac{x_k - x_k^0}{|\boldsymbol{x} - \boldsymbol{x}^0|} \quad (k = 1, 2, 3). \tag{10.212}$$

From the Pauli–Dirac representation (10.177) of A_k $(k = 1, 2, 3)$ we have

$$\sum_{k=1}^{3} a_k A_k = \begin{pmatrix} 0 & \sum_{k=1}^{3} a_k \sigma_k \\ \sum_{k=1}^{3} a_k \sigma_k & 0 \end{pmatrix}. \tag{10.213}$$

Let λ be the eigenvalue of matrix $\sum_{k=1}^{3} a_k A_k$. Since $\sum_{k=1}^{3} a_k A_k$ is a Hermitian matrix, λ must be real. In addition, λ^2 must be the eigenvalue of the matrix

$$\left(\sum_{k=1}^{3} a_k A_k \right)^2 = \begin{pmatrix} \left(\sum_{k=1}^{3} a_k \sigma_k \right)^2 & 0 \\ 0 & \left(\sum_{k=1}^{3} a_k \sigma_k \right)^2 \end{pmatrix};$$

then λ^2 is also the eigenvalue of the matrix $\left(\sum_{k=1}^{3} a_k \sigma_k \right)^2$. But from (10.188) and (10.189) we can easily show that the eigenvalues of $\sum_{k=1}^{3} a_k \sigma_k$ are $\pm\sqrt{a_1^2 + a_2^2 + a_3^2} = \pm 1$, so the eigenvalue of $\left(\sum_{k=1}^{3} a_k \sigma_k \right)^2$ is $\lambda^2 = 1$. This implies that the eigenvalue λ of the matrix $\sum_{k=1}^{3} \alpha_k A_k$ satisfies

$$|\lambda| = 1, \quad \lambda \in \mathbb{R}. \tag{10.214}$$

It follows immediately from (10.214) that the matrix $I + \sum_{k=1}^{3} a_k A_k$ is positive semidefinite; i.e.,

$$\psi^* \left(I + \sum_{k=1}^{3} a_k A_k \right) \psi \geqslant 0, \quad \forall \psi. \tag{10.215}$$

Thus, using (10.211), it follows from (10.210) that

$$\int_{\Omega_\tau} \rho(\tau, \boldsymbol{x}) \mathrm{d}x - \int_{\Omega_0} \rho(0, \boldsymbol{x}) \mathrm{d}x \leqslant 0.$$

This finishes the proof of the theorem. □

We can see immediately from Theorem 10.1 that if the initial value $\psi^0(\boldsymbol{x})$ is zero in the bottom Ω_0 of the light cone $\Gamma(t^0)$, then the wave function $\psi(t, \boldsymbol{x})$ determined by it is identically equal to zero in the whole light cone. This indicates that for the Dirac system, the propagation speed of perturbation cannot exceed the speed of light c.

As a consequence, the uniqueness of solutions to the Cauchy problem (10.203)–(10.204) of the Dirac system can be inferred from Theorem 10.1.

10.4.6 Lorentz Invariance of the Dirac System

In this subsection, we discuss the invariance of the Dirac system under the Lorentz transformation. For this, first we need to examine whether or not the Dirac wave function $\psi = (\psi_1, \psi_2, \psi_3, \psi_4)^{\mathrm{T}}$ is a vector in Minkowski four-space-time. If not, what will this wave function be under the Lorentz transformation?

First we notice that ρe is just the charge density in space and $e\boldsymbol{j}$ is just the current density vector, where e is the charge of the electron and ρ and $\boldsymbol{j}$ are given by (10.156) and (10.168), respectively. $(c\rho e, e\boldsymbol{j})$ is called the four-current density in the theory of electromagnetic fields (see section 9.7). Therefore,

$$(j^0, j^1, j^2, j^3) = (c\rho, j_1, j_2, j_3) \tag{10.216}$$

is a vector in Minkowski four-space-time (see section 9.4 in Chapter 9), and the conservation equation (10.167) can be written in the four-divergence form

$$\frac{\partial j^\alpha}{\partial x^\alpha} = 0, \tag{10.217}$$

where $(x^0, x^1, x^2, x^3) = (ct, x_1, x_2, x_3)$, and repeated Greek letters stand for summation from 0 to 3.

Now we discuss only the normal Lorentz transformation

$$\widetilde{x}^\alpha = a^\alpha_\beta x^\beta; \tag{10.218}$$

i.e., we assume in the above transformation that $\det(a^\alpha_\beta) = 1$ and $a^0_0 > 0$. This actually excludes reflection (with respect to time and space). From Theorem 9.1, any given normal Lorentz transformation can be compounded by a Lorentz transformation of special form given by (9.19) and a spatial rotation transformation.

First, we look at the Lorentz transformation of special form given by (9.19). It can be written in the following form:

$$\begin{cases} \widetilde{x}^0 = (\mathrm{ch}\,\theta)x^0 - (\mathrm{sh}\,\theta)x^1, \\ \widetilde{x}^1 = -(\mathrm{sh}\,\theta)x^0 + (\mathrm{ch}\,\theta)x^1, \\ \widetilde{x}^2 = x^2, \\ \widetilde{x}^3 = x^3, \end{cases} \tag{10.219}$$

where

$$\mathrm{ch}\,\theta = \frac{c}{\sqrt{c^2 - V^2}}, \quad \mathrm{sh}\,\theta = \frac{V}{\sqrt{c^2 - V^2}}. \tag{10.220}$$

Suppose that the Dirac wave function ψ is changed into $\widetilde{\psi}$ under the above Lorentz transformation. Noting that $(c\rho, j_1, j_2, j_3)$ is a four-vector, according to the change law of vectors in Minkowski four-space-time (see section 9.4), we have

$$\widetilde{\psi}^* \widetilde{\psi} = \psi^*((\mathrm{ch}\,\theta)I - (\mathrm{sh}\,\theta)A_1)\psi, \tag{10.221}$$

$$\widetilde{\psi}^* A_1 \widetilde{\psi} = \psi^*((\mathrm{ch}\,\theta)A_1 - (\mathrm{sh}\,\theta)I)\psi, \tag{10.222}$$

$$\widetilde{\psi}^* A_{2,3} \widetilde{\psi} = \psi^* A_{2,3} \psi. \tag{10.223}$$

Because we need them in further discussion, we give the following two lemmas.

Lemma 10.2. *Suppose that the square matrix A satisfies $A^2 = I$; then*

$$\mathrm{e}^{\theta A} = (\mathrm{ch}\,\theta)I + (\mathrm{sh}\,\theta)A, \quad \forall \theta \in \mathbb{R}. \tag{10.224}$$

Proof. From $A^2 = I$ we can easily see that for any given natural number n we have

$$A^{2n} = I, \quad A^{2n+1} = A;$$

then

$$\mathrm{ch}(\theta A) = \frac{1}{2}(\mathrm{e}^{\theta A} + \mathrm{e}^{-\theta A}) = (\mathrm{ch}\,\theta)I,$$

$$\mathrm{sh}(\theta A) = \frac{1}{2}(\mathrm{e}^{\theta A} - \mathrm{e}^{-\theta A}) = (\mathrm{sh}\,\theta)A.$$

Noting that $\mathrm{e}^{\theta A} = \mathrm{ch}(\theta A) + \mathrm{sh}(\theta A)$, the lemma follows immediately. □

Lemma 10.3. *If square matrices A and B are commutative, i.e., $AB = BA$, then*

$$B\,\mathrm{e}^{\theta A} = \mathrm{e}^{\theta A} B, \quad \forall \theta \in \mathbb{R}, \tag{10.225}$$

while if A and B are anticommutative, i.e., $AB = -BA$, then

$$B\,\mathrm{e}^{\theta A} = \mathrm{e}^{-\theta A} B, \quad \forall \theta \in \mathbb{R}. \tag{10.226}$$

Proof. The proof is obvious, so it is omitted here. □

Using Lemma 10.2 and noting $A_1^2 = I$, transformation formulas (10.221) and (10.222) can be rewritten as

$$\widetilde{\psi}^* \widetilde{\psi} = \psi^* \mathrm{e}^{-\theta A_1} \psi, \tag{10.227}$$
$$\widetilde{\psi}^* A_1 \widetilde{\psi} = \psi^* A_1 \mathrm{e}^{-\theta A_1} \psi. \tag{10.228}$$

In order for the above two relations to be satisfied, the transformation of the wave function can be taken as follows:

$$\widetilde{\psi} = \mathrm{e}^{-\frac{\theta}{2} A_1} \psi. \tag{10.229}$$

Since A_1 is a Hermitian matrix, obviously we have

$$\widetilde{\psi}^* = \psi^* \mathrm{e}^{-\frac{\theta}{2} A_1}; \tag{10.230}$$

then (10.227) is certainly satisfied. According to Lemma 10.3, we also have

$$\mathrm{e}^{-\frac{\theta}{2} A_1} A_1 = A_1 \mathrm{e}^{-\frac{\theta}{2} A_1},$$
$$\mathrm{e}^{-\frac{\theta}{2} A_1} A_{2,3} = A_{2,3} \mathrm{e}^{\frac{\theta}{2} A_1}.$$

From this we can see that for transformation (10.229), formulas (10.228) and (10.223) are also satisfied. Thus, (10.229) is exactly the transformation formula of the wave function under the Lorentz transformation (10.219).

Now we look at the rotation transformation of the space coordinate system. Without loss of generality, suppose that the rotation is around the x_3 axis, and the corresponding transformation is

$$\begin{cases} \widetilde{x}^0 = x^0, \\ \widetilde{x}^1 = (\cos\phi)x^1 + (\sin\phi)x^2, \\ \widetilde{x}^2 = -(\sin\phi)x^1 + (\cos\phi)x^2, \\ \widetilde{x}^3 = x^3. \end{cases} \tag{10.231}$$

According to the change law of vectors in Minkowski four-space-time, under the above transformation we have

$$\widetilde{\psi}^* \widetilde{\psi} = \psi^* \psi, \tag{10.232}$$
$$\widetilde{\psi}^* A_1 \widetilde{\psi} = \psi^*((\cos\phi)A_1 + (\sin\phi)A_2)\psi, \tag{10.233}$$
$$\widetilde{\psi}^* A_2 \widetilde{\psi} = \psi^*(-(\sin\phi)A_1 + (\cos\phi)A_2)\psi, \tag{10.234}$$
$$\widetilde{\psi}^* A_3 \widetilde{\psi} = \psi^* A_3 \psi. \tag{10.235}$$

Similarly to Lemmas 10.2 and 10.3, we can prove the following lemmas.

Lemma 10.4. *Suppose that the square matrix A satisfies $A^2 = I$; then*

$$\mathrm{e}^{\mathrm{i}\phi A} = (\cos\phi)I + \mathrm{i}(\sin\phi)A, \quad \forall \phi \in \mathbb{R}.$$

Lemma 10.5. *If square matrices A and B are commutative, then*

$$B\,\mathrm{e}^{\mathrm{i}\phi A} = \mathrm{e}^{\mathrm{i}\phi A} B, \quad \forall \phi \in \mathbb{R},$$

while if A and B are anticommutative, then

$$B\,\mathrm{e}^{\mathrm{i}\phi A} = \mathrm{e}^{-\mathrm{i}\phi A}\,B, \quad \forall\phi\in\mathbb{R}.$$

Denote

$$\Sigma_3 = \begin{pmatrix} \sigma_3 & 0 \\ 0 & \sigma_3 \end{pmatrix}, \tag{10.236}$$

where σ_3 is a Pauli matrix. It is easy to verify that

$$A_2 = \mathrm{i}\,A_1\Sigma_3. \tag{10.237}$$

Plugging (10.237) into (10.233) and (10.234), noting $\Sigma_3^2 = I$, and using Lemma 10.4, we obtain that transformation formulas (10.233) and (10.234) are equivalent to

$$\widetilde{\psi}^* A_1 \widetilde{\psi} = \psi^* A_1 \mathrm{e}^{\mathrm{i}\phi\Sigma_3}\,\psi, \tag{10.238}$$

$$\widetilde{\psi}^* A_1 \Sigma_3 \widetilde{\psi} = \psi^* A_1 \Sigma_3 \mathrm{e}^{\mathrm{i}\phi\Sigma_3}\,\psi. \tag{10.239}$$

In order for the above two formulas to hold, we take the transformation of the wave function as

$$\widetilde{\psi} = \mathrm{e}^{\mathrm{i}\frac{\phi}{2}\Sigma_3}\,\psi. \tag{10.240}$$

Obviously, we have

$$\widetilde{\psi}^* = \psi^*\,\mathrm{e}^{-\mathrm{i}\frac{\phi}{2}\Sigma_3}. \tag{10.241}$$

Since A_1 and Σ_3 are anticommutative, from Lemma 10.5 we have

$$\mathrm{e}^{-\mathrm{i}\frac{\phi}{2}\Sigma_3}\,A_1 = A_1\,\mathrm{e}^{\mathrm{i}\frac{\phi}{2}\Sigma_3}.$$

From this we can immediately see that (10.238) is satisfied for the transformation given by (10.240). In addition to the above formula, from Lemma 10.5 we also have

$$\Sigma_3\,\mathrm{e}^{\mathrm{i}\frac{\phi}{2}\Sigma_3} = \mathrm{e}^{\mathrm{i}\frac{\phi}{2}\Sigma_3}\,\Sigma_3,$$

so (10.239) is also satisfied.

Since A_3 and Σ_3 are commutative, we can also obtain (10.235) from Lemma 10.5. It is obvious that (10.232) is correct. Therefore, (10.240) is exactly the transformation formula of the wave function under the Lorentz transformation (10.231).

It can be shown from (10.229) and (10.240) that the change law of the Dirac wave function is different from that of vectors in Minkowski four-space-time. As stated above, the change law of vectors is given by angle θ or ϕ. It can also be shown that the change law of second-order tensors is given by the double angle 2θ or 2ϕ. However, the change law of the Dirac wave function is represented by the half angle $\theta/2$ or $\phi/2$. This kind of quantity transformed by (10.229) and/or (10.240) is called the *spinor* or the *half-order tensor*.

To show that the Dirac system (10.160) remains unchanged under the above transformation of the wave function, we write it in the following form:

$$\frac{\partial}{\partial x^\alpha}(\gamma^\alpha \psi) - \frac{m_0 c}{\hbar}\psi = 0, \tag{10.242}$$

where

$$\gamma^0 = (-\mathrm{i}\,B)^{-1}, \quad \gamma^k = (-\mathrm{i}\,B)^{-1} A_k \quad (k = 1,2,3). \tag{10.243}$$

Denoting the above transformation of the wave function by

$$\widetilde{\psi} = M\psi, \tag{10.244}$$

and plugging it into system (10.242), we get

$$\frac{\partial}{\partial x^\alpha}(\gamma^\alpha M^{-1}\widetilde{\psi}) - \frac{m_0 c}{\hbar} M^{-1}\widetilde{\psi} = 0. \tag{10.245}$$

Noting furthermore that under the Lorentz transformation (10.218) we have

$$\frac{\partial}{\partial x^\alpha} = a_\alpha^\beta \frac{\partial}{\partial \widetilde{x}^\beta},$$

system (10.245) can be written as

$$a_\alpha^\beta \gamma^\alpha M^{-1} \frac{\partial \widetilde{\psi}}{\partial \widetilde{x}^\beta} - \frac{m_0 c}{\hbar} M^{-1}\widetilde{\psi} = 0.$$

Multiplying both sides of the above formula by M from the left, we have

$$\left(a_\alpha^\beta M \gamma^\alpha M^{-1} \frac{\partial}{\partial \widetilde{x}^\beta} - \frac{m_0 c}{\hbar} \right) \widetilde{\psi} = 0. \tag{10.246}$$

Therefore, if the transformation (10.244) of the wave function is such that

$$a_\alpha^\beta M \gamma^\alpha M^{-1} = \gamma^\beta \quad (\beta = 0,1,2,3), \tag{10.247}$$

then the Dirac system remains unchanged under the Lorentz transformation.

As to the transformation of the wave function given by (10.229) and (10.240), we now verify the corresponding condition (10.247).

(1) For Lorentz transformation (10.219), we have

$$\begin{aligned}
&a_0^0 = \mathrm{ch}\,\theta, \quad a_1^0 = -\,\mathrm{sh}\,\theta,\\
&a_0^1 = -\,\mathrm{sh}\,\theta, \quad a_1^1 = \mathrm{ch}\,\theta,\\
&a_2^2 = 1, \quad a_3^3 = 1,
\end{aligned}$$

and the other components of (a_α^β) are all zero. Now condition (10.247) is reduced to

$$M((\operatorname{ch}\theta)\gamma^0-(\operatorname{sh}\theta)\gamma^1)=\gamma^0 M, \tag{10.248}$$

$$M((-\operatorname{sh}\theta)\gamma^0+(\operatorname{ch}\theta)\gamma^1)=\gamma^1 M, \tag{10.249}$$

$$M\gamma^2=\gamma^2 M, \tag{10.250}$$

and

$$M\gamma^3=\gamma^3 M, \tag{10.251}$$

where

$$M=\mathrm{e}^{-\frac{\theta}{2}A_1}. \tag{10.252}$$

We only verify the correctness of (10.248) here. Others can be verified similarly, which is left to the reader as an exercise.

Using (10.252), (10.243), and Lemma 10.2, we have

$$M((\operatorname{ch}\theta)\gamma^0-(\operatorname{sh}\theta)\gamma^1)=\mathrm{e}^{-\frac{\theta}{2}A_1}(-\mathrm{i}\,B)^{-1}\mathrm{e}^{-\theta A_1}.$$

From the anticommutativity of A_1 and B we can obtain the anticommutativity of A_1 and B^{-1}; then Lemma 10.3 yields

$$B^{-1}\mathrm{e}^{\frac{\theta}{2}A_1}=\mathrm{e}^{-\frac{\theta}{2}A_1}B^{-1}.$$

Thus,

$$\begin{aligned}&M((\operatorname{ch}\theta)\gamma^0-(\operatorname{sh}\theta)\gamma^1)\\&=(-\mathrm{i}\,B)^{-1}\mathrm{e}^{\frac{\theta}{2}A_1}\mathrm{e}^{-\theta A_1}\\&=(-\mathrm{i}\,B)^{-1}\mathrm{e}^{-\frac{\theta}{2}A_1}=\gamma^0 M.\end{aligned}$$

This proves (10.248).

(2) For the transformation (10.231), we have

$$\begin{aligned}&a_0^0=1,\\&a_1^1=\cos\phi,\quad a_2^1=\sin\phi,\\&a_1^2=-\sin\phi,\quad a_2^2=\cos\phi,\\&a_3^3=1,\end{aligned}$$

and the other components of (a_α^β) are all zero. Now condition (10.247) is reduced to

$$M\gamma^0=\gamma^0 M, \tag{10.253}$$

$$M((\cos\phi)\gamma^1+(\sin\phi)\gamma^2)=\gamma^1 M, \tag{10.254}$$

$$M(-(\sin\phi)\gamma^1+(\cos\phi)\gamma^2)=\gamma^2 M, \tag{10.255}$$

and

$$M\gamma^3=\gamma^3 M, \tag{10.256}$$

where

$$M=\mathrm{e}^{\mathrm{i}\frac{\phi}{2}\Sigma_3}. \tag{10.257}$$

Using (10.243), (10.237), and Lemma 10.4, we get

$$\begin{aligned}
&M((\cos\phi)\gamma^1+(\sin\phi)\gamma^2)\\
&=\mathrm{e}^{\mathrm{i}\frac{\phi}{2}\Sigma_3}(-\mathrm{i}B)^{-1}((\cos\phi)A_1+(\sin\phi)A_2)\\
&=\mathrm{e}^{\mathrm{i}\frac{\phi}{2}\Sigma_3}(-\mathrm{i}B)^{-1}A_1\,\mathrm{e}^{\mathrm{i}\phi\Sigma_3}.
\end{aligned}$$

It is easy to verify directly that A_1 and Σ_3 are anticommutative, while B and Σ_3 are commutative; then B^{-1} and Σ_3 are commutative. Hence, using Lemma 10.5 we have

$$\begin{aligned}
\mathrm{e}^{\mathrm{i}\frac{\phi}{2}\Sigma_3}A_1&=A_1\,\mathrm{e}^{-\mathrm{i}\frac{\phi}{2}\Sigma_3},\\
\mathrm{e}^{\mathrm{i}\frac{\phi}{2}\Sigma_3}B^{-1}&=B^{-1}\,\mathrm{e}^{\mathrm{i}\frac{\phi}{2}\Sigma_3}.
\end{aligned}$$

Thus we obtain

$$\begin{aligned}
&M((\cos\phi)\gamma^1+(\sin\phi)\gamma^2)\\
&=(-\mathrm{i}B)^{-1}\mathrm{e}^{\mathrm{i}\frac{\phi}{2}\Sigma_3}A_1\,\mathrm{e}^{\mathrm{i}\phi\Sigma_3}\\
&=(-\mathrm{i}B)^{-1}A_1\,\mathrm{e}^{-\mathrm{i}\frac{\phi}{2}\Sigma_3}\mathrm{e}^{\mathrm{i}\phi\Sigma_3}\\
&=\gamma^1M.
\end{aligned}$$

This proves (10.254). Similarly, (10.255)–(10.256) and (10.253) can be verified, which is left to the reader as an exercise.

Exercises

1. Suppose that $\psi_1(t,\boldsymbol{x})$ and $\psi_2(t,\boldsymbol{x})$ are two solutions of the Schrödinger equation

$$\mathrm{i}\hbar\frac{\partial\psi}{\partial t}=-\frac{\hbar^2}{2m}\Delta\psi+V(\boldsymbol{x})\psi,$$

and they and their first-order derivatives with respect to the spatial variables go to zero at a sufficiently rapid rate as $|\boldsymbol{x}|\to\infty$, and $V(\boldsymbol{x})$ is a real-valued function. Prove that

$$\frac{\mathrm{d}}{\mathrm{d}t}\int_{\mathbb{R}^3}\overline{\psi_1(t,\boldsymbol{x})}\psi_2(t,\boldsymbol{x})\mathrm{d}x=0,\quad\forall t\in\mathbb{R}.$$

2. Suppose that $\Omega\subset\mathbb{R}^3$ is a bounded domain and the potential field $V(\boldsymbol{x})$ is a real-valued function defined in Ω. Prove the uniqueness of solutions to the following initial-boundary value problem:

$$\begin{cases}
\mathrm{i}\hbar\dfrac{\partial\psi}{\partial t}=-\dfrac{\hbar^2}{2m}\Delta\psi+V(\boldsymbol{x})\psi,\quad t>0,\boldsymbol{x}\in\Omega,\\
\psi=0,\quad t>0,\boldsymbol{x}\in\partial\Omega\ (\partial\Omega\text{ is the boundary of }\Omega),\\
\psi=\psi_0(\boldsymbol{x}),\quad t=0,\boldsymbol{x}\in\Omega,
\end{cases}$$

where $\psi_0(\boldsymbol{x})$ is a given function.

3. Change the boundary condition in Exercise 2 to

$$\frac{\partial \psi}{\partial n} = 0, \quad t > 0, \boldsymbol{x} \in \partial\Omega,$$

where $\boldsymbol{n}$ stands for the unit outward normal vector on $\partial\Omega$. Prove the uniqueness of solutions to the corresponding initial-boundary value problem.

4. Prove that

$$E(t,x) = H(t)\frac{1-\mathrm{i}}{2a\sqrt{2\pi t}}\mathrm{e}^{\mathrm{i}\frac{x^2}{4a^2 t}}$$

is the fundamental solution of the Schrödinger equation

$$\frac{\partial \psi}{\partial t} - \mathrm{i}a^2\frac{\partial^2 \psi}{\partial x^2} = 0,$$

where $a^2 > 0$ is a real number.

5. Prove that the velocity field of a particle determined by the Schrödinger equation of one particle is irrotational; that is, if we assume that

$$\boldsymbol{v} = \boldsymbol{j}/\rho,$$

where ρ and $\boldsymbol{j}$ are defined by (10.58) and (10.59), respectively, prove that

$$\mathrm{rot}\, \boldsymbol{v} = 0.$$

6. Suppose that a particle with energy $E > 0$ enters from the left in the following potential well:

$$V(x) = \begin{cases} -V_0, & x < 0, \\ 0, & x > 0, \end{cases}$$

where $V_0 > 0$ is a constant. Find the reflection coefficient of this particle on the wall $x = 0$.

7. Suppose that a particle moves in the potential field $V(\boldsymbol{x})$, and its wave function ψ and the first-order space derivatives of the wave function go to zero as $|\boldsymbol{x}| \to \infty$ at a sufficiently rapid rate. Prove that its average energy is

$$\langle H \rangle = \int_{\mathbb{R}^3} W \mathrm{d}x,$$

where

$$W = \frac{\hbar^2}{2m}|\nabla \psi|^2 + \overline{\psi} V \psi$$

is the *energy density*.

8. For the energy density defined in Exercise 7, prove the following conservation law of energy:

$$\frac{\partial W}{\partial t} + \mathrm{div}\, \boldsymbol{Q} = 0,$$

where

$$\boldsymbol{Q} = -\frac{\hbar^2}{2m}\left(\frac{\partial \overline{\psi}}{\partial t}\nabla\psi + \frac{\partial \psi}{\partial t}\nabla\overline{\psi}\right)$$

is the *energy flux density vector*.

9. Suppose that a particle moves in the potential field $V(\boldsymbol{x})$, and denote by $\langle \boldsymbol{x} \rangle$ and $\langle \boldsymbol{p} \rangle$ the average of the particle position and momentum, respectively. Prove

$$\langle \boldsymbol{p} \rangle = m \frac{\mathrm{d}\langle \boldsymbol{x} \rangle}{\mathrm{d}t},$$

where m is the mass of the particle.

10. Suppose that a particle moves in the smooth potential field $V(\boldsymbol{x})$. Denote

$$\boldsymbol{f} = -\nabla V(\boldsymbol{x}),$$
$$\langle \boldsymbol{f} \rangle = \int \overline{\psi} \boldsymbol{f} \psi \mathrm{d}x.$$

Prove that

$$\frac{\mathrm{d}\langle \boldsymbol{p} \rangle}{\mathrm{d}t} = \langle \boldsymbol{f} \rangle,$$

where $\langle \boldsymbol{p} \rangle$ is the average of the particle momentum. This shows that replacing quantities in quantum mechanics with their average values will result in classical mechanics.

11. Prove that any given second-order complex matrix can be represented by the complex linear combination of Pauli matrices σ_1, σ_2, σ_3 and the second-order unit matrix I.

12. Suppose that under the Galilean transformation

$$\begin{cases} t' = t, \\ \boldsymbol{x}' = \boldsymbol{x} - \boldsymbol{v}t \end{cases}$$

the potential field $V(\boldsymbol{x})$ and the wave function ψ change according to the following transformation laws:

$$V'(\boldsymbol{x}') = V(\boldsymbol{x}),$$
$$\psi'(t', \boldsymbol{x}') = \mathrm{e}^{-\mathrm{i}m(\boldsymbol{v}\cdot\boldsymbol{x} + \frac{|\boldsymbol{v}|^2}{2}t')/\hbar} \psi(t', \boldsymbol{x}' + \boldsymbol{v}t').$$

Prove that the Schrödinger equation is invariant under the Galilean transformation.

13. For the transformation of the wave function given by (10.252), verify (10.249)–(10.251).

14. For the transformation of the wave function given by (10.257), verify (10.255)–(10.256) and (10.253).

Bibliography

[1] Gu C., Li T. et al. *Mathematical Physics Equations* (in Chinese), Second Edition. Beijing: Higher Education Press, 1979.

[2] Gu C., Li T. et al. *Mathematical Physics Equations* (in Chinese). Shanghai: Shanghai Science and Technology Press, 1987.

[3] Guo D. *Quantum Mechanics: A Preliminary* (in Chinese). Beijing: Higher Education Press, 1978.

[4] Hawking S. *A Brief History of Time: From the Big Bang to Black Holes*. New York: Bantam Books, 1988.

[5] Jackson J. D. *Classical Electrodynamics*, Third Edition. New York: John Wiley & Sons, 1998.

[6] Landau L. D., Lifshitz E. M. *Quantum Mechanics (Non-Relativistic Theory)*, Third Edition. Amsterdam: Elsevier Ltd., 1977.

[7] Lu H. *Remarks on Quantum Theory of Copenhagen School* (in Chinese). Shanghai: Fudan University Press, 1984.

[8] Pazy A. *Semigroups of Linear Operators and Applications to Partial Differential Equations*. New York: Springer–Verlag, 1983.

[9] Penrose R. *Emperor's New Mind: Concerning Computers, Minds, and the Laws of Physics*. Oxford: Oxford University Press, 1989.

[10] Qi M. *Talking about Mathematics at the Turn of the Century*. Wuhan: Hubei Education Press, 2000.

[11] Treves F. *Basic Linear Partial Differential Equations*. Orlando, FL: Academic Press, 1975.

[12] You G. *How Did Einstein Establish the Theory of Relativity* (in Chinese). Changsha: Hunan Education Press, 1993.

[13] Zeng J. *Quantum Mechanics* (in Chinese), Vol. I. Beijing: Science Press, 1990.

[14] Zeng J. *Quantum Mechanics: An Introduction* (in Chinese). Beijing: Peking University Press, 1992.

[15] Zeng J. *Quantum Mechanics* (in Chinese), Vol. II. Beijing: Science Press, 1993.

[16] Zhou S. *Quantum Mechanics* (in Chinese). Shanghai: Shanghai Science and Technology Press, 1961.

Appendix C

Tensors in Minkowski Four-Space-Time

C.1 Minkowski Four-Space-Time and the Lorentz Transformation

Definition C.1. *If we define the following inner product in a four-dimensional real linear space* $V = \{x = (x^0, x^1, x^2, x^3) \in \mathbb{R}^4\}$*:*

$$\langle x, y \rangle = x^0 y^0 - x^1 y^1 - x^2 y^2 - x^3 y^3, \quad \forall x, y \in V, \tag{C.1}$$

then this linear space is called a Minkowski four-space-time*, denoted by* M.

For $x, y \in M$, if

$$\langle x, y \rangle = 0, \tag{C.2}$$

then we say that x and y are mutually *orthogonal*. If $x \in M$ and $\langle x, x \rangle = \pm 1$, then x is called a *unit vector*.

Notice that the quadratic form used to define the inner product (C.1) is not positive definite.

The set of vectors

$$e_0 = \begin{pmatrix} 1 \\ 0 \\ 0 \\ 0 \end{pmatrix}, \quad e_1 = \begin{pmatrix} 0 \\ 1 \\ 0 \\ 0 \end{pmatrix}, \quad e_2 = \begin{pmatrix} 0 \\ 0 \\ 1 \\ 0 \end{pmatrix}, \quad e_3 = \begin{pmatrix} 0 \\ 0 \\ 0 \\ 1 \end{pmatrix} \tag{C.3}$$

can obviously be regarded as a basis of M satisfying

$$\langle e_\alpha, e_\beta \rangle = g_{\alpha\beta} \quad (\alpha, \beta = 0, 1, 2, 3), \tag{C.4}$$

where

$$g_{\alpha\beta} = \begin{cases} 1 & \text{if } \alpha = \beta = 0, \\ -1 & \text{if } \alpha = \beta = 1, 2, 3, \\ 0 & \text{if } \alpha \neq \beta. \end{cases} \tag{C.5}$$

A basis of the Minkowski four-space-time M satisfying condition (C.4) is called an *orthonormal basis*.

Definition C.2. *Let $L: M \to M$ be a linear transformation. If the inner product between any two given vectors in M remains unchanged under this transformation, namely,*

$$\langle Lx, Ly \rangle = \langle x, y \rangle, \quad x, y \in M, \tag{C.6}$$

then L is called an orthogonal transformation *in M, also called a* Lorentz transformation.

Just like the role of the orthogonal transformation in defining the Cartesian tensor in Euclidean space, the Lorentz transformation plays a fundamental role in defining the tensor in Minkowski four-space-time.

It is easy to directly verify that any given orthonormal basis in M is still an orthonormal basis after a Lorentz transformation.

Let $\{e_0, e_1, e_2, e_3\}$ be an orthonormal basis in M and let L be a Lorentz transformation. Denote

$$e'_\alpha = Le_\alpha \quad (\alpha = 0, \dots, 3). \tag{C.7}$$

Denote furthermore

$$e'_\alpha = a^\gamma_\alpha e_\gamma \quad (\alpha = 0, \dots, 3), \tag{C.8}$$

where a^γ_α are real numbers, and the repeated superscript and subscript represented by Greek letters imply the summation over this index from 0 to 3. Since $\{e_0, \dots, e_3\}$ is an orthonormal basis in M, from (C.8) and (C.4) we get

$$\begin{aligned} \langle e'_\alpha, e'_\beta \rangle &= a^\gamma_\alpha a^\delta_\beta \langle e_\gamma, e_\delta \rangle \\ &= a^\gamma_\alpha a^\delta_\beta g_{\gamma\delta}. \end{aligned} \tag{C.9}$$

Noticing furthermore that $\{e'_0, \dots, e'_3\}$ is also an orthonormal basis in M, from (C.9) and (C.4) we immediately get

$$a^\gamma_\alpha a^\delta_\beta g_{\gamma\delta} = g_{\alpha\beta}. \tag{C.10}$$

Denote

$$A = \begin{pmatrix} a^0_0 & a^1_0 & a^2_0 & a^3_0 \\ a^0_1 & a^1_1 & a^2_1 & a^3_1 \\ a^0_2 & a^1_2 & a^2_2 & a^3_2 \\ a^0_3 & a^1_3 & a^2_3 & a^3_3 \end{pmatrix}. \tag{C.11}$$

Noting the definition (C.5) of $g_{\alpha\beta}$, it is not hard to show from (C.10) that the linear transformation L given by (C.8) is a Lorentz transformation if and only if the row vectors of its corresponding transformation matrix A constitute an orthonormal basis of M.

It is not difficult to prove (left as an exercise) that (C.10) is equivalent to

$$a^\gamma_\alpha a^\delta_\beta g^{\alpha\beta} = g^{\gamma\delta}, \tag{C.12}$$

where $g^{\alpha\beta} = g_{\alpha\beta}$. This also implies that the linear transformation L given by (C.8) is a Lorentz transformation if and only if the column vectors of matrix A constitute an orthonormal basis of M. Equation (C.12) can also be written equivalently in matrix form as

$$A^{\mathrm{T}}\begin{pmatrix}1&&&\\&-1&&\\&&-1&\\&&&-1\end{pmatrix}A=\begin{pmatrix}1&&&\\&-1&&\\&&-1&\\&&&-1\end{pmatrix}, \tag{C.13}$$

where A^{T} stands for the transpose of A.

From (C.13) we have

$$\begin{aligned}A^{-1}&=\begin{pmatrix}1&&&\\&-1&&\\&&-1&\\&&&-1\end{pmatrix}A^{\mathrm{T}}\begin{pmatrix}1&&&\\&-1&&\\&&-1&\\&&&-1\end{pmatrix}\\&=\begin{pmatrix}a_0^0&-a_1^0&-a_2^0&-a_3^0\\-a_0^1&a_1^1&a_2^1&a_3^1\\-a_0^2&a_1^2&a_2^2&a_3^2\\-a_0^3&a_1^3&a_2^3&a_3^3\end{pmatrix}.\end{aligned} \tag{C.14}$$

Denoting by b_α^β the elements of $B=A^{-1}$, we obtain from (C.8) the form of the orthonormal basis $\{e_0,\ldots,e_3\}$ expressed by $\{e'_0,\ldots,e'_3\}$:

$$e_\alpha=b_\alpha^\beta e'_\beta \quad (\alpha=0,\ldots,3). \tag{C.15}$$

C.2 Tensors in Minkowski Four-Space-Time

We first show how a vector in M changes under the Lorentz transformation.

Let $\{e_0,\ldots,e_3\}$ be an orthonormal basis of M. Any given vector x in M can be expressed linearly by this basis:

$$x=u^\alpha e_\alpha. \tag{C.16}$$

Assume that L is the Lorentz transformation given by (C.8), and under the new orthonormal basis $\{e'_0,\ldots,e'_3\}$, x can be expressed as

$$x=u'^\beta e'_\beta. \tag{C.17}$$

Plugging e_α expressed by (C.15) into (C.16), we get

$$x=b_\alpha^\beta u^\alpha e'_\beta. \tag{C.18}$$

Comparing this with (C.17), we obtain

$$u'^\beta=b_\alpha^\beta u^\alpha \quad (\beta=0,1,2,3). \tag{C.19}$$

In Euclidean space, the transformation formula of vector components is the same as that of orthonormal bases (see Appendix A in Volume I). But for Minkowski four-space-time, the transformation formula (C.19) of vector components is different from the transformation formula (C.8) of orthonormal bases. The transformation formula (C.8) of bases is given by matrix A, while the transformation formula (C.19) of vector components is given by the transpose of matrix A.

Definition C.3. *Suppose that a quantity can be represented by four real numbers* (u^α) *under an orthonormal basis of* M*, and also by four real numbers* (u'^α) *under a new orthonormal basis after the Lorentz transformation* (C.8). *If this quantity satisfies the relations*

$$u'^\beta = b^\beta_\alpha u^\alpha \quad (\beta = 0,\dots,3), \tag{C.20}$$

then (u^α) *is called a* contravariant vector *or a* first-order contravariant tensor.

Similarly, we can define the second-order tensor.

Definition C.4. *Suppose that a quantity can be represented by* 16 *real numbers* $(T^{\alpha\beta})$ *under an orthonormal basis of* M*, and also by* 16 *real numbers* $(T'^{\alpha\beta})$ *under a new orthonormal basis after the Lorentz transformation* (C.8). *If this quantity satisfies*

$$T'^{\alpha\beta} = b^\alpha_\gamma b^\beta_\delta T^{\gamma\delta} \quad (\alpha,\beta = 0,\dots,3), \tag{C.21}$$

then $(T^{\alpha\beta})$ *is called a* second-order contravariant tensor.

Definitions for higher-order contravariant tensors in Minkowski space-time can be given similarly; we omit the details here.

For a given vector $x \in M$ and an orthonormal basis of M, using (u^α), namely, (C.16), is not the only way to express this vector. In fact, this vector can also be expressed by components (u_α) defined as follows:

$$u_\alpha = \langle x, e_\alpha \rangle \quad (\alpha = 0,\dots,3). \tag{C.22}$$

Just like (u^α), (u_α) has one-to-one correspondence with x.

Under a new orthonormal basis $\{e'_\alpha\}$ after the Lorentz transformation (C.8), the corresponding components turn into

$$u'_\alpha = \langle x, e'_\alpha \rangle \quad (\alpha = 0,\dots,3). \tag{C.23}$$

Plugging (C.8) into the above formula, we get

$$u'_\alpha = a^\gamma_\alpha \langle x, e_\gamma \rangle = a^\gamma_\alpha u_\gamma \quad (\alpha = 0,1,2,3). \tag{C.24}$$

This implies that for such components of a vector, the transformation law under the Lorentz transformation is exactly the same as the transformation law (C.8) of orthonormal bases.

Definition C.5. *Suppose that a quantity can be represented by four real numbers* (u_α) *under an orthonormal basis of* M*, and also by four real numbers* (u'_α) *under a new orthonormal basis after the Lorentz transformation* (C.8). *If this quantity satisfies*

$$u'_\alpha = a^\gamma_\alpha u_\gamma \quad (\alpha = 0,\dots,3), \tag{C.25}$$

then (u_α) *is called a* covariant vector *or a* first-order covariant tensor. *A* second-order covariant tensor $(T_{\alpha\beta})$ *can be defined by the following transformation law:*

$$T'_{\alpha\beta} = a^\gamma_\alpha a^\delta_\beta T_{\gamma\delta} \quad (\alpha,\beta = 0,\ldots,3). \tag{C.26}$$

Similarly, we can define a second-order mixed tensor (T^β_α), *whose transformation law is*

$$T'^{\beta}_{\alpha} = a^\gamma_\alpha b^\beta_\delta T^\delta_\gamma. \tag{C.27}$$

We can similarly define the higher-order mixed tensors. *A* zero-order tensor *is just a* scalar; *namely, this quantity can be expressed by a real number that remains unchanged under the Lorentz transformation.*

It is easy to verify that

$$(g^{\alpha\beta}) = (g_{\alpha\beta}) = \begin{pmatrix} 1 & & & \\ & -1 & & \\ & & -1 & \\ & & & -1 \end{pmatrix} \tag{C.28}$$

is a second-order contravariant tensor as well as a second-order covariant tensor. It remains unchanged under all coordinates. Hereafter, for narrative convenience, "under different coordinates" means "for different orthonormal bases."

C.3 Operations of Tensors

The linear operations of tensors go without saying; we now focus on indicating the product and the contraction of tensors.

(1) *Product of tensors.* We only take the following case as an example to show this. Suppose that (u^α) is a first-order contravariant tensor and (T^γ_β) is a second-order mixed tensor; then their tensor product $(u^\alpha T^\gamma_\beta)$ is a third-order mixed tensor $(S^{\alpha\gamma}_\beta)$, that is, a mixed tensor with two contravariant indices and one covariant index. As a matter of fact, under a new coordinate system, we have

$$S'^{\alpha\gamma}_{\ \ \beta} = u'^\alpha T'^\gamma_{\ \beta} = b^\alpha_\lambda u^\lambda a^\mu_\beta b^\gamma_\nu T^\nu_\mu = b^\alpha_\lambda a^\mu_\beta b^\gamma_\nu S^{\lambda\nu}_\mu.$$

(2) *Contraction of a tensor.* For components of a given mixed tensor, the operation of taking a superscript-subscript pair to be the same and summing over this index from 0 to 3 is called the *contraction of a tensor.* A tensor after contraction is still a tensor with order two less than the original. For instance, suppose that $(S^{\alpha\beta}_\gamma)$ is a third-order mixed tensor; then $u^\alpha = (S^{\alpha\beta}_\beta)$ is a first-order contravariant tensor. In fact, under a new coordinate system, we have

$$u'^\alpha = S'^{\alpha\beta}_{\ \ \beta} = b^\alpha_\lambda b^\beta_\mu a^\nu_\beta S^{\lambda\mu}_\nu. \tag{C.29}$$

Since B is the inverse matrix of A, we have

$$b^{\beta}_{\mu}a^{\nu}_{\beta} = \begin{cases} 1 & \text{if } \mu = \nu, \\ 0 & \text{if } \mu \neq \nu. \end{cases}$$

Thus, (C.29) can be written as

$$u'^{\alpha} = b^{\alpha}_{\lambda} S^{\lambda\mu}_{\mu} = b^{\alpha}_{\lambda} u^{\lambda}.$$

This implies that $(S^{\alpha\beta}_{\beta})$ is a first-order contravariant tensor.

Example C.1. Suppose that (u^{α}) and (w_{β}) are a first-order contravariant tensor and a first-order covariant tensor, respectively; then $(u^{\alpha}w_{\alpha})$ is a scalar. In fact, $(u^{\alpha}w_{\alpha})$ results from the contraction of the second-order mixed tensor $(u^{\alpha}w_{\beta})$ obtained by the product of two tensors. ■

Example C.2. Suppose that $(T_{\alpha\beta})$ is a second-order covariant tensor and (u^{γ}) is a first-order contravariant tensor; then $(T_{\alpha\beta}u^{\beta})$ is a first-order covariant tensor. ■

Example C.3. Let (u^{α}) be a first-order contravariant tensor. Since $(g_{\alpha\beta})$ is a second-order covariant tensor, then

$$(u_{\alpha}) = (g_{\alpha\beta}u^{\beta})$$

is a first-order covariant tensor. u_{α} is called the covariant component of the first-order contravariant tensor (u^{α}). On the other hand, we have

$$u^{\alpha} = g^{\alpha\beta}u_{\beta}.$$

In fact, it is easy to see from the definition (C.28) of $g_{\alpha\beta}$ and $g^{\alpha\beta}$ that the contravariant component u^{α} and the covariant component u_{α} of a vector have the following simple relations:

$$u_0 = u^0, \quad u_1 = -u^1, \quad u_2 = -u^2, \quad u_3 = -u^3. \tag{C.30}$$

■

C.4 Covariant Derivatives of Tensors

Suppose that the components of a tensor are functions of the point coordinates (x^{α}) under a certain coordinate system. Now we investigate the derivatives of this tensor with respect to (x^{α}). We take the second-order contravariant tensor $(T^{\alpha\beta})$ as an example to make this clear. Denote

$$S^{\alpha\beta}_{\gamma} = \frac{\partial T^{\alpha\beta}}{\partial x^{\gamma}}; \tag{C.31}$$

then $(S^{\alpha\beta}_{\gamma})$ is a third-order mixed tensor. That is, taking the gradient once to the components of the tensor with respect to (x^{α}) increases the covariant index by one as well as the order by one. In fact, under the new coordinates, we have

$$\begin{aligned} S'^{\alpha\beta}_{\gamma} &= \frac{\partial T'^{\alpha\beta}}{\partial x'^{\gamma}} \\ &= \frac{\partial x^{\delta}}{\partial x'^{\gamma}} \frac{\partial}{\partial x^{\delta}} (b^{\alpha}_{\lambda} b^{\beta}_{\mu} T^{\lambda\mu}) \\ &= a^{\delta}_{\gamma} b^{\alpha}_{\lambda} b^{\beta}_{\mu} \frac{\partial T^{\lambda\mu}}{\partial x^{\delta}} \\ &= a^{\delta}_{\gamma} b^{\alpha}_{\lambda} b^{\beta}_{\mu} S^{\lambda\mu}_{\delta}. \end{aligned} \tag{C.32}$$

Here we used

$$x^{\delta} = a^{\delta}_{\gamma} x'^{\gamma}, \tag{C.33}$$

which is easy to obtain from (C.19). Equation (C.32) shows that $(S^{\alpha\beta}_{\gamma})$ is a third-order mixed tensor with one covariant index and two contravariant indices.

Index